Rainer Brüggemann
Lars Carlsen

Partial Order in Environmental Sciences and Chemistry

Rainer Brüggemann
Lars Carlsen
(Editors)

Partial Order in Environmental Sciences and Chemistry

With 140 Figures and 50 Tables

Springer

DR. RAINER BRÜGGEMANN
Leibniz Institute
of Freshwater Ecology and Inland Fisheries
Dept. Ecohydrology
Müggelseedamm 310
12587 Berlin-Friedrichshagen
Germany

E-mail:
brg@igb-Berlin.de

PROF. DR. LARS CARLSEN
Awareness-Center
Hyldeholm 4
4000 Roskilde-Veddelev
Denmark

E-mail:
LC@AwarenessCenter.dk

ISBN-10 3-642-42833-9 Springer Berlin Heidelberg New York
ISBN-13 978-3-642-42833-3 Springer Berlin Heidelberg New York

Springer is a part of Springer Science+Business Media
springer.com

Softcover re-print of the Hardcover 1st edition 2006

Cover design: Erich Kirchner
Typesetting: camera-ready by the editors
Production: Christine Jacobi
Printing: Krips bv, Meppel
Binding: Stürtz AG, Würzburg

Printed on acid-free paper 30/2133/cj 5 4 3 2 1 0

Preface

When you edit a book, the editors should ask themselves, why are we doing this and whom are we doing this for? To whom could this book be valuable as a source of information and possibly inspiration and of course are there other books with similar topics on the market? Indeed the mathematical structure 'partial order' is explained in many mathematical textbooks, which require different degrees of mathematical skills to comprehend. Thus, as far as we can tell, all these books are dedicated directly towards mathematician working in the area of Discrete Mathematics and Theoretical Informatics. Although partial order is very well known in quantum mechanics, especially within the context of Young-diagrams, literature stressing the application aspect of partial order seems to be not available. However, an increasing number of publications in scientific journals have in recent years appeared, applying partial order to various fields of chemistry and environmental sciences. A recent summary can be found in a special issue of the journal Match - Commun.Math.Comput. Chem. 2000, edited by Klein and Brickmann. However, we believe that this journal possibly is too specific and as such it may not reach scientists actually applying partial order in various fields of research. Hence, we dared to initiate the editing of this book in order to address a broader audience and we were happy to convincing distinguished scientists working with different aspects of partial order theory to contribute to this book. We are indeed indebted to all of them.

What is a partial order? A general explanation can be found just in the first chapters of this book and according to the different application aspects, correspondingly adopted definitions can be found in many other chapters; however, it might be useful briefly to explain the concept here by a simple example. Thus, if a chemical is toxic and is bioaccumulating then obviously the chemical may exert an environmental risk. If there are two other chemicals, one exhibiting a lower toxicity but a higher bioaccumulation potential and another with a much higher toxicity but a lower bioaccumulation potential, we may have a problem to assess their individual environmental risks. This kind of problems can be analyzed with partial order. The only mathematical operation needed is the comparison, i.e. is a larger or smaller than b. Hence, partial order in its various application aspects is the science of comparisons! Comparisons of chemical properties, comparisons of environmental systems, and even comparisons of strategies or management options are all topic that advantageously may be analyzed using partial order theory. Our objective with this book is to demonstrate how to use partial order in the field of pure chemistry, in substance property estimations, and in environmental sciences. Some chapters will show how partial order can be applied in field monitoring studies, in deriving decisions and in judging the quality of databases in the context of environmental systems and chemistry. The charming aspect of partial order is just that by comparison we learn something about the objects, which are to be compared!

Most of the readers will probably be trained within differential calculus, with linear algebra, or with statistics. All the mathematical operations needed in these disciplines are by far more complex than that single one needed in partial order. The point is that operating without numbers may appear somewhat strange. The book aims to reduce this uncomfortable strange feeling.

Thus, we hope that this book will broaden the circle of scientists, which find partial order as a useful tool for their work. The theoretical and practical aspects of partial order are discussed in, e.g., the INDO-US-workshop on Mathematical Chemistry, a series of scientific symposia initialized by Basak and Sinha, 1998, and in specific workshops about partial order in chemistry and environmental systems. We urge scientist, newcomers as well as established partial order users to contribute to these workshops, contacts can be found by our E-Mail-addresses (brg@igb-Berlin.de or brg_home@t-online.de (Brueggemann) or LC@AwarenessCenter.dk (Carlsen)).

April 2006

Rainer Brüggemann and Lars Carlsen

Acknowledgement

This book could not have been reality without the enthusiasm of all our contributing authors. We are truly grateful and thank each of them cordially. We thank Alexandra Sakowsky for her help and her patience in rewriting texts in the correct layout, Dagmar Schwamm, Grit Siegert, Barbara Kobisch and Dr. Torsten Strube for helping us. Last not least we thank the Leibniz-Institute of Freshwater Ecology and Inland Fisheries for supporting this work.

We thank the publishing house 'Springer' for its patience.

Contents

List of Contributors

BEHRENDT, H.
Leibniz-Institute of Freshwater Ecology and Inland Fisheries
Müggelseedamm 310, D-12587 Berlin, Germany
e-mail: behrendt@igb-berlin.de

BRÜGGEMANN, R.
Leibniz-Institute of Freshwater Ecology and Inland Fisheries
Müggelseedamm 310, D-12587 Berlin, Germany
e-mail: brg@igb-berlin.de or brg_home@t-online.de

CAI, Y.
Department of Statistics, The Pennsylvania State University
Univ. Park, PA 16802, USA
e-mail: yzc102@psu.edu

CARLSEN, L.
Awareness Center
Veddelev, Hyldeholm 4, 4000 Roskilde, Denmark
e-mail: LC@AwarenessCenter.dk

CONSONNI, V.
Milano Chemometrics and QSAR Research Group
Dept. of Environmental Sciences, University of Milano-Bicocca
P.za della Szienza, I-20126 Milano, Italy
e-mail: viviana.consonni@unimib.it

EL-BASIL, S.
Faculty of Pharmacy, University of Cairo
Kasr Al-Aini st. Cairo 11562, Egypt
e-mail: sherifbasil@hotmail.com

GRAMATICA, P.
QSAR and Environmental Chemistry Research Unit
Dept. of Structural and Functional Biology, University of Insubria
via Dunant 3, I-21100 Varese, Italy
e-mail: paola.gramatica@uninsubria.it

HALFON, E.
Burlington, Ontario, 4481 Concord Place, Canada L7L1J5
e-mail: info@butx.com

HEFFERLIN, R.
Southern Adventist University, Collegedale, Tennessee 37315, USA
e-mail: hefferln@southern.edu

HEININGER, P.
Federal Institute of Hydrology (BfG), Dept. Qualitative Hydrology
P.O. Box 200253, D-56002 Koblenz, Germany
e-mail: heininger@bafg.de

HELM, D.
Robert Koch-Institute, Seestr. 10, D-13353 Berlin, Germany
e-mail: helmd@rki.de

IVANCIUC, T.
Texas A&M University, Galveston, Texas, USA
e-mail: oiivanci@utmb.edu

KERBER, A.
Department of Mathematics, University of Bayreuth, Germany
e-mail: Adalbert.Kerber@uni-bayreuth.de

KLEIN, D. J.
Texas A&M University, Galveston, Texas, USA
e-mail: kleind@tamug.tamu.edu

LERCHE, D. B.
The National Environmental Research Institute, Department of Policy Analysis, Frederiksborgvej 399, DK-4000 Roskilde, Denmark
e-mail: dortelerche@hotmail.com

MYERS, W. L.
124 Land & Water Research Bildg, The Pennsylvania State University, Univ. Park, PA 16802, USA
e-mail: wlm@psu.edu

PATIL, G. P.
Department of Statistics, The Pennsylvania State University, Univ. Park, PA 16802, USA
e-mail: gpp@stat.psu.edu

PAVAN, M.
Milano Chemometrics and QSAR Research Group, Dept. of Environmental Sciences, University of Milano-Bicocca, P.za della Szienza, I-20126 Milano, Italy.
e-mail: manuela.pavan@unimib.it (recently: manuela.pavan@jrc.it)

PUDENZ, S.
Criterion-Evaluation & Information Management
Mariannenstr. 33, D-10999 Berlin, Germany
e-mail: stefan.pudenz@criteri-on.de

SEITZ, W.
Department of Marine Sciences, University at Galveston, Texas 77539, P.O. Box 1675, USA
e-mail: seitzw@tamug.edu

SIMON, U.
Leibniz-Institute of Freshwater Ecology and Inland Fisheries
Müggelseedamm 310, D-12587 Berlin, Germany
e-mail: ute.simon@geo.hu-berlin.de

SØRENSEN, P. B.
Department of Policy Analysis, National Environmental Research Institute, Vejlsoevej 25, DK-8600 Silkeborg, Denmark
e-mail: pbs@dmu.dk

THOMSEN, M.
The National Environmental Research Institute, Department of Policy Analysis, Frederiksborgvej 399, DK-4000 Roskilde, Denmark
e-mail: mth@dmu.dk

TODESCHINI, R.
Milano Chemometrics and QSAR Research Group, Dept. of Environmental Sciences, University of Milano-Bicocca, P.za della Szienza, I-20126 Milano, Italy
e-mail: roberto.todeschini@unimib.it

VOIGT, K.
GSF-Research Centre for Environment and Health,
Institute for Biomathematics and Biometry,
Ingolstädter Landstr. 1, D-85758 Oberschleissheim, Germany
e-mail: kvoigt@gsf.de

WALKER, J. D.
TSCA Interagency Testing Committee (ITC), Office of Pollution Prevention and Toxics (7401), Washington, D.C. 20460, USA
e-mail: Walker.Johnd@epamail.epa.gov

1 Chemistry and Partial Order

In this section the fundamentals of partial orders are introduced in three chapters, which are rather different, albeit they point to the same item: partial order in chemistry. The reader will learn basic concepts and a manifold how to derive a partial order from chemical concepts.

In the first chapter, by El-Basil, the main terms and concepts of partial order are explained. It shows that there are many different ways to apply the axioms of partial order. Especially the important theorem of Muirhead and its generalization are broadly discussed. The reader may learn how to develop Young diagrams and how to extract useful results form the partially ordered set of Young diagrams. The examples are mainly following the chemistry of aromatics. Hence, the reader will become familiar with the broad topic of Kekulé structures and counting them.

The detection of the periodic system of chemical elements was a break through in the theoretical understanding of chemistry. Hefferlin discusses periodicities of chemical elements and small molecules. He shows how general the concept of posets is. Why not explore the properties of small molecules by means of a Hasse diagram? Hefferlin shows by the example of Phosphorus oxides how this may be done.

The first two chapters are devoted to a static presentation of chemical concepts. However, chemistry is the science of reactions and interactions. In the third chapter Klein and Ivanciuc show, how partial order can be applied within the context of substitution patterns. The authors demonstrate for example that partial order relations and an order based on environmental toxicities match very well and how a parameter free approach to QSAR can be found (see also topic 3). Methodologically the reader will learn how chemical structures and partially ordered sets can be related and how interpolation schemes are working. Finally, the important idea to extend the field of chemical property estimations by the concept or quantitative super-structure activity relationships is discussed.

Partial Ordering of Properties: The Young Diagram Lattice and Related Chemical Systems

Sherif El-Basil

Faculty of Pharmacy, University of Cairo, Kasr Al-Aini st. Cairo 11562, Egypt

e-mail: sherifbasil@hotmail.com

Abstract

The basic definitions related to the general topic of ordering are reviewed and exemplified including: partial ordering, posets, Hasse diagrams, majorization of structures and comparable / incomparable structures.

Young Diagram lattice (of Ruch) and the ordering scheme of tree graphs (of Gutman and Randić) are described and it is shown, how the two schemes coincide with each other, i.e. generate identical orders.

The role of Young diagrams in the ordering of chemical structures is explained by their relation to alkane hydrocarbons and unbranched cata-condensed benzenoid systems.

The Basic Terms: Examples of Posets, The Hasse Diagram

The concept of a partial order appears to be very useful in environmental science when evaluation and comparative study of properties are required. The object to be studied form an object set and the partially order set (≡ poset) depends on the ≤ , (greater than- or equal to-) relation (Luther et al. (2000). We now introduce some of the popular definitions in an intuitive approach, which avoids the "dryness" of mathematical rigor.

Partially ordered set (poset)

It may be helpful to consider the following graph and analyze some parts of it: (cf. Fig. 1)

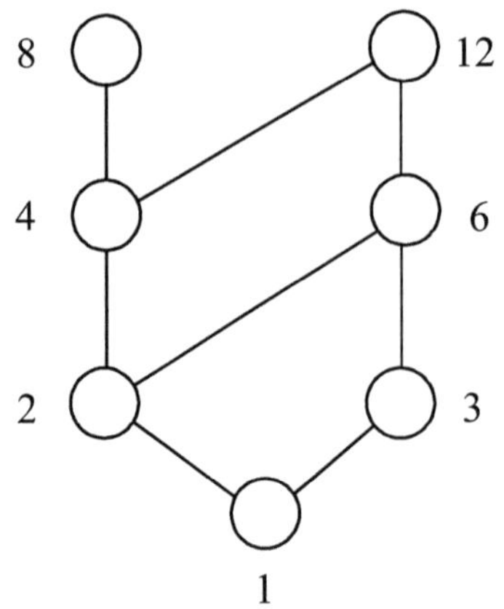

Fig. 1. A labelled graph, which corresponds to a relation on a set of numbers

Obviously, the above graph describes some sort of a relation, R, on the components of the set of integers:

$$S = \{1, 2, 3, 4, 6, 8, 12\} \quad (1)$$

We consider S as ground set (object set), whose elements are labelled vertices of a graph. The relation among the vertices, graphically displayed by lines (called "edges") depends on the questions one has. For example: One observes that numbers, which divide others are connected, those that do not divide each other are not. One, then, says that the above graph represents some sort of ordering relation expressed as.

$$\{(a,b) \mid a \text{ divides } b\} \text{ on } S = \{1, 2, 3, 4, 6, 8, 12\} \quad (2)$$

The relations among integers are described as follows:

a) Because every element of S is related to itself, i. e., $(a, a) \in R$; R is said to be reflexive.
b) While, e.g., 2 divides 4, 4 does not divide 2 and so on. Such a relation is said to be anti-symmetric.

c) The last property may be exemplified on the subset {2, 4, 8}: 2 divides 4; 4 divides 8 hence 2 divides 8, which is true for other components, i. e.: if (a, b) ∈ R and (b, c) ∈ R then (a, b, c) ∈ R.

The above property is called the transitive character of R. A poset may then be defined as a relation R, on a set S if R is reflexive, anti-symmetric and transitive.

The graph, which describes a particular poset, is called a Hasse diagram after the 20th century German mathematician Helmut Hasse (1898-1979) (Rosen 1991). See also chapter by Halfon p. 385.

A word on Hasse diagrams:
Actually the object shown in Fig. 1 is just a graph (not a diagram!): perhaps the word diagram is associated to it from the way it is used to be drawn. In fact all self-evident edges are now removed such as all loops, which describe the reflexive relation and also which result from the transitive character, e.g., edges (2, 8), (3, 12) and (1, all other vertices) are removed. Also arrows that indicate relative positions of components are no longer indicated, yet the "old name": diagram, (instead of graph) remained.

The Hasse diagram can be drawn in different ways maintaining the main information, the order relations. Such Hasse diagrams are isomorphic to each other.

Majorization of Structures: Relative Importance

Sometimes in (partial) ordering problems one may be interested in the relative importance of the components of a set. This situation reminds us with the relation $A \leq B$ i.e., "A is a descendent of B" or that: "B majorizes A". A popular example is the partial ordering $\{(A, B) \mid A \subseteq B\}$ on the power set S = {a, b, c} where $A \subseteq B$ means that A is a subset of B. Whenever this relation exists one says that B majorizes A. The power set S contains $2^3 = 8$ elements, viz., {a}, {b}, {a, b}, {a, c}, {b, c}, {a, b, c} and ∅, where ∅ is the empty set.

For this particular case the Hasse diagram is simply a cube, labelled as shown in Fig. 2.

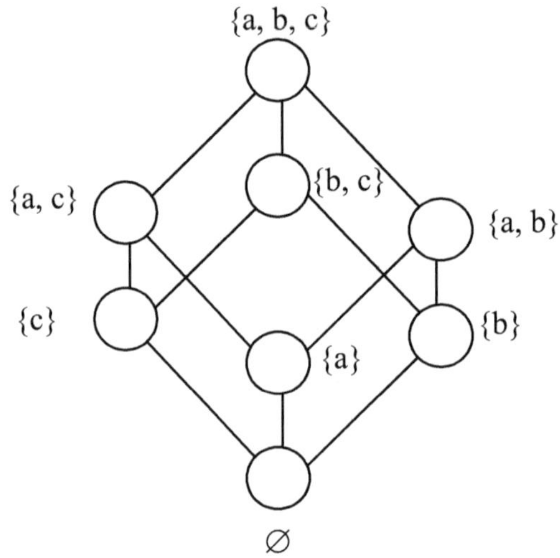

Fig. 2. The Hasse diagram of S = {a, b, c}. Each subset is attached to its direct off-spring, so that the descendant (less important components) lies in lower levels

One observes that {a, b}, {a, c} and {b, c} are subsets of {a, b, c} and therefore of lower relative importance and analogously for the single-component subsets {a}, {b}, {c}. The above example represents one of the simplest cases of relative importance ordering problems, which finds chemical applications (section 'Relative importance of Kekulé Structures of Benzenoid Hydrocarbons: Chain ordering').

Comparable and incomparable elements: Chain and Anti-chain

The elements a and b of a poset (S,<) are called comparable if either $a \leq b$ or $b \leq a$. When a and b are elements of S such that neither $a \leq b$ nor $b \leq a$, a and b are called incomparable. For example the subsets {a,c}, {b,c} and {a, b} are incomparable with each other: (they are not directly connected (= adjacent) to each other, cf. Fig. 2). On the other hand, because {a, b, c} majorizes {a, c}, e.g., they are comparable components of S.

Partial ordering may, then, be viewed as first weakening (≡ relaxation) of the usual total ordering which is required for every pair of elements, $a,b \in S$, that it must be $a \leq b$ or $b \leq a$ or $a = b$. Of course the standard total ordering is that of "greater than or equal to" on the set of real members. In Fig. 2, the subset of vertices, labelled {{a, b, c}, {a, c}, {c}, ∅} is called a chain because every two elements of this subset are comparable. On the other hand the subset {{a, b, c}, {b, c}, {a}, ∅}} is called an anti-chain because every two elements are incomparable.

One can immediately see the advantage of mathematical (graph-theoretical) techniques over quantum-chemical calculations in fields, which requires analysis of structure-property relation (such as environmental sciences). We quote the following paragraph from a paper by Randić et al. (1985).

"Quantum Chemistry appears to be preoccupied with evaluation of the wave function and potential surfaces, a worthy goal- but of limited use when one considers whole families of molecules and when one is concerned with structure-property relationships".

We quote further (Randić et al. 1985):

"Graph Theory is concerned with relations, and, in chemistry, the relationships between molecular structure and molecular properties are of particular interest".

Namely using graph-theoretical techniques, a structure is "replaced", so-to-speak, by a collection of its mathematical properties (≡ graph-invariants) (Randić et al. 1985) and whence allows the generation of various types of posets as we shall see the next section.

Some Posets of Chemical Interest

In this section we show some of the posets produced by researchers in mathematical chemistry over the past quarter of a century:

Relative importance of Kekulé Structures of Benzenoid Hydrocarbons: Chain ordering

Individual formal valence structures of conjugated hydrocarbons are excellent "substrates" for research in chemical graph theory, whereby many of the concepts of discrete mathematics and combinatorics may be applied to chemical problems. The lecture note published by Cyvin and Gutman (Cyvin, Gutman 1988)) outlines the main features of this type of research mostly from enumeration viewpoint. In addition to their combinatorial properties, chemists were also interested in relative importance of Kekulé valence-bond structures of benzenoid hydrocarbons. In fact, as early as 1973, Graovac et al. (1973) published their Kekulé index, which seems to be one of the earliest results on the ordering of Kekulé structures: These authors used ideas from molecular orbital theory to calculate their indices but the resulting ordering is not partial: it is a chain-type (also called total or linear order).

Graph-theoretical Ordering of Kekulé structures

A few years later, Randić (1977) analyzed a valence-bond Kekulé structure into conjugated circuits of π-electrons: For benzenoid systems R_n implies (4n+2) π-electrons. Randić, then, parameterized his R_n's and ordered them as $R_1 > R_2 > R_3 > \ldots$ where he studied both the relative importance of Kekulé structures as well as the stabilities of their benzenoid hydrocarbons (compare Fig. 6, section 'Partial-Ordering of Kekulé Structures').

Partial Ordering of Kekulé Structures

A decade ago El-Basil (1993) generated vertex-transitive graphs (i.e., 2-cube (≡ square), 3-cubes (≡ cube), 4-cube (≡ tesseract), etc.) using terminal R_1 circuits in a (sub)-set of Kekulé structures by defining two Kekulé structures as "adjacent" if one can be obtained from the other by sextet rotation in only one terminal R_1 through 60°. Formally, this is an operation on a power set composed of n terminal R_1 conjugated circuits (≡ terminal sextets). When n=2 one obtains a square (2-cube, because $2^2 = 4$), n=3 generates a cube ($2^3 = 8$) while a tesseract requires 4 terminal circuits ($2^4 = 16$) and so on. The base 2 originates from the fundamental fact that there are only two ways in which the double bonds are arranged in a hexagon, viz., proper, (+1) and improper (-1): (Fig. 3)

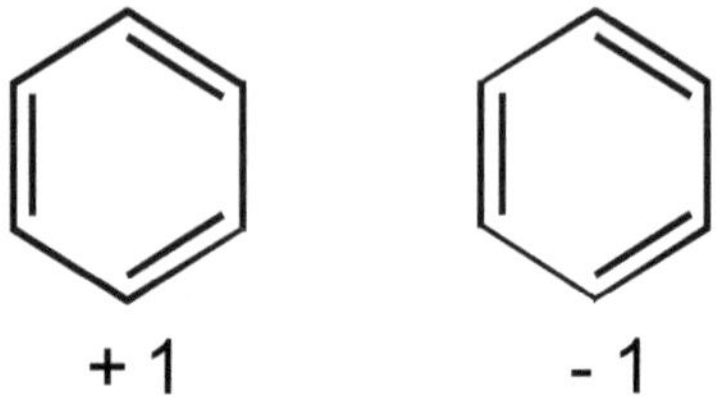

Fig. 3. The two orientations of π-electrons in a hexagon

The sextet rotation operation defines our adjacency relation among the set of Kekulé structures and the vertex-transitive graphs generated are nothing else but posets of Kekulé structures.

Coding Kekulé structures of catacondensed benzenoids

1. Arrange the skeleton of the benzenoid hydrocarbon so that some of its edges are vertical
2. Starting from the top left corner of the benzenoid graph, assign +1 or -1 to terminal rings according to the orientation of their aromatic sextets (Fig. 3).
3. Two Kekulé structures X, Y are defined to be adjacent (El-Basil (1993) if their codes differ in the sign of only one position: A skeleton $X \geq Y$ iff $q_i(X) \geq q_i(Y)$ with q being a sequence of +1 and -1.

An example is shown in Fig. 4

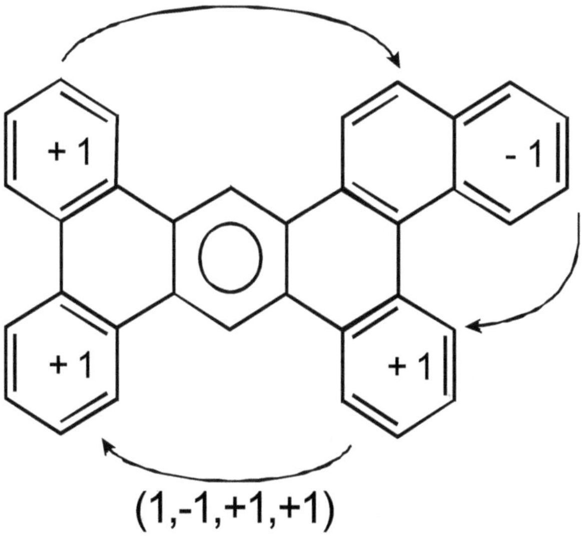

Fig. 4. The code of Kekulé structure of catacondensed benzenoid system containing 4 terminal hexagons. Compare also Fig. 3

In Fig. 5 we show the posets (Hasse diagrams), which correspond to hydrocarbons containing 3, 4 and 5 terminal hexagons.

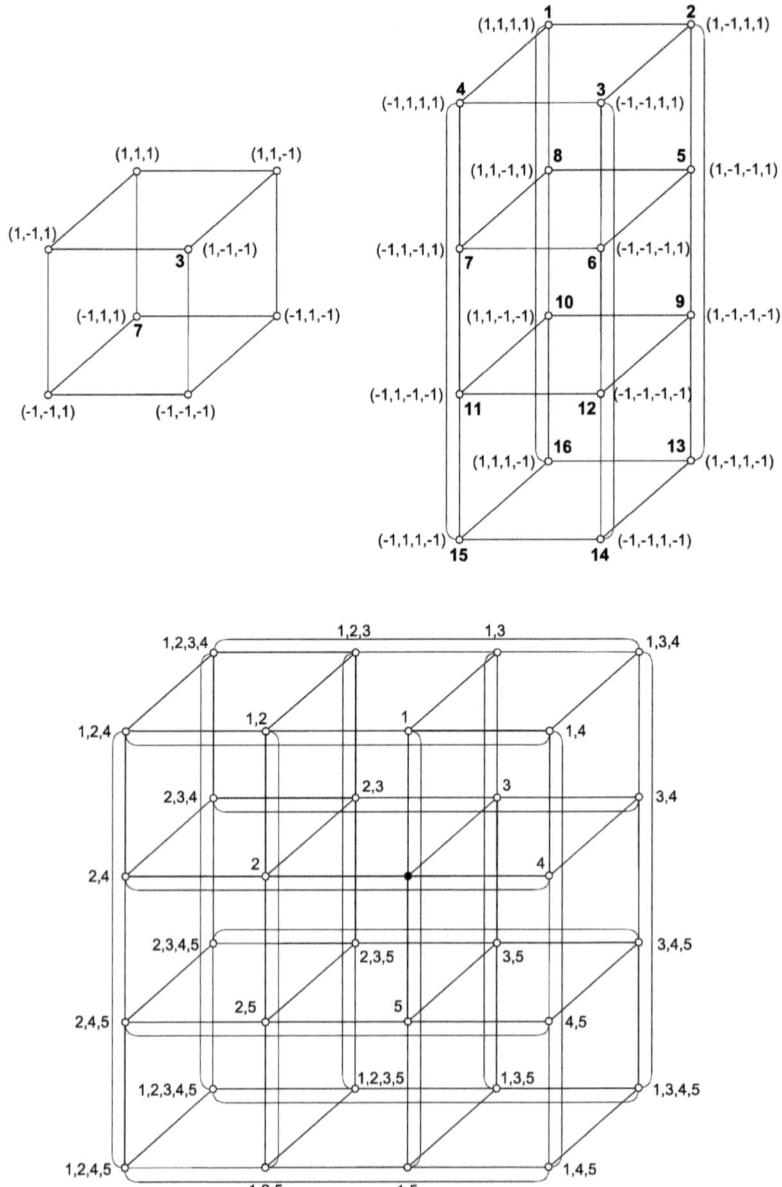

Fig. 5. 3-, 4- and 5-cubes, which represents posets generated from sets of Kekulé structures

Note that in Fig. 5 benzenoid hydrocarbons are shown, having 3, 4 and 5 terminal hexagons. Codes of Kekulé structures are indicated. For the 5-cube only places of negative signs of the code are written

A) The cube poset:

Fig. 6 shows the cube which results when 8 (out of the 9) structures of triphenylene are ordered according to their adjacency relations of their codes (El-Basil (1993).

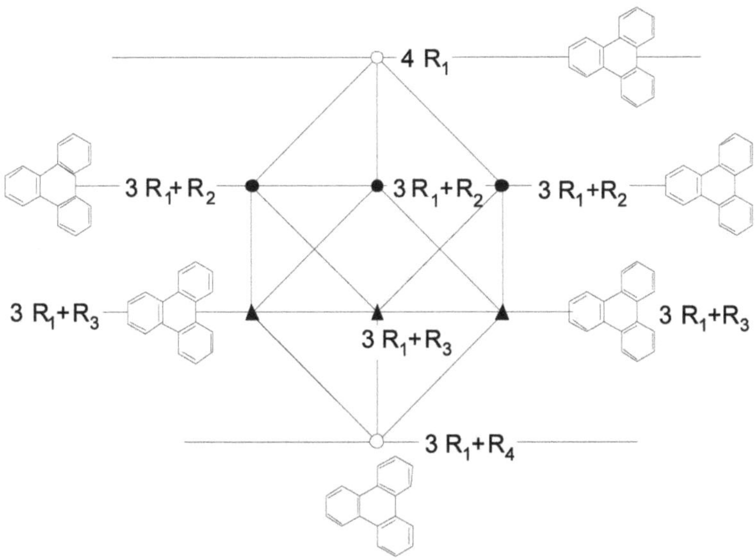

Fig. 6. The Hasse diagram of ordering 8 (out of 9) Kekulé structures of triphenylene

In Fig. 6 incomparable structures are indicated by solid circles and by solid triangles. Dotted lines indicate levels of stability of the Kekulé structures: $4R_1 > 3R_1+R_2 > 3R_1+R_3 > 3R_1+R_4$

Counts of conjugated circuits are shown, from which we see several chain orders, e.g., one of which leads to the following relative stabilities:

$$4R_1 > 3R_1+R_2 > 3\,R_1+R_3 > 3R1+R_4 \qquad (3)$$

All the resulting partial orders are consistent with the conjugated-circuits model of Randić (1977).

Kekulé structures, which correspond to identical circuit counts, e.g. vertices labelled by ($3R_1+R_2$) and by ($3R_1+R_3$) in Fig. 6, are incomparable. They represent vertices, which are not connected on the poset. Some Kekulé structures are shown as representative examples along with their conjugated - circuit counts.

B) The tesseract (The 4-dimensional cube)

This 16 (= 2^4) vertex-transitive graph may be generated using a catacondensed benzenoid system with 4 terminal hexagons. Again, the individual Kekulé structures are partially ordered in accord with their conjugated-circuits counts (Randić (1977).

Fig. 7 shows the resulting poset.

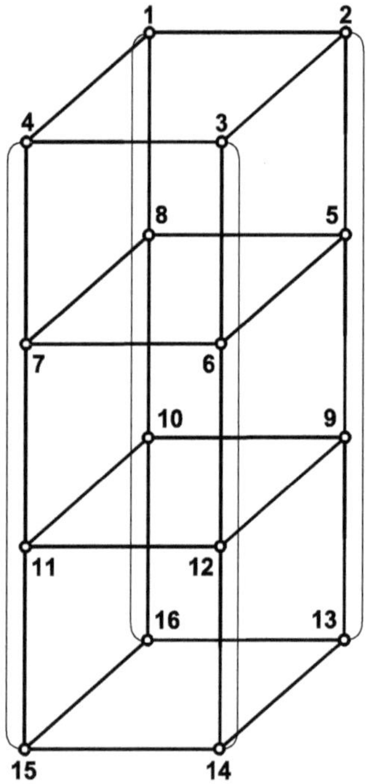

Fig. 7. Four-dimensional cube generated from a benzenoid system containing 4 terminal hexagons

In Fig. 7 the conjugated circuits correspond to Kekulé structure-positions as follows: 1 = ($7R_1$); 2 = ($6R_1 + R_2$); 3 = ($6R_1 + R_3$); 4 = ($6R_1 + R_2$); 5 = ($5R_1 + 2R_2$); 6 = ($5R_1 + R_2 + R_3$); 7 = ($5R_1 + 2R_2$); 8 = ($6R_1 + R_2$); 9 = ($5R_1 + R_2 + R_3$); 10 = ($6R_1 + R_3$); 11 = ($5R_1 + R_2 + R_3$); 12 = ($5R_1 + 2R_3$); 13 = ($5R_1 + 2R_2$); 14 = ($5R_1 + R_2 + R_3$); 15 = ($5R_1 + 2R_2$); 16 = ($6R_1 + R_2$).

The Young-Diagram Lattice, Ordering of Muirhead and generalization of Karamata

Ordering implies a comparison, and instead of actual structures, one normally compares sequences of numbers characterizing a molecular graph of a chemical structure. Frequently the required sequences are derived from an enumeration of selected graph invariants. If the selected invariants lead to integers, then the ordering theory of Muirhead (1903) is most suited for these special cases:

At the beginning of last century Muirhead (1903) introduced a theory of ordering and comparing sequences of integers. Muirhead's method calls for the construction of partial sums derived from integral sequences. If for every entry in two such sequences of partial sums, members of one structure are larger or equal (but not smaller) than the corresponding entries in other sequence, the structures can be ordered with the first structure preceding the second. If these conditions are not satisfied, the structures are not comparable leading to a partial ordering.

$$(a_1 \geq a_2 \geq \ldots \geq a_n > 0) \text{ and } (b_1 \geq b_2 \geq \ldots \geq b_n > 0) \quad (4)$$

Be two sequences of integers. Then, Muirhead's method states that: $(a_1, a_2, \ldots \geq a_n)$ majorizes $(b_1, b_2, \ldots \geq b_n)$, if a series of statement holds: (Table 1):

Table 1. Muirhead's method (Muirhead (1903)

$(a_1,a_2,\ldots,a_n)$	majorizes	$(b_1,b_2,\ldots,b_n)$	if
a_1	$\geq$	b_1	
a_1+a_2	$\geq$	b_1+b_2	
…	$\geq$	…	
$a_1+a_2+\ldots+a_n$	$\geq$	$b_1+b_2+\ldots+b_n$	

Restrictions to integral entries have subsequently been removed and for these more general situations, Karamata (Beckenbach, Bellmann 1961) derived an important theorem, which allows definite conclusions to be drawn from properties of the structures to be studied, if graph invariants are not integral quantities.

More recently Ruch (1975) used ideas of Muirhead in connection with representations of the symmetric group and generated a partial ordering of partitions of integers. For each partition one associates a row of an equal number of dots or boxes so that the rows are arranged in a non-increasing order. For example there are 5 partitions of 4, represented in Fig. 8.

(4)	4		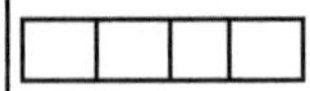
(3,1)	3+1	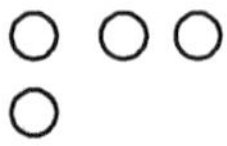	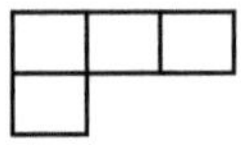
(2^2)	2+2	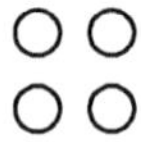	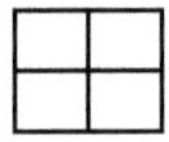
$(2,1^2)$	2+1+1	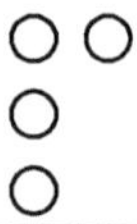	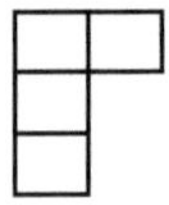
(1^4)	1+1+1+1		

Fig. 8. Partitions of 4 and the corresponding Ferrers graphs and Young diagrams. Both to be read horizontally

Sometimes, the diagrams of dots are referred to as Ferrers graphs (Coleman 1968), after the English mathematician of the latter part of the nineteenth century. However, Young, employed similar devices now known as Young diagrams where he replaces dots with small squares. We let:

$$\Gamma_a = a_1 + a_2 + \ldots + a_n\,,$$
$$\Gamma_b = b_1 + b_2 + \ldots + b_n \qquad (5)$$

be two partitions of the same integer (i.e., $a_a + a_2 + \ldots + a_n = b_1 + b_2 + \ldots + b_n$). Then Γ_a is said to dominate or majorize Γ_b if equation (4) is satisfied, otherwise, the two partitions (or the corresponding graphs) are not comparable. As an illustration, we form the partial sums of the partitions of 4:

4	$\to (4, 0, 0, 0) \to (4, 4, 4, 4)$
3+1	$\to (3, 1, 0, 0) \to (3, 4, 4, 4)$
2+2	$\to (2, 2, 0, 0) \to (2, 4, 4, 4)$
2+1+1	$\to (2, 1, 1, 0) \to (2, 3, 4, 4)$

$$1{+}1{+}1{+}1 \quad \rightarrow (1, 1, 1, 1) \rightarrow (1, 2, 3, 4) \qquad (6)$$

Muirhead's ordering conditions in the above case are a chain on five vertices because all successive sequences (≡ graphs) are comparable.

For n = 5 there are 7 partitions, the ordering of which also leads to a chain. But starting with n = 6, one observes two pairs of non-comparable graphs (leading to two bifurcations). (This poset is shown in Fig. 10 (section 'The Young-Diagram Lattice, Ordering of Muirhead and generalization of Karamata')).

The first bifurcation is generated at $\Gamma_4 = 4 + 1 + 1$ and $\Gamma_5 = 3 + 3$, which correspond to the partial sums:

$$(4, 5, 6, 6, 6, 6) \quad , \quad (3, 6, 6, 6, 6, 6) \qquad (7)$$

We observe that the first component of the first sequence is greater than that of the second sequence, but the reverse order for the second components, i.e. $4 > 3$ but $5 < 6$. Hence the sequences (7) are an incomparable pair. Both Γ_4 and Γ_5 are comparable with $\Gamma_3 = 4 + 2$ whose partial sum is (4, 6, 6, 6, 6, 6) which majorizes both partial sums shown in eqn. (7).

Ordering of Tree graphs

In Gutman and Randić (1977) published their work on the algebraic characterization of skeletal branching of tree graphs (cf. Harary (1972). (A graph is viewed as an abstract representation of a molecule where vertices replace atoms and edges replace chemical bonds. The degree of a vertex equals the valence of the corresponding atom. Often in hydrocarbons a H-suppressed graph is useful, where the hydrogen atoms are neglected.) The steps involved in the scheme of Gutman and Randić are outlined as follows:

a) List the valences of a tree graph in a non-increasing way.
b) Form the partial sums of the above sequence.
c) Order a set of partial sums according to eqn. (4).

Remarkably the poset obtained (by Ruch) for Young diagrams were also obtained (by Gutman and Randić) for the trees! In fact a set of Young diagram containing n vertices is isomorphic with a set of trees containing (n + 2) vertices. As an illustration we show how a set of Young diagrams on 6 boxes and a set of trees on 8 vertices generate the same poset (Fig. 9).

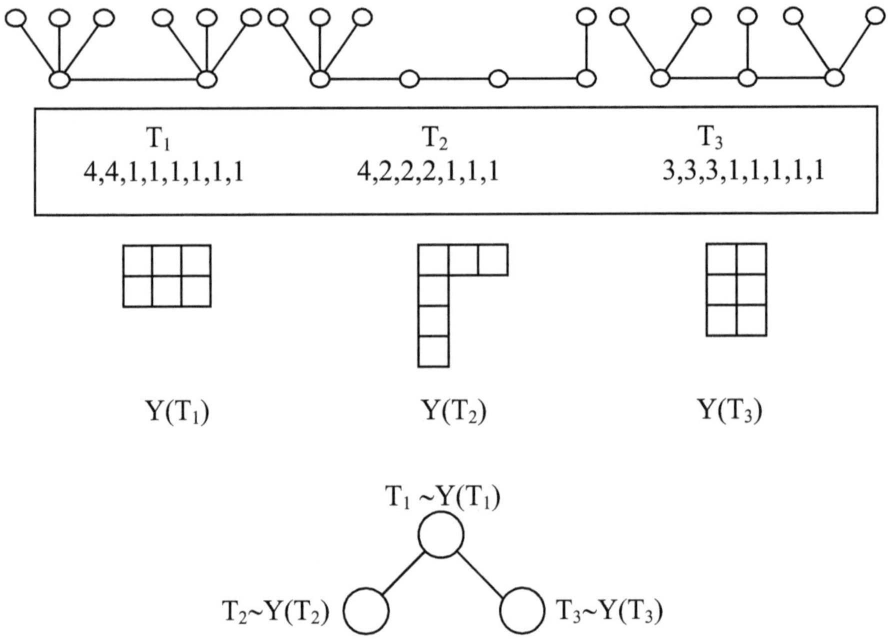

Fig. 9. A set of trees $\{T_1, T_2, T_3\}$ and the corresponding set of Young diagrams $\{Y (T_1), Y (T_2), Y (T_3)\}$, their valence sums and partial sums leading to a poset with one bifurcation which defines a non-comparable pair $\{T_2, T_3\}$, cf. eqn. (5)

The valences of these trees are listed in a non-increasing way together with the corresponding partial sums. When rules of Muirhead (eqn. 4) are applied to these partial sums one obtains the poset shown in Fig. 10. We observe that T_1 majorizes (i.e., more important than ≡ dominates) T_2 and T_3 but T_2 and T_3 cannot be ordered: Muirhead's theory describes T_2 and T_3 (or their corresponding Young diagrams) as being incomparable. Pairs of incomparable objects generate sites of bifurcations.

In Fig. 10 we show the ordering of the set of Young diagrams on 6 boxes according to the rules of Ruch (1975). The corresponding tree graphs are also shown (see also chapter by Seitz p. 367, where a set of Young diagrams with 10 boxes is represented and discussed with respect to complexity measures).

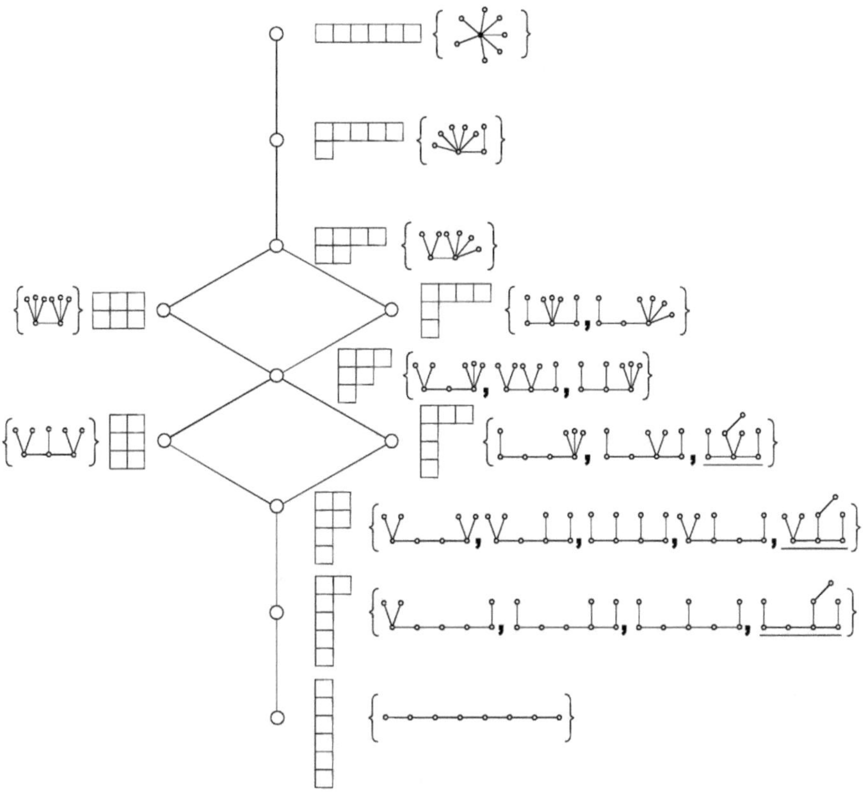

Fig. 10. Ordering of the set of Young diagrams containing 6 boxes. The corresponding tree graphs are shown. Underlined graphs are non-caterpillar trees

The Fig. 10 illustrates how the ordering theory of Ruch (1975) coincides with that of Gutman and Randić (1977) (See also Fig. 12).

The overlap between the ordering schemes of Ruch and that of Gutman & Randić

These two ordering schemes may be made to overlap (i.e., generate the same poset) for a set of trees containing (n = 2) vertices and a set of Young diagrams containing n boxes as follows:

a) Suppress information on terminal vertices (of valence = 1)
b) Reduce valence of each vertex by one
c) The resulting sequence of integers (from left to right) represents rows of boxes from top to bottom.

Example:
T_2 (Fig. 9) generates the following sequence of integers representing valences of vertices arranged in a non-descending order:

4,2,2,2,1,1,1 (8)

We adopt steps a-c:

a) Suppressing information of terminal vertices leads to the sequence:

4,2,2,2 (9)

b) Reduction of valence of each vertex by one leads to:

3,1,1,1 (10)

c) The above sequence corresponds to the following Young diagram Fig. 11.

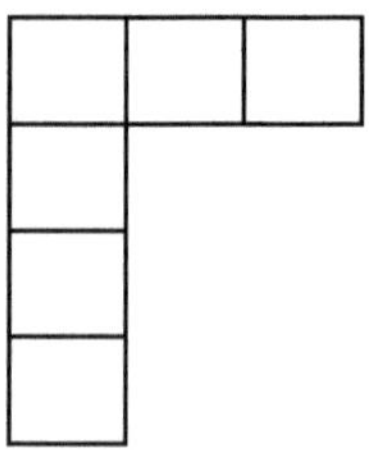

$= Y(T_2)$

Fig. 11. A specific Young diagram, corresponding to the sequence 3,1,1,1

Correlation of Young Diagrams with Alkanes and benzenoid Hydrocarbons

A remarkable type of tree graph is called a Caterpillar El-Basil (1987) (or a caterpillar tree): P_j (m_1, m_2, … , m_j) which is obtained by the addition of m_1 monovalent vertices to the first vertex υ_1 of path P_j , m_2 monovalent vertices to υ_2 of P_j and so on. The three tree graphs shown in Fig. 10 are all caterpillar trees and may be designated respectively as:

P_2 (3,3) ; P_4 (3,0,0,1) ; P_3 (2,1,2)

An example of a non-caterpillar tree is shown in Fig. 12.

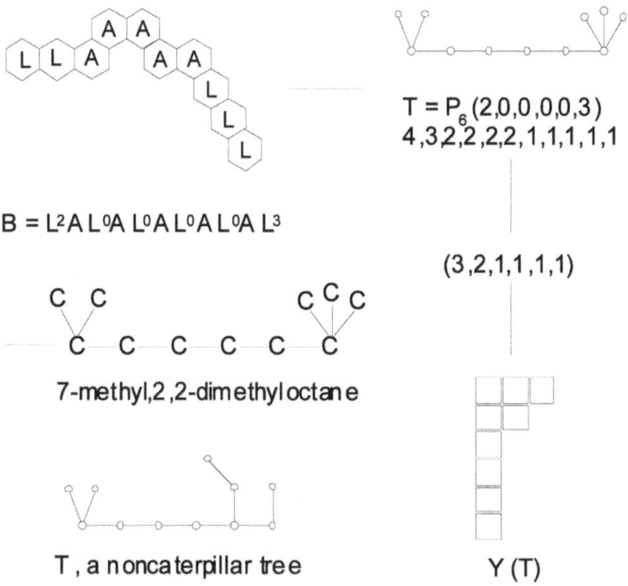

Fig. 12. An unbranched benzenoid hydrocarbon, the corresponding caterpillar tree T, alkane hydrocarbon skeleton and Young diagram. T' is a non-caterpillar tree

Caterpillar trees are related to other combinatorial objects of chemistry and physics (such as rook boards, Clar graphs, and King polyomino graphs) (El-Basil, Randić (1992) but most importantly, caterpillar trees represent in fact unbranched catacondensed benzenoid hydrocarbons (El-Basil 1987), El-Basil, Randić 1992). To envisage this important connection we distinguish two types of annellation of hexagons (Cyvin, Gutman (1988), viz., linear, L, and angular, A, modes (Fig. 13):

L A

Fig. 13. Linear and angular modes of annelating hexagons

An unbranched benzenoid may thus be "coded" by its LA – sequence written, say, from left to right, as LA – units, viz.

$$L^{m_1} A L^{m_2} A \ldots L^{j}$$

The corresponding caterpillar tree is composed by the addition of m_1 monovalent vertices to υ_1, m_2 monovalent vertices to υ_2 , … , j monovalent vertices to $j^{\underline{th}}$ vertex of path P_j (on j vertices).

In Fig. 12 we illustrate these concepts.

Ordering of Unbranched Benzenoid Hydrocarbons

We have seen in the previous section (cf. Fig. 12) that a caterpillar tree can be made to overlap with an unbranched catacondensed benzenoid hydrocarbon. I.e., the modes of hexagon annellation are, in fact, "stored" so-to-speak in the distribution of the terminal vertices of a caterpillar tree. One is, then, tempted to go back to posets such as the one shown in Fig. 10 and replaces the caterpillar trees by their corresponding benzenoids. In this way a set of benzenoid hydrocarbons has been partially ordered according to the theory of Ruch (1975), using Muirhead's rules (Muirhead (1903) (eqn. 4) or equivalently according to the scheme of Gutman and Randić (1978) using valences of vertices of tree graphs as input for eqn. (4). The question now becomes: does the resulting (purely structural) partial ordering reflect the (chemical) properties of benzenoids? The answer is quite encouraging: In Fig. 12 the corresponding benzenoids are represented as their respective LA-sequences and their stabilities are measured by the set of Herndon's permutation integrals (Herndorn, Ellzey Jr (1974) (γ_1, γ_2, γ_3, γ_4) where γ_i involves permutation of (4i+2) π-electrons. Observing that twice these integrals are numbers of conjugated circuits (cf. Randić (1977), listed, respectively, as (R_1, R_2, R_3, R_4), we can use this poset to order a set of hydrocarbons according to their stability.

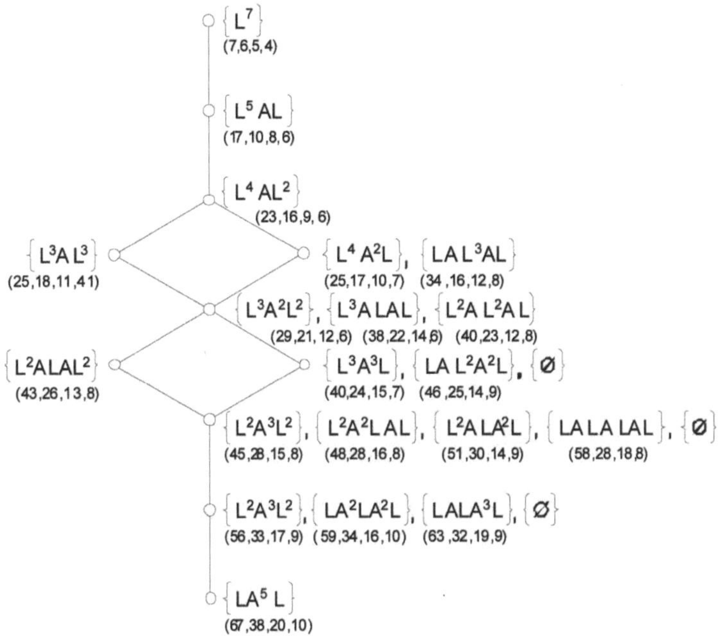

Fig. 14. Partial ordering of the set of unbranched benzenoids containing 7 hexagons. Each benzenoid is coded by its L-A sequence of hexagon annelations

In Figure 14, numbers in parentheses are permutation integrals (Herndorn, Ellzey Jr (1974). The poset is isomorphic with the one shown in Fig. 10. Chemical stability goes up as one goes down along the poset.

In the present case, stability increases as one goes down the edges of the poset. The limits are defined as $\{L^7\}$ and $\{LA^5L\}$ representing a linear acene (heptacene) and a single zigzag chain, all- benzenoid, system. Linear acenes are known to be coloured unstable hydrocarbons while angular annellations of hexagons leads to colourless stable systems (Clar 1972). In Fig. 15 this situation is illustrated with a few examples of unbranched benzenoids for which UV data are available (Clar 1972), which serves to illustrate general features.

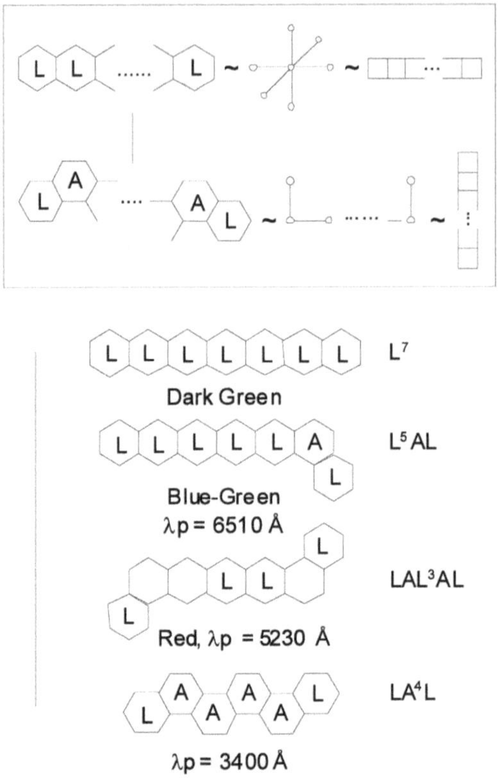

Fig. 15. Limits of stability of a set of catacondensed benzenoids

In Fig. 15 the linear acene represents the most unstable system while the all-kinked acene being the most stable. These limits may be modelled respectively with a star tree (or a row of boxes) and a path (or a column of boxes). The para bands of UV spectra are indicated for some cases for which data are available (Clar (1972). It is interesting to observe that sequences of numbers which represent (γ_1, γ_2, γ_3, γ_4) lead to bifurcations (i.e. incomparable pairs) when the sequences, which correspond to Young diagrams, are also incomparable! For example at the first bifurcation (Fig.'s 10, 12) one finds the following pair of sequences of γ's.

$$(25, 18, 11, 4)\,, (25, 17, 10, 7), \tag{11}$$

which leads to the following non-comparable partial sums:

$$(25, 43, 54, 58)\,, (25, 42, 52, 59) \tag{12}$$

i.e., $54 > 52$ while $58 < 59$ leading to a bifurcation!

An observation regarding Young diagrams and tree graphs
While there is a unique Young diagram for every tree graph, the opposite is not true, viz., several trees may occupy the same position characterizing a single Young diagram on a given poset. Namely, several trees lead to the same partition of vertex-valences and whence the same Young diagram. As an illustration one may observe in Fig. 10, that each of position 4, 6, 7, 9, 10 of the poset shown characterizes a single Young diagram but several tree graphs! Take, e.g. the sequence of vertex-degree 3,3,2,2,1,1,1,1, then, the ordering rules of Gutman and Randić (1977) lead to rows of boxes (from top to bottom) of lengths 2,2,1,1 which define a unique Young diagram, but the vertex-degrees in this case generate four caterpillar and one non-caterpillar trees. (position 9 of the poset shown in Fig. 10).

Grid Graphs Based on Molecular Path Codes of Lengths 2 and 3: Relation to Ordering of Young Diagrams

In a series of publications Randić et al. (Randić, Wilkins (1979) generated grid graphs of molecular graphs of classes of compounds based on their path codes of lengths 2 and 3. Such periodic tables are reminiscent of the Hasse diagrams of partial orderings and may be viewed as multiposets, the nodes of which represent the partial ordering of a given property. Several properties were studied, which include enthalpies, heat capacities, critical volumes, index of refraction, entropy changes and several others.

Here, we observe how these grids are related to Young diagrams. As an example we consider in Fig. 14, the diagram that shows positions of the set of octane isomers in the coordinate system (P_2, P_3). We also indicate the corresponding Young diagrams associated with the tree graphs representing the octanes. Interestingly the resulting grid successfully orders subsets of Young diagrams in accord with rules of Ruch (1975) as well as the scheme of Gutman and Randić.

Example (Fig. 16):

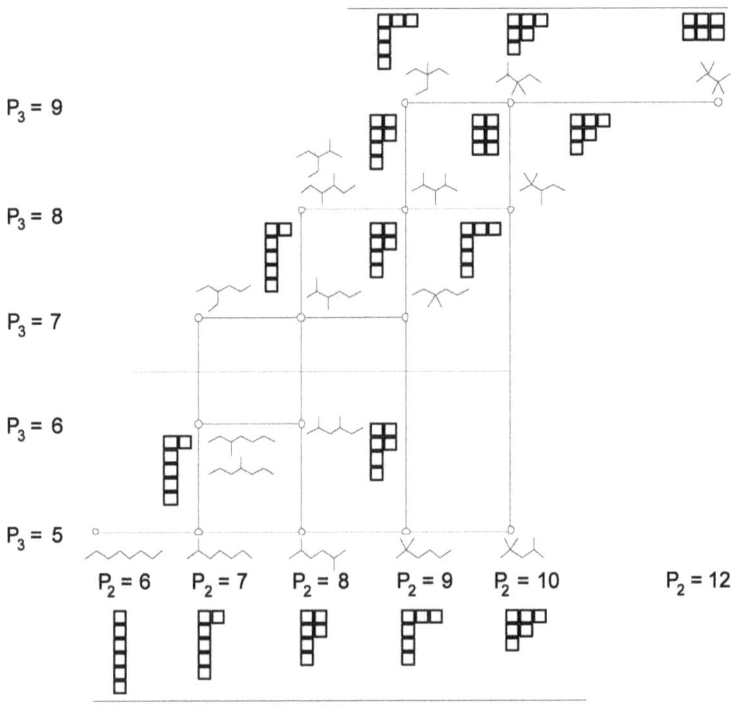

Fig. 16. The diagram showing positions of various octane isomers in the coordinate system (P_2, P_3)

In Fig. 16 the corresponding Young diagrams are correctly ordered in horizontal lines (dashed arrows) in accord with rules of Ruch (1975) as well as Gutman and Randić (1977).

References

Beckenbach EF, Bellmann R (1961) Inequalities, Ergebnisse der Mathematik und Ihre Grenzgebiete. Neue Folge, Heft 30, Springer Berlin
see also:
Randić M (1978) On Comparability of Structures. Chem Phys Lett 55:547-551

Clar E (1972) The Aromatic Sextet. John Wiely & sons, London

Coleman AJ (1968) The Symmetric Group Made Easy, in Advances in Quantum Chemistry 4. Ed. Löwdin PO, Academic Press Inc New York p 86

Cyvin SJ, Gutman I (1988) Kekulé Structures in Benzenoid Hydrocarbons, Lecture Notes in Chemistry 46. Springer-Verlag Berlin

El-Basil S (1987) Applications of Caterpillar Trees in Chemistry and Physics. J Math Chem 1:153-174
See also:
El-Basil S (1986) Gutman Trees, Combinatorial-Recursive Relations of counting Polynomial: Data Reduction using Chemical Graphs. J Chem Soc Faraday Trans 2, 82:299-316
El-Basil S (1990) Caterpillar (Gutman) Trees in Chemical Graph Theory. Topics in Current Chemistry, Springer-Verlag Berlin pp 273-289

El-Basil S (1993) Graph Generation: Quasicrystal-like Benzenoid System. J Am Chem Soc Faraday Trans 89(6):909-920
see also:
El-Basil S (1982) On a Novel Graph-Theoretical Basis of Ordering Kekulé Structures. Mat Ch 13:183-197
El-Basil S (1993) Generation of Lattice Graphs: An Equivalence Relation on Kekulé Counts of Catacondensed Benzenoid Hydrocarbons. J Mol Struct (Theochem) 288:67-84
El-Basil S (1993) Kekulé Structures as Graph Generators. J Math Chem 14:305-318
El-Basil S (2000) Benzenoid Hydrocarbons and Modern Frontiers of Science. Mat Ch 42:233-259
El-Basil S, Botros S, Ismail M (1985) A Revisit to the Relative Importance of Kekulé Structures; On a Novel non-numerical Ordering Scheme. Mat Ch 17:45-74

El-Basil S, Randić M (1992) Equivalence of Mathematical Objects of Interest in Chemistry and Physics. Adv Quant Chem 24:239-290

Graovac A, Gutman I, Randić M, Trinajstić N (1973) Kekulé Index for Valence Bond Structures of Conjugated Polycyclic System. J Am Chem Soc 95:6267-6273

Gutman I, Randić M (1977) Algebraic Characterization of Skeletal Branching. Chem Phys Lett 47:15-19

Harary F (1972) Graph Theory. Addison-Wesley Publishing Company, Reading, Massachusetts, Chapter 4

Herndorn WC, Ellzey Jr, ML (1974) Resonance Theory; Resonance Energies of Benzenoid and Nonbenzenoid Systems. J Am Chem Soc 96:6631-6642

Luther B, Brüggemann R, Pudenz S (2000) An Approach to Combine Cluster Analysis with Order Theoretical Tools in Problems of Environmental Pollution. Mat Ch 42:119-143 and references cited therein

Muirhead RF (1903) Some methods applicable to identities and inequalities of symmetric algebraic function of n letters. Proc Edinburg Math Soc 21:144-157

Randić M (1977) A graph Theoretical Approach to Conjugation and Resonance Energies of Hydrocarbons. Tetrahedron 33:1905-1920

Randić M (1977) Aromaticity and Conjugation. J Am Chem Soc 99:444-450

Randić M, Trinajstić N, Knop JV, Jericević Z (1985) Aromatic Stability of Heterocyclic Conjugated Systems. J Am Chem Soc 107:849-859

Randić M, Wilkins CL (1979) Graph Theoretical Ordering of Structures as a Basis for Systematic Searches for Regularities in Molecular Data. J Phys Chem 83:1525-1540
see also:
Randić M, Trinajstić N (1982) On Isomeric Variations in Decanes. Mat Ch 13:271-290
Randić M, Wilkins CL (1979) On a Graph-Theoretical Basis for the Ordering of Structures. Chem Phys Lett 63:332-336
Randić M, Wilkins CL (1980) Graph Theoretical Analysis of Molecular Properties: Isomeric Variation in nonanes. Int J Quant Chem 18:1005-1027

Rosen KH (1991) Discrete Mathematics and its Applications. McGraw-Hill, Inc, New York, p 391

Ruch E (1975) The Diagram Lattice as Structural Principle. Theor Chim Acta 38:167-183

Hasse Diagrams and their Relation to Molecular Periodicity

Ray Hefferlin

Southern Adventist University, Collegedale, Tennessee 37315, USA

e-mail: hefferln@southern.edu

Abstract

Hasse diagrams are applied to molecules and radiation phenomena. Then the relation of these diagrams to periodicity in atoms is noted. The possibility is raised that Hasse diagrams can also be related to the growing body of evidence that periodicity exists in molecules with two, three, and four atoms; in binary inorganic molecules; and in some organic molecules.

Individual molecules

The definition and partially ordered set theory behind the Hasse diagram are given in the chapter by El-Basil, p. 3 and are not repeated; we proceed immediately to discuss some interesting Hasse diagrams.

Would it be possible to construct a molecule or a decay scheme that is its own Hasse diagram? Fig. 1 shows such a hypothetical species. The first number in each box is the period number in which the atom is found; the second is the number of valence electrons. The lone pair on carbon is not shown and the left-hand line is actually a double bond. Fig. 2 shows such a possible atomic decay scheme; similar radiation pathways occur in molecules and also in nuclei.

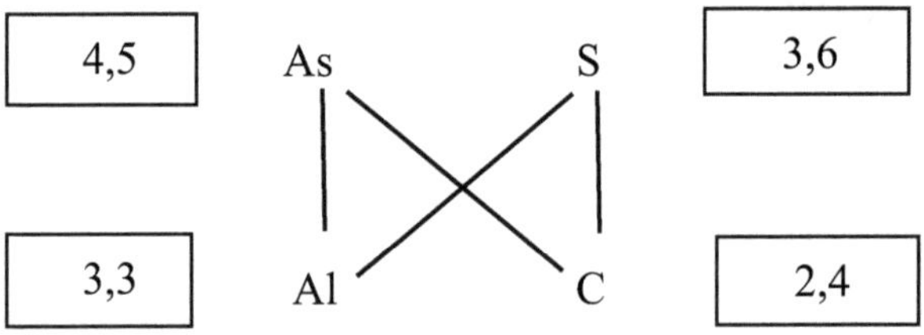

Fig. 1. The hypothetical molecule serves as its own Hasse diagram

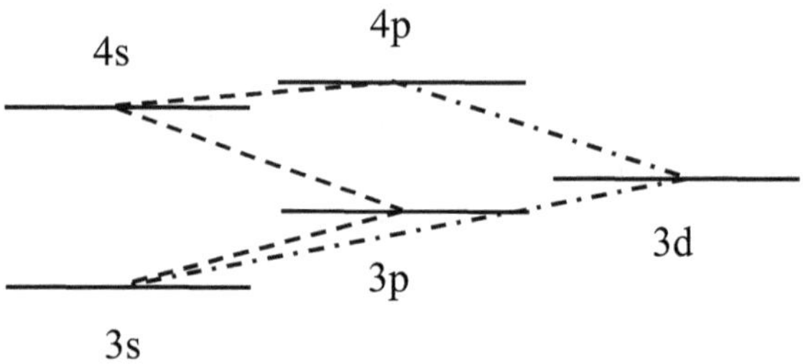

Fig. 2. A possible atomic radiation scheme is shown

Inorganic molecules

Klein has already represented the chart of the elements as the Hasse diagram of a multi-poset (Klein 1995). Over many decades the periodicity of many properties of main-group and transition metal diatomic molecules, in various phases, and main-group anions has been exhaustively documented (Clark 1935; Cornish 1959; Hefferlin et al. 1979; Hefferlin and Kutzner 1981; Kong 1982; Hefferlin, 1989; Boldyrev et al. 1994). Merging these two trains of thought make it seem reasonable that a somewhat more complex multiposet should pertain to them. The same is true for main-group triatomic and tetra-atomic molecules, since their periodicity has been demonstrated (Kong 1989, 1993).

A massive research project on the periodic properties of molecules has been pursued by generations of Russian chemists during the Soviet period and, to a lesser extent, subsequently. This research was to a large degree led by Shchukarev of Leningrad State University. He considered a general periodic system as being one immense "supermatrix" connecting innumerable "matrices" of elemental atomic states, of the compounds that they may form, of their properties, of their functional dependencies on the external conditions, and so on. The term "supermatrix" was used to represent the chemical element periodic chart and the immense molecular spaces accessed by stepping through any one or more of its compartments. The word "matrices" was also used loosely, referring in many cases to graphical representations of data; "databases" and "subsystems" would do as well.

The transformation from one matrix to any other, within the supermatrix, was to be accomplished with advanced mathematics such as group theory. Knowing that such transformations were beyond his capability at the time, Shchukarev and his colleagues collected data in preparation for the day when the capability would materialize. The data are represented in scores of graphs such as Fig. 3.

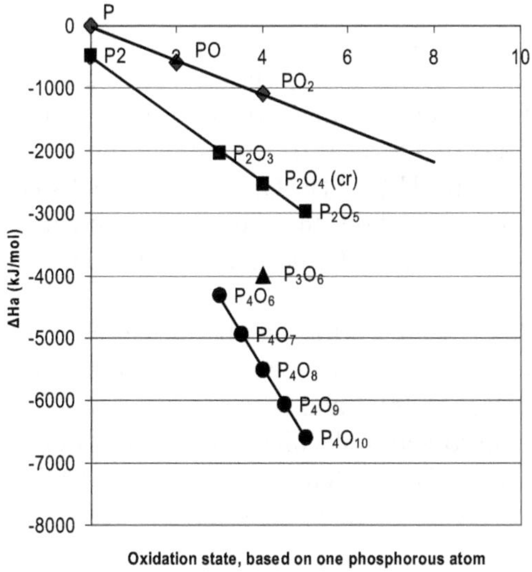

Fig. 3. Heats of atomization for oxides of phosphorus plotted against the oxidation state, normalized to one P atom, of phosphorus. The graphs as plotted by Shchukarev also normalize the data vertically to one P atom

The vertical axis shows the heats of formation (Gurvich 1978, 1979, 1981, 1982) of the molecules from free atoms; all of the molecules are in the gas phase (except for P_2O_4, which is in the crystal phase). The Hasse diagram shown in Fig. 4 represents the upper half of Fig. 3; it is easily seen how the lower half would be appended.

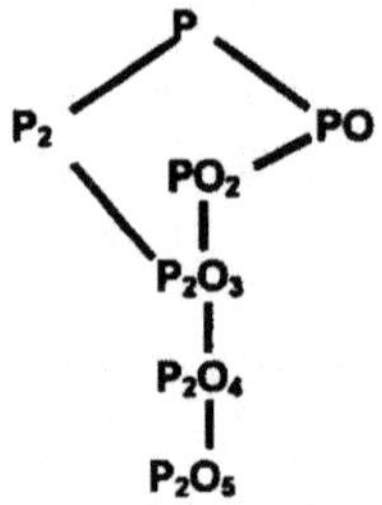

Fig. 4. An organization of phosphorus and its oxides up to P_2O_5 (higher oxides exist). Each line indicates the addition of one or more atoms to the species at its beginning

Molecular periodicity is clearly evident in perusing the graphs in Shchukarev's books (1970 and 1974); they contains graphs of ΔH_a (heat of atomization) like Fig. 3 (except for vertical normalization) for group 1 oxides (O), Mg and Ca (O), 3 to 12 (O), 13 (halides and H), 14 (O and H), 15 (O and H), 16 (O, S, and H), 17 (F, O, and H), and 18 (F). These graphs are amazingly similar, and show that heats of atomization of binary and ternary inorganic species echo the periodicity of atoms in the Mendeleev chart.

This extent of this research effort can be measured by noting that it includes work on how molecules are changed by going into water solution (Latysheva and Hefferlin 2004) and how the water solution is changed by the presence of the dissolved molecules (Lilitch and Mogilev 1954; Lilitch 1964; Burkov et al. 1977; Lilitch and Chernykh 1977; Khripun et al. 1983). A detailed review is given by Latysheva (1998).

Organic molecules

Morozov explored periodicity among alkanes nearly a century ago (Morozov 1907). Beautiful periodic tables have been constructed for polycyclic aromatic hydrocarbons by Dias (1996), for acyclic hydrocarbons by Bytautas et al (2000), and for fullerenes by Torrens (2004). So it is clear that periodicity has been observed in organic molecules.

Klein and Bytautas (2000) have given the Hasse diagram—technically, the partially ordered set reaction diagram—for halogenated benzenes. It starts with benzene at the top, concludes with fully chlorinated benzene at the bottom, and has the partially chlorobenzenes in between at ordinates approximating the boiling point scale. An almost identical diagram, with boiling points that progress in a similar way, pertains to methylbenzenes. From the point of view of periodicity, it is significant that the methyl functional group is isovalent to the halides (Haas 1982).

The most extreme diagram that comes to mind is one for polychlorinated biphenyls. Presumably it would have 1,1-biphenyl at the top, decachloro 1,1-biphenyl at the bottom, and the 207 remaining isomers between, ranked according to data for some property. Unfortunately, there are nowhere near enough data to begin the diagram now. There are, however, enough data to suggest periodicity in halogenated biphenyls (Fig. 5). The suggestion would be reinforced if the point for 4-fluoro 1,1'-biphenyl is shown to be in error to be at about 540K.

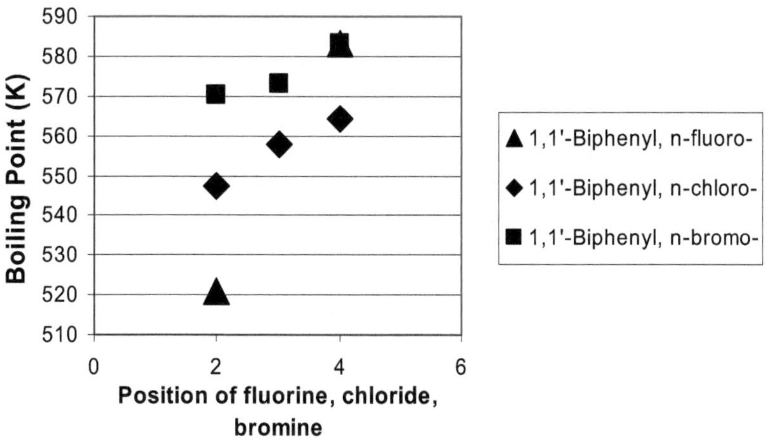

Fig. 5. Boiling points of 2-, 3-, and 4- fluoro-, chloro-, and bromo- 1,1'-biphenyls. The x = 2 (F,Cl,Br) arrangement, which would exist also at x = 4 if one point were moved, suggests periodicity

References

Boldyrev AI, Gonzales N, Simons J (1994) Periodicity and Peculiarity in 120 First- and Second-Row Diatomic Molecules. J Phys Chem 98:9931-9944
Burkov KA, Busko EA, Lilitch LS (1977) Termodinamicheskie kharakteristiki reaktzii gidroliza i obrazovaniya gidrokompleksov. Khimiya i termodinamika rastvorov, Leningrad State University, St. Petersburg, Russia pp 15-43
Bytautas L, Klein DJ, and Schmalz TG (2000) All Acyclic Hydrocarbons: Formula Periodic Tables and Property Overlap Plots vs. Chemical Combinatorics. New J Chem 24:329-336
Clark CHD (1935) The Periodic Groups of Non-hydride Di-atoms. Trans Faraday Soc 31:1017-1036
Cornish AJ (1959) Arrays of Inorganic Semiconducting Compounds. J Electrochem Soc 106:685-689
Gurvich LV, editor (1978, 1979, 1981, 1982) Thermodinamicheskie svoista individual'nikh veschestv. Volumes 1-4 Nauka, Moscow
Dias JR (1996) Formula Periodic Tables—Their Construction and Related Symmetries. J Chem Inf Comput Sci 36:361-366
Haas A (1982) The Element Displacement Principle: A New Guide in p-block Element Chemistry. Adv Inorgan Chem Radiochem 28:167-202
Hefferlin R (1989) Periodic Systems and their Relation to the Systematic Analysis of Molecular Data. Edwin Mellen, Lewiston, New York
Hefferlin R, Campbell R, Gimbel D, Kuhlman H, Cayton T (1979) The Periodic Table of Diatomic Molecules—I. An Algorithm for Retrieval and Prediction of Spectrophysical Properties. J Quant Spectrosc Radiat Transfer 21:315-336
Hefferlin R, Kutzner M (1981) Systematics of Ground-State Potential Minima between two Main-Group Atoms or Ions. Astrophys J 75:1035-1036
Khripun MK, Lilitch LS, Efimov AYu, Bulgakov SA (1983) Rasvitie strukturnodinamicheskikh predstavlenii o koncentrirovannykh rastvorakh elektrolitov. Problemy sovremennoi khimii koordinatzionnykh soedinenii, Leningrad State University, St. Petersburg, Russia pp 58-101
Klein DJ (1995) Similarity and Dissimilarity in Posets. J Math Chem 18:321-348
Klein DJ, Babic D (1997) Partial Orderings in Chemistry. J Chem Inf Comp Sci 37:656-671
Klein DJ, Bytautas L (2000) Directed Reaction Graphs as Posets. Comm Math Comput Chem 42:261-290
Kong FA (1982) The Periodicity of Diatomic Molecules. J Mol Struct THEOCHEM 90:17-28
Kong FA (1989) An Alternative Periodic Table for Triatomic Molecules. In: Hefferlin R Periodic Systems and their Relation to the Systematic Analysis of Molecular Data. Edwin Mellen, Lewiston, New York
Kong FA (1993) unpublished communication
Latysheva VA (1998) Vodno-solevye rastvory. Systemnyi podkhod. Saint Petersburg University, Saint Petersburg, Russia p 342

Latysheva VA, Hefferlin R (2004) Periodic Systems of Molecules as Elements of Shchukarev's "Supermatrix," i.e. the Chemical Element Periodic System. J Chem Inf Comput Sci 44:1202-1209

Lilitch LS (1964) Periodicheskii zakon i svoistva rastvorov. Khimiya i termodinamika rastvorov. Leningrad State University pp 5-13

Lilitch LS, Chernykh LV (1977) Zakonomernosti ismeneniya davleniya parov vody v binarnykh rastovrakh elektrolitov. Khimiya i termodinamika rastvorov, Leningrad State University, St. Petersburg, Russia pp 43-54

Morozov N (1907) Periodicheskia sistemy. Stroenia Veshchestva. I. D. Svitina, Moscow

Shchukarev SA (1970, 1974) Neorganicheskaya khimia. Vol 1, 2. Vyshchaya shkola, Moscow

Torrens F (2004) Table of Periodic Properties of Fullerenes Based on Structural Parameters. J Chem Inf Comput Sci 44:60-67

Directed Reaction Graphs as Posets

D. J. Klein and T. Ivanciuc

Texas A&M University @ Galveston, Galveston, Texas 77553-1675

e-mail: kleind@tamug.tamu.edu

Abstract

Reaction diagrams are considered especially for the circumstance of progressive substitution (or addition) on a fixed molecular skeleton, and it is noted that these naturally form Hasse diagrams for a partially ordered set (or poset) of the substituted structures. The possibility that different properties are similarly ordered is a further natural consideration, and is here illustrated for several different properties for (methyl & chloro) substituted benzenes.

This posetic approach thence provides a novel approach to structure/property and structure/bioactivity correlations, with focus in some sense beyond simple molecular structure, in that this approach attends to how a structure fits into a systematic (reaction) network of structures. Different manners for fitting and prediction of properties are noted, with illustration of an especially simple "poset-average" scheme. Some numerical evidence indicates that such approaches are quite reasonable. It is emphasized that such directed reaction graphs admitting posetic treatment are widespread.

Introduction

Reaction graphs occur throughout chemistry. For instance, directed reaction graphs occur in (directed) syntheses, as reviewed in Corey & Cheng's (1989) seminal book 'The Logic of Chemical Synthesis'. The structures of synthesis graphs are crucial in differentiating between "linear" and "convergent" synthetic approaches (Hendrickson 1977), though there has been only a little formal graph-theoretic work (e.g., in Hendrickson 1977, and

Bertz 1984, 1986) on such graphs, so that there could perhaps be some further investigation incorporating more fully the directed aspect of their edges (and consequent possibility for being interpretable as posets). In many cases it seems that these synthesis graphs are simply linear chains or nearly so, whence the posetic structure is especially simple (a total order). The seemingly less common synthesis graphs for so-called "convergent" syntheses (Hendrickson 1977, Bertz 1984, and Bertz 1986) feature branching, but still very often these graphs are trees (even when the directions are eliminated from the edges), though this is not a logical necessity. Another commonly occurring and important type of directed reaction graph is found in molecular biological applications, often showing cycles (or "hypercycles") say in the area of "enzyme kinetics", again with directions on the edges. These reaction schemes have been much considered, even in a formal sense, e.g., as in Eigen (1971, 1977) and Hill (1977). But in these cases a posetic interpretation is complicated by the general occurrence of cycles. Reaction graphs for degenerate rearrangements have been considered in a general graph-theoretic network (as reviewed by Balaban 1994), though here not only are there cycles, but the edges are best viewed as undirected. General complex sets of chemical reactions (as considered by Temkin et al. 1996) typically similarly exhibit cycles.

Still there is a fairly general class of directed reaction networks which can be neatly viewed as partially ordered sets, or posets. That is, in some cases there is an intrinsic natural order, say as for the possible results of substitutional chlorination of benzene, as illustrated in Fig. 1. There only the hexagon of carbons is shown, and the Cl-substituted carbon vertices are shown as (larger) black dots. An arrow is directed from one structure α to a second β, if β can be obtained from α by the replacement of one H-atom by one Cl-atom (without moving around any other Cl-atoms which might already be attached). Note that in general not all the n-substituted isomers so arise from a particular isomer with $n-1$ substituents. Thence there is more information than just the degree of substitution as with para-dichloro-benzene which gives only one of the three trichloro-benzenes. The arrow in these diagrams represents a single minimal step of chlorination. And the diagram as a whole is (Klein and Bytautas 2000) the Hasse diagram associated to a poset of different chloro-benzenes. The ordering relation $\alpha \succ \gamma$ then means that there is some non-negative number of Cl-atoms which can be (substitutionally) added to structure α so as to obtain structure γ.

The general class of posetic reaction diagrams are (Klein and Bytautas 2000) then those for which there is a progressive degree of reaction, via substitution, addition, dissociation, or local rearrangement. The progressive substitution as in Fig. 1 is just one example. The possibility for the use of such posetic networks in structure/property correlations is a natural consideration, and is illustrated here.

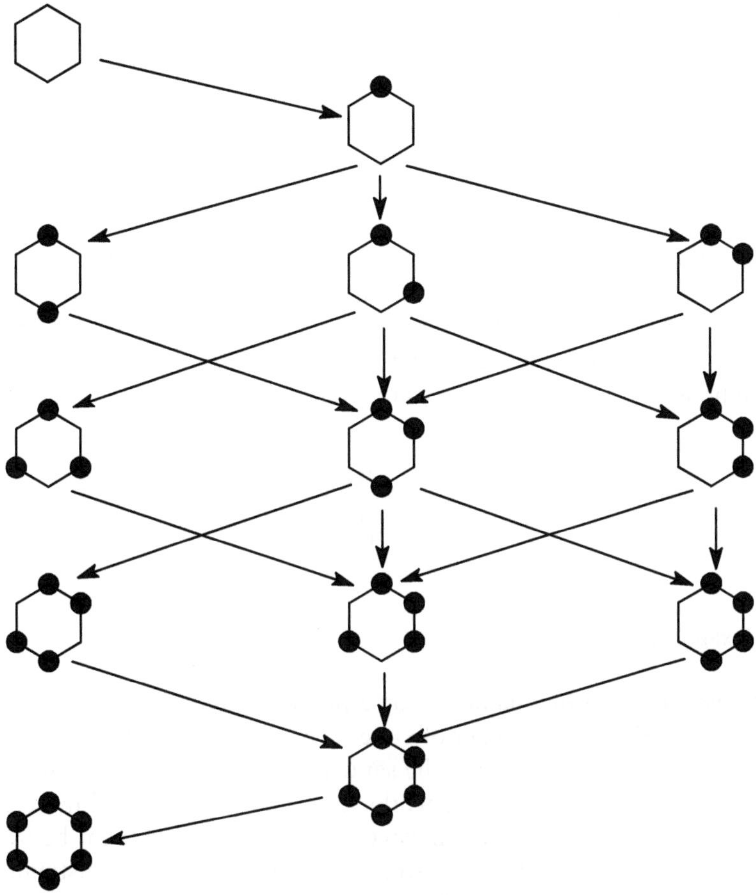

Fig. 1. The posetic reaction diagram for successive chlorination of the benzene skeleton, with black dots identifying Cl-substituted sites

Posetic Substitution Reaction Networks & Chemical Structure

An interesting question concerns how much the posetic diagram might determine about the molecular structure of the species involved. In particular, what does the substitution reaction diagram for benzene imply about the skeleton being a regular hexagon? In fact the overall question of the structure of benzene was of great historical relevance, with Kekulé arguing (1865, 1866, and 1872) for a regular hexagonal skeleton, as based on several points including:

- first, the atom-count formula;
- second, the various possible isomeric substituents; and
- third, substitution reaction graphs ordered as in our posetic reaction diagrams.

Particularly Körner in 1874 investigated permutational questions for the case of benzene using (at least implicitly) substitution reaction diagrams with an effort to avoid inadvertent geometrical assumptions. As an alternative to the hexagonal structure for benzene Ladenburg (1869) proposed a triagonal prismatic structure, as indicated in Fig. 2.

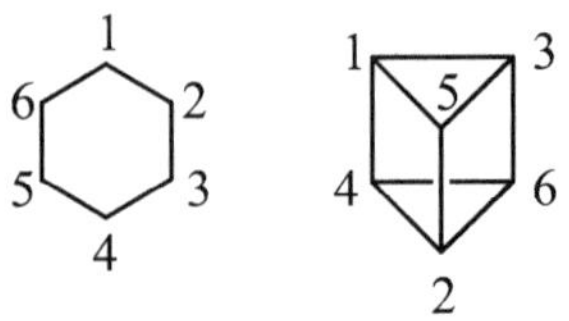

Fig. 2. Hexagonal and prismatic benzene

There we indicate the "natural" corresponding numbering for the two C_6H_6 skeletons: that is, an isomer of either skeleton with n substituents at a given common set S of sites gives the same n+1 substituted sets S_+ of sites. Thus their substitution reaction networks are isomorphic, and all these criteria do not distinguish Kekulé's hexagonal skeletal structure for benzene from that of Ladenburg's (1869) prismane. At least they do not distinguish the two so long as the chirality of some of the substituted prismane structures is ignored – the chirality of molecular structures only being proposed a few years later by Van't Hoff (1874) & LeBel (1874), and then for a slightly different circumstance for which there was already evidence in hand. Also to distinguish the two choices it does not help to further consider multiple substitutions (of some other substituent in addition to Cl). Evidently further criteria to choose the structure needed to be invoked, and were, involving a sort of chemical consistency for a wide range of chemical phenomena, with geometric considerations entering in different subtle manners. Of course there were complications: the absence of two ortho disubstituted derivatives (across single or double bonds), often imagined to be accounted for by oscillations between the two classical benzene structures; the fact that the properties of benzene were not like that of (a mixture of two) conformations with localized double bonds; and the difficulties with the (Claus 1866) centric formula in that reliance solely on the connectivity pattern would indicate that ortho & para disubstituted derivatives should be equivalent. The fully unambiguous verification of the hexagonal skeleton would await the development of x-ray diffraction, and a full understanding of the behaviour would await quantum mechanical insight. This is discussed to some degree elsewhere (Rocke 1985, Klein and Bytautas 2000).

Different examples of substitution (or other sorts of progressive) reaction networks appear occasionally in the literature. Though the benzene-substitution network has a long history, as already noted, it is occasionally rediscovered (e.g., as in Dolfing and Harrison 1993). Another nice network is that for a skeleton of dimethyl- bicyclo[1.1.1]pentane-1,3-dicarboxylate as considered by Shtarev et al. (2001). Some networks can be quite complex, as must be the one for the addition of H-atoms to buckminsterfullerene, where even when only the substitution patterns admitting a Kekulé structure are retained (as in Babić et al. 2004) there remain $\sim 10^{14}$ members in the consequent network. Several further examples of different progressive reaction networks are given in Klein and Bytautas (2000). Beyond substitution and addition, such networks can also involve coagulation and other constructive processes. The posetic reaction networks need not correspond to typically realized reactions, but might be contemplated as an organizational principle. And the reaction networks need not be finite but can be without an end. For example one may consider the alkane poset for which the beginning part is shown in Fig. 3. Here the contemplated reaction is the replacement of an H-atom by a methyl $-CH_3$ group.

One sees that the structure of a posetic reaction diagram retains some information about molecular structure. Certainly the investigation of the information inherent in a reaction diagram should be developed in a theoretical format, and different possible uses should be explored. To recognize the attention beyond the local structure to placement in a whole network, one might then speak of quantitative super-structural/activity relationships, or QSSAR. Or in dealing with more conventional physical or chemical properties one could speak of QSSPR.

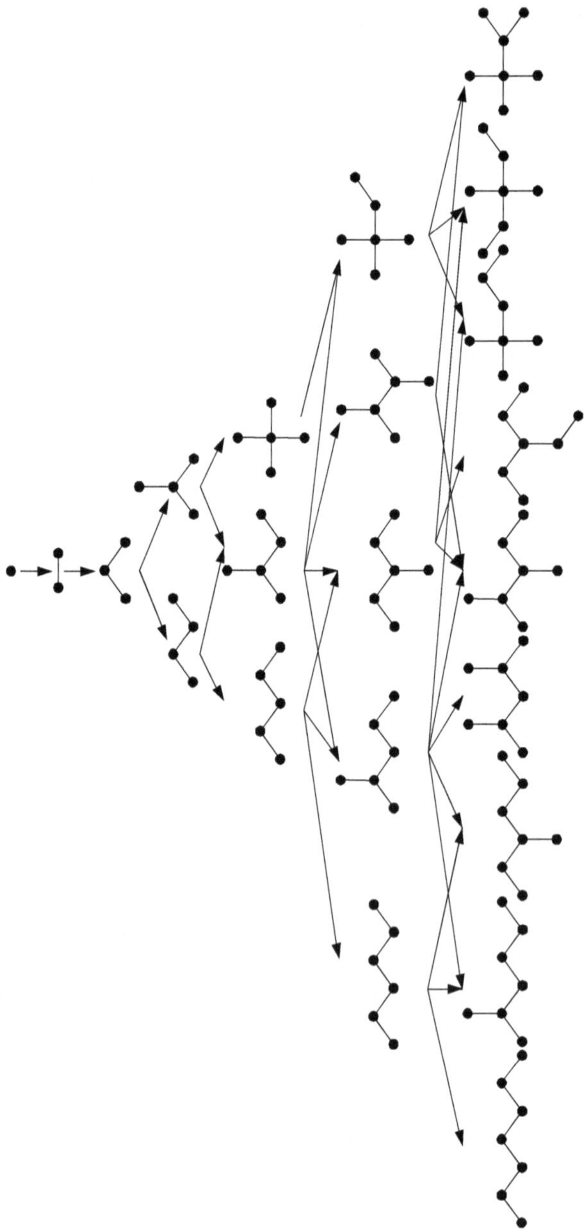

Fig. 3. The posetic reaction network for alkanes

Molecular Properties & the Reaction Poset

It is rather natural to consider that different chemical properties might be ordered in concert with the partial ordering of a progressive reaction poset. That is, for general structures A & B in the reaction poset such that B follows A, then property *X* might be ordered such that $X(A)<X(B)$. If this is true for every such pair, then the ordering for *X* is consonant with the partial ordering of the reaction network. Even if whenever B follows A, it turns out that $X(A)>X(B)$, the property ordering is still "consonant" with the partial ordering of the reaction network – the sense of the correspondence between the two orderings is just reversed. Such consonance is more technically termed isotonicity. This indeed does happen, e.g., as illustrated in Fig. 4, for toxicities of the chloro-benzenes, and in Fig. 5 for chromatographic retention indices (on squalene at 96°C) of methylbenzenes.

Beyond perfect isotonicity for a property *X* there is also the possibility of "imperfect isotonicity". Such sometimes occurs, e.g., as in Fig. 6, for density (in g/cm^3) of methylbenzenes. Here there are several neighbour pairs (indicated in the figure with dashed arrows) in the diagram such as to be ordered differently than the other neighbour pairs. Generally to each property one might associate an isotonicity score, defined as the ratio of the number of correctly ordered neighbour pairs to the total number of neighbour pairs. Thus for our benzene substitution poset, as of Fig. 1, one sees 20 neighbour pairs, and in Fig. 5 a perfect isotonicity score of 20/20 is attained, while for Fig. 6 with a misordering the (rather poor) score is 15/20 as indicated by the five broken lines.

In some cases it might be that property values for all species in a reaction network are not available. Even in such cases one can make an isotonicity score, just by taking into account neighbour pairs (or equivalently the number of covering relations) of species such that property values for both are known. Thus for instance, if the property value for the hexa-substituted species (ϕCl_6) were not available, then the remaining available number of neighbour pairs would be 19, as in Fig. 4. Thence the property score for this figure is seen to be 19/19.

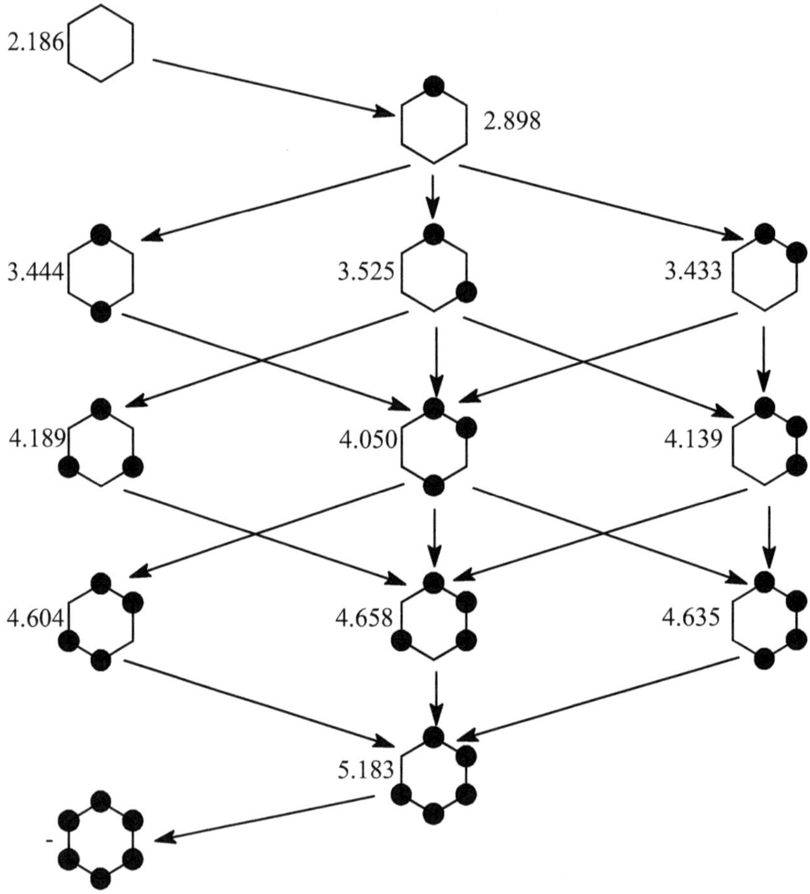

Fig. 4. The posetic presentation of acute aquatic toxicity to guppy (*Poecilia reticulata*) for chloro-benzenes (as $\log LC_{50}$, [μmol/l])

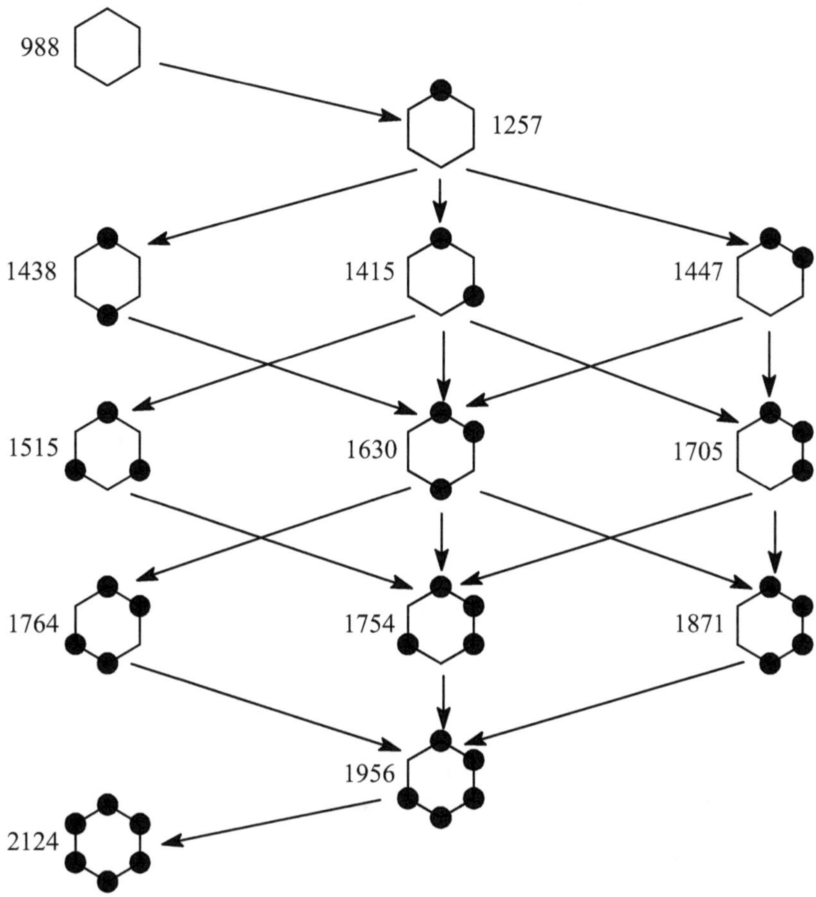

Fig. 5. The posetic presentation of chromatographic retention indices for chlorobenzenes (retention indices on CARBOWAX 20M column, polar stationary phase, at 140°C)

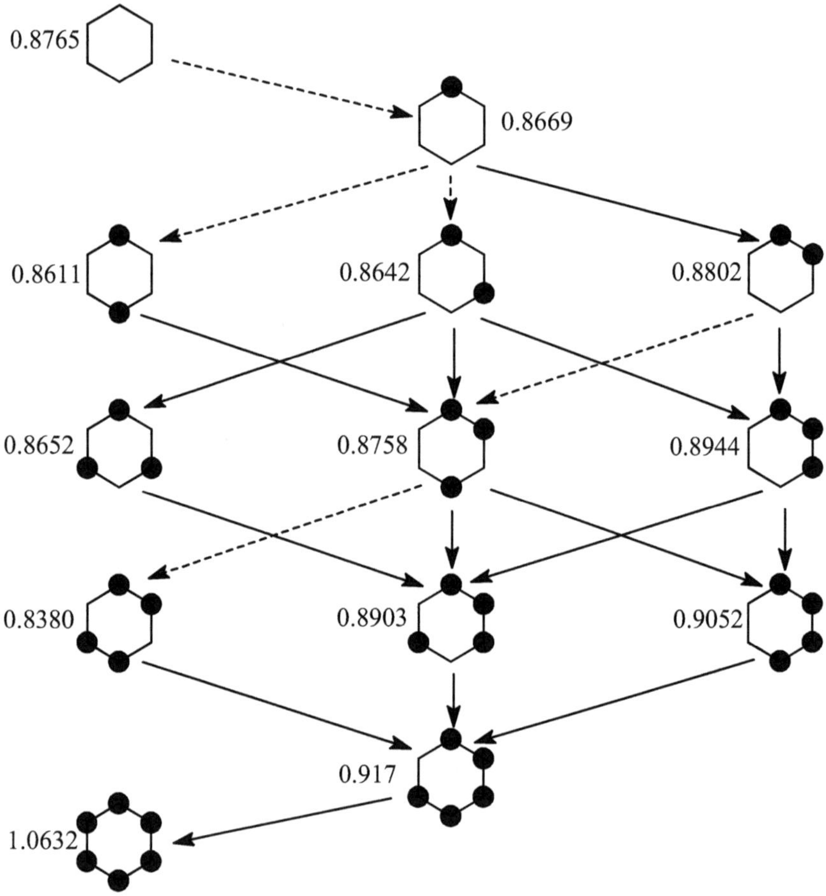

Fig. 6. The posetic presentation of densities (in g/cm^3) of the methylbenzenes. Dashed arrows indicate the incorrectly ordered pairs of the reaction poset

Our article (Ivanciuc and Klein 2004) reports an examination of the literature to find reasonably complete tabulations of over two dozen properties for chloro- and methylbenzenes.

Table 1. Properties and Their Isotonicity Scores for Chloro-Benzenes

No.	Property			Score ϕCl_n	Degree indifference	$\sigma_{\phi Cln}$
1	logP	partition coefficients octanol/water		20/20	6/6	0.089
					6/6	0.065
					6/6	0.142
2	RI	retention indices		20/20	6/6	23.299[a]
				20/20	6/6	42.949[b]
					6/6	69.299[c]
					6/6	16.627[d]
3	k	capacity factor		20/20		2.182[d]
						0.876[e]
4	t_R	retention time	min	20/20		1.064
				19/19	6/6	0.713
5	$\Delta_f H^o_{gas}$	enthalpy of formation of gas	[kJ/mol]	20/20	6/6	6.992
6	log S	water solubility	[mol/l]	20/20	6/6	0.438
					6/6	0.380
7	$\log V_p$	vapor pressure at 25°C	[Pa]	20/20	6/6	0.156
8	$\Delta_{vap} H^o$	enthalpy of vaporization	[kJ/mol]	20/20	6/6	0.534
					6/6	1.514
9	$\ln(\gamma M_s)$	molecular activities on SE-30 stationary phase		19/20	6/6	0.515
						0.289
10	ΔH_f	enthalpy of fusion	[kJ/mol]	17/20	3/6	3.020
11	T_f	temperature of fusion	[K]	16/20	4/6	47.770
12	ΔS_f	entropy of fusion	[J/molK]	15/20	4/6	3.120
13	log LC_{50}	aqueous toxicity for Poecilia reticulata		19/19	6/6	0.164
14	E_{2d}	reduction potential	[V]	19/19	6/6	0.017
15	$\ln\gamma^{\infty}$	infinite dilution activity coefficients in water at 298.15K		15/19	-	2.031
16	p	solute polarity parameter		16/16	-	0.101
17	IP	ionization potential	[eV]	12/16	-	0.087
18	ρ	density	[g/cm^3]	12/12	-	0.102
19	n_D	refractive index at 20°C		10/10	-	0.010
20	p_c	critical pressure	[bar]	9/9	-	0.357
21	T_c	critical temperature	[K]	8/9	-	11.079
22	V_c	critical volume	[ml/g]	8/9	-	36.53
23	ρ_c	critical density	[g/ml]	6/6	-	0.015
24	$\Delta_f H^o_{liquid}$	enthalpy of formation of liquid	[kJ/mol]	5/5	-	2.443

[a] retention indices on SE-30 column, non-polar stationary phase, at 120°C. [b] retention indices on CARBOWAX 20M column, polar stationary phase, at 140°C. [c] retention indices on SE-30 column, non-polar stationary phase, at 160°C. [d] retention indices and capacity factor on HP-5 column at 120°C. [e] capacity factor on HP-5 column at 140°C; where $\boldsymbol{\sigma_{\phi Cln}}$ represents standard deviations.

Table 1a. References of Table 1

No.	References
1	Chemosphere 2000, 20, 457-512. J. Moleculare Structure 2003, 622, 127-145.
2	J. Chromatography A 2003, 1002, 155-168. J. Chromatography 1985, 319, 1-8.
3	J. Chromatography A 1996, 734, 277-287. J. Chromatography A 2003, 1002, 155-168.
4	J. Chromatography A 1999, 835, 19-27. Electroanalytical Chemistry and Interfacial Electrochemistry 1975, 61, 303-313.
5	www.nist.gov/chemistry NIST Chemistry Webbook J. Phys.Chem. A 2002, 106, 6618-6627.
6	J. Chem. Inf. Comput. Sci. 1996, 36, 100-107. J. Chem. Inf. Comput. Sci. 1998, 38, 283-292. Chemosphere 1997, 34, 275-298.
7	J. Chem. Inf. Comput. Sci. 1999, 39, 1081-1089. http://cas.org./SCIFINDER/SCHOLAR
8	Thermochimica Acta 1991, 179, 81-88. http://cas.org./SCIFINDER/SCHOLAR
9	J. Chromatography A 2003, 1002, 155-168.
10	Thermochimica Acta 1991, 179, 81-88.
11	www.nist.gov/chemistry NIST Chemistry Webbook Thermochimica Acta 1991, 179, 81-88.
12	Thermochimica Acta 1991, 179, 81-88.
13	J. Chem. Inf. Comput. Sci. 2001, 41, 1162-1176.
14	Electroanalytical Chemistry and Interfacial Electrochemistry 1975, 61, 303-313.
15	Fluid Phase Equilibra 2003, 205, 303-316. Fluid Phase Equilibra 1997, 131, 145-179.
16	J. Chem. Inf. Comput. Sci. 2003, 43, 1240-1247.
17	J. Phys.Chem. A 2003, 34, 6580-6586.
18	CRC Handbook of Chemistry and Physics 1997-1998 Ed. 8, 5-43.
19	Physical Properties of Chemical Compounds, American Chemical Society, Washington D.C 1995, 1-87. J. Chem. Inf. Comput. Sci. 1998, 38, 840-844.
20	http:/159.226.63.177/scripts/opes/properties Engineering Chemistry Database Computers and Chemistry 2002, 26, 159-169.
21	http:/159.226.63.177/scripts/opes/properties Engineering Chemistry Database
22	Physical Properties of Chemical Compounds, American Chemical Society, Washington D.C 1995, 1-87. http:/159.226.63.177/scripts/opes/properties Engineering Chemistry Database
23	Physical Properties of Chemical Compounds, American Chemical Society, Washington D.C 1995, 1-87.
24	www.nist.gov/chemistry NIST Chemistry Webbook CRC Handbook of Chemistry and Physics 1997-1998 Ed. 8, 5-43.

These properties, their posetic manifestation, and their isotonicity scores were reported, and the results for chloro-benzene properties are reproduced in Table 1. For most properties isotonicity scores were fairly high, with several being perfect. There are a few properties which exhibit lower isotonicity scores – these properties include melting point, heat of fusion, and entropy of fusion. Also in these tables one finds a "degree of indifference" score, such as has to do with a comparison to another somewhat natural partial ordering, wherein every ϕCl_n is viewed to precede every ϕCl_{n+1} (ignoring the "incomparability" of some of the different substitution patterns). That is, one might imagine that some properties are ordered such that the values of every n-substituted species are similarly ordered with respect to every n+1 – substituted species – one especially simple such property being the molecular weight. This alternative more complete partial ordering has an additional 6 pairs in the Hasse diagram and the degree of indifference is defined as the fraction of these additional relations which would also be satisfied by a considered property. E.g., from Fig. 4 one sees that the toxicity values simply increase with the number of substituted Cl-atoms, so that the degree of indifference is 6/6. It is seen that the degree of indifference is often high – but not always, more so for the case of methylbenzenes.

Molecular Property Interpolation

There is an especially simple sort of scheme by which to use the reaction poset to interpolatively predict property values, in suitable cases. By this scheme a prediction $\overline{X}(\mathrm{B})$ is made for a property X at a position B in the poset if the property values are available for all other positions immediately adjacent to A in the Hasse diagram. We take $\overline{X}(\mathrm{B})$ as the average of two values: first the average of all X(A) with A just above B, and second the average of all X(C) with C just below B (see similar ideas in Klein & Bytautas 2000, Brüggemann et al. 2001, Carlsen et al. 2002 and 2004). Thence for ortho-dichoro-benzene the predicted toxicity from the data for chlorobenzene and for 1,2,3- & 1,2,4-trichloro-benzenes is

$$\begin{aligned}\overline{X}(o\mathrm{Cl}_2\phi) &= \frac{1}{2}\{X(\mathrm{Cl}\phi)+\frac{1}{2}[X(123\mathrm{Cl}_3\phi)+X(124\mathrm{Cl}_3\phi)]\}\\ &= \frac{1}{2}\{2.23+\frac{1}{2}[1.17+1.11]\}\\ &= 1.685\end{aligned}$$

This scheme is notably rather like the neighbour-averaging scheme originally used by Mendeleev to make predictions for selected properties of (then) missing elements (e.g. eka-silicon, later named germanium). Indeed this scheme in application to Mendeleev's periodic table is often described in texts for introductory chemistry courses. Notably Mendeleev's scheme evidently appears to have even more in common with our present work when Mendeleev's periodic table is viewed as a poset (such as indeed has been suggested). Errors (as $\mathrm{BP(A)} - \overline{\mathrm{BP}}(\mathrm{A})$) for the so predicted boiling points of the methylbenzenes are exhibited in Fig. 7.

Significantly beyond predicting already known property values so as to make comparisons, one may also make predictions for property values such as have not been experimentally reported. Such predictions are reported in our article (Ivanciuc and Klein 2004). But there are further methods to use posets to make predictions.

First, there is posetic splinoid fitting described in Došlić and Klein (2004), which is more widely applicable than the posetic averaging already described. This procedure imagines a function defined on each of the line segments shown in the Hasse diagram (each such segment corresponding to a single step of the reaction here considered), the function is taken to be a cubic polynomial, such that: first, its values (corresponding to the property values) at the end points are the same for each incident line segment; second, it is smooth in the direction of change of the poset; and otherwise the net effective "curvature" of the function is minimized. Such a fitting for the special case of a completely ordered poset yields the long standard and highly successful spline fit – see, e.g., de Boor (1978) or Ruitishauser (1990). For more general posets such as the reaction network posets here, the technique is new – but we believe, promising. The consequent linear-algebraic formulas as developed in Došlić and Klein 2004 (and summarized in Ivanciuc, Ivanciuc and Klein 2005), involve standard matrix inversions of matrices with sizes given by the number of members of the poset.

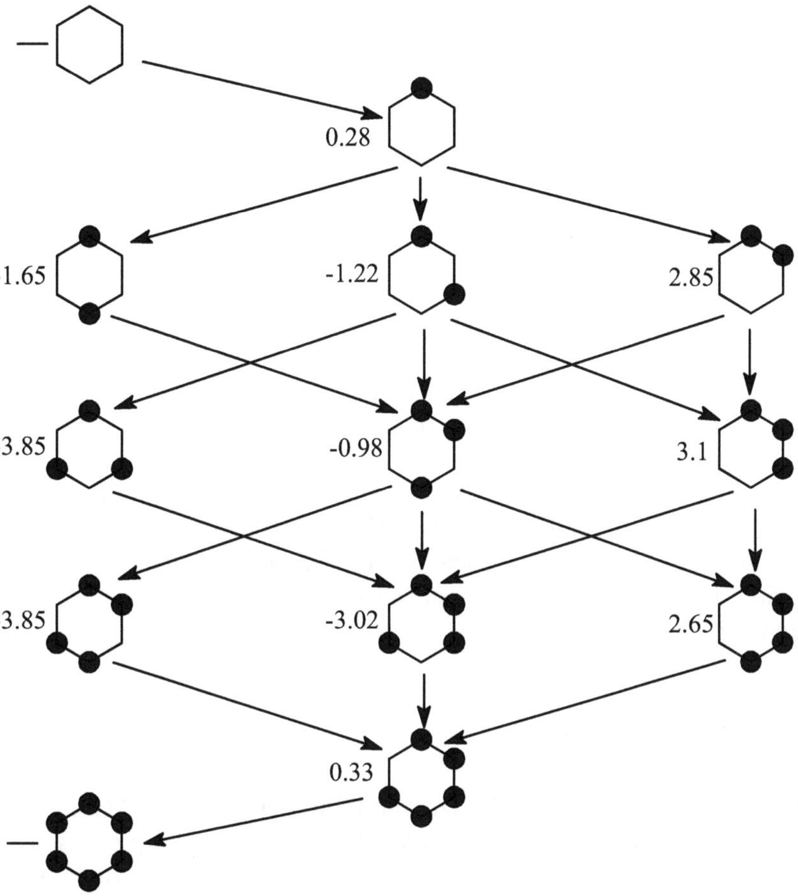

Fig. 7. The posetic presentation of the errors for the boiling points (in °C) of the methylbenzenes

With this approach even when structures at adjacent positions in a reaction network have property values which are unknown and are wished to be predicted, the spline-fitting method is applicable. Indeed there is initial evidence that the scheme is robust under the circumstance that even sizable fractions of the property values are missing – perhaps even a majority of the values might be missing. The errors may of course be anticipated generally to increase with increasing deficits of known property values. As an example this spline-fitting scheme is especially useful to deal with toxicities of polychlorinated biphenyls (PCBs). Here there are about 40 toxicities available (Ivanciuc, Ivanciuc and Klein 2006), though there are 210 members of the substitution reaction poset (presuming that one does not distinguish stereostructures associated with lack of rotation of the two benzene rings about the interconnecting bond between the two).

A further fitting method is to use a cluster expansion for the reaction poset. This gives a property X for species S as a linear combination of a corresponding "constituent" property x for earlier members of the poset

$$\mathrm{X(S)} = \sum_{\mathrm{R} \prec S} c_R x(R) \tag{1}$$

with c_{R} determined from the poset and the x(R) determined either by "fitting" or by "inversion". Usually a simplifying approximation is invoked wherein all but a few of the very earliest R of the poset have $x(\mathrm{R}) = 0$. This also is developed to recent work (Ivanciuc, Klein, Ivanciuc 2005), and a summary is given in Ivanciuc, Ivanciuc and Klein 2005. In fact this method is intimately related to more standard substructural cluster expansions discussed at length in reviews (Klein 1986 and Klein et al. 1999). The substructural cluster expansions are based upon the subgraph partial ordering, but the general idea applies to rather general posets as emphasized by Rota (1964).

General Discussion

It may be worthwhile to emphasize a little further the generality and ubiquity of the general framework considered here. The idea of progressive reaction posets can be argued to underlie the general idea of periodic tables, with Mendeleev's periodic table of the elements providing naught but the most conspicuous example of such a periodic table. E.g., there also is Randić and Wilkins "periodic table of alkanes" (such as appears on a cover of the Journal of Chemical Education Randić 1992), Dias' "formula periodic table of benzenoids", and "periodic table of all acyclics" (Bytautas et al. 2000). All these periodic tables fall into two-dimensional arrangements with one type of reaction down columns and another type along rows. Most of these periodic table examples end up with more than one chemical species at each position (or node) of what then is recognized as a reaction network. More detail about such an interpretation (for all these periodic tables as well as several others) may be found elsewhere (Klein 2005). Especially in Randić & Wilkins periodic table poset for alkanes the type of property interpolation techniques developed here should be rather directly applicable. In the progressive reaction poset for substituted benzenes it may even be mentioned that our diagram of Fig. 1 may be viewed to anticipate a second type of reaction – namely a rearrangement reaction going across rows of the diagram. Here it is understood that we start with the most symmetric species (of a given overall composition) and then proceed to increasingly asymmetric species with each horizontal (rearrangement) step to the right, so that, e.g., the molecular dipole moment should increase from left to right. Thence the present poset of substituted benzenes might be viewed as a (simple) "periodic table". Overall it seems that the general ideas underlying our approach are widespread, and perhaps many applications of the general theory of posets may be made, maybe with some of the techniques specifically adapted to the evidently ubiquitous case of progressive reaction posets.

Note that the currently noted interpolative fitting techniques (poset-averaging, splinoid fitting, cluster expansion) go beyond simple ordering and ranking. This may be contrasted with the work of Carlsen (and also in that of Brüggemann 2001, Carlsen et al. 2002 and 2004) the considered poset is not determined a priori but rather is defined in terms of the considered property values, so that at least these property values are strictly isotonic. What is different in the current approach is that: miss-orderings of the experimental values are allowed, quantitative predictions are entertained, and consequent standard deviations are obtained. That is, with predicted values $\overline{X}(\mathrm{B})$ comparisons with measured values and consequent standard deviations are naturally made, as in the preceding section. Both the isotonicity score and the standard deviation provide measures of the degree of organization achieved by the posetic classification. The overall results as reported in Table I seem encouraging. A further indication of the utility of the results is made upon making plots of predicted *vs.* measured property values as indicated: in Fig. 8 for aquatic toxicities of chlorobenzenes; in Fig. 9 for octanol-water partition coefficients for chlorobenzenes; in Fig. 10 for octanol-water partition coefficients for methylbenzenes; and in Fig. 11 for aqueous solubility of methylbenzenes. Evidently even this simple parameter free poset-averaging can sometimes work quite well. Further work with the splinoid fit and the cluster expansion indicate that they also work quite well, while also being more robustly tolerant of missing information. Especially with larger posets this robustness is often especially relevant, e.g., with the chloro-substitution reaction for biphenyl, where out of 210 members one finds only about 40 toxicities reported in the literature.

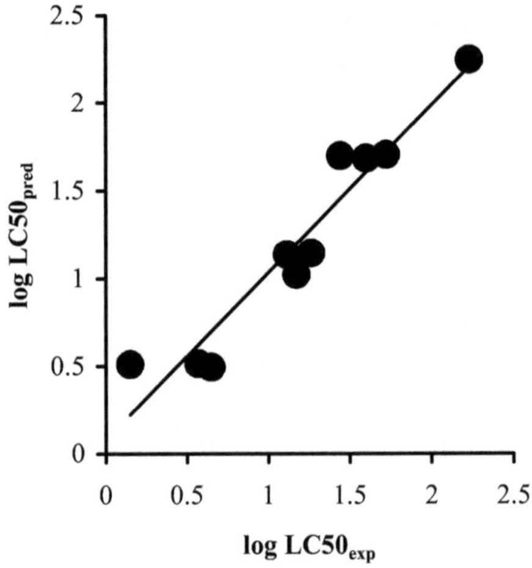

Fig. 8. Comparison of predicted and measured toxicities for chloro-benzenes

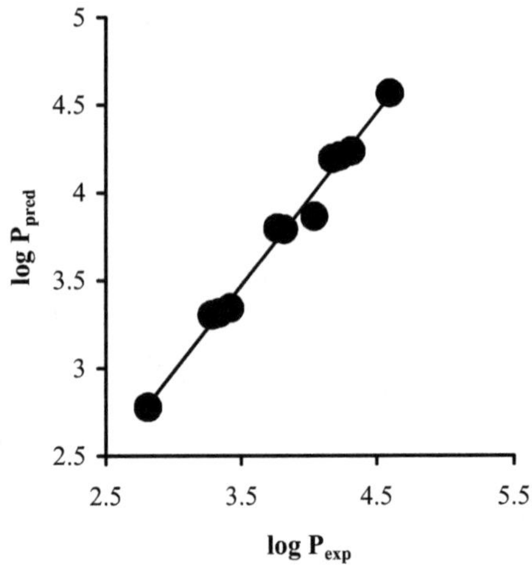

Fig. 9. Comparison of predicted and measured partition coefficients octanol-water for chloro-benzenes (as logP)

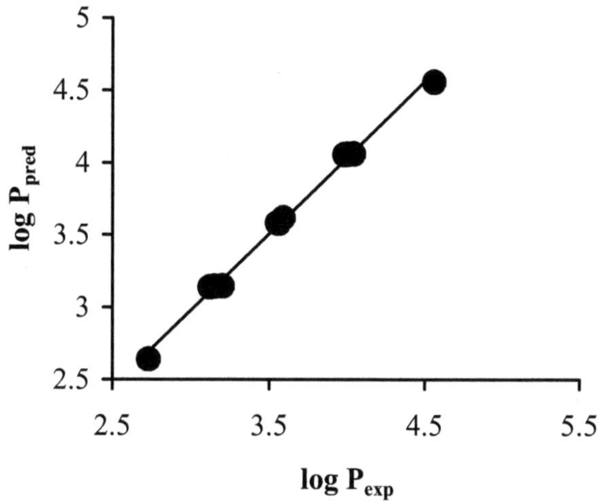

Fig. 10. Comparison of predicted and measured partition coefficients octanol-water for methylbenzenes (as logP)

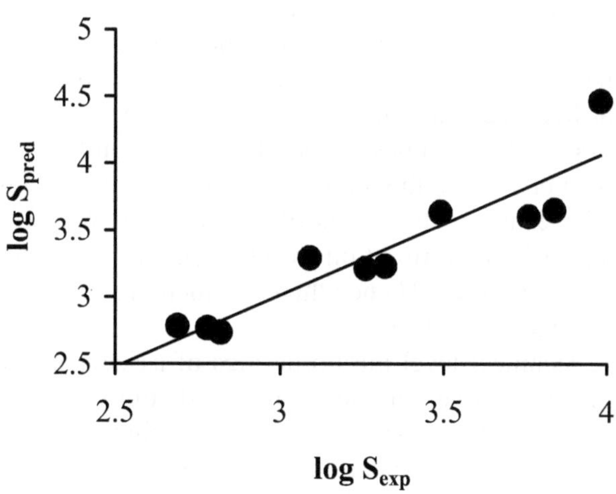

Fig. 11. Comparison of predicted and measured aqueous solubilities (mol/L) for methylbenzenes (as logS)

Prognosis

Evidently progressive reaction network are in fact posets. And properties for species appearing in such a network can be fit in a poset-attentive manner to understand the pattern within the poset and to make useful predictions. Some of the fitting procedures are of a notably different nature than commonly used QSAR fitting procedures. Indeed the "poset-average" and splinoid fitting procedures are in essence parameter free. The cluster-expansion approach yields fitting parameters in a more conventional manner, and the possibility of their transfer between different reaction posets is a problem for future study. In the (few) cases so far studied to make fits and predictions of properties, the various techniques seem to work quite reasonably. The "poset-average" procedure recommends itself with its great simplicity. The splinoid and cluster-expansion approaches however have an advantage of being tolerant to larger deficits of information, while still seemingly making reasonable predictions. The novelty of the approach in comparison to typical QSAR approaches, and in particular the attention beyond structure to placements in reaction networks, suggests the designation of one approach to be that of QSSAR (quantitative super-structure/activity relationships).

Further it is emphasized that such directed reactions graphs are wide spread in chemistry conceivably even in other sciences as well. Sometimes the posets may be quite complex as with the briefly noted addition - reaction poset for hydrogenation of buckmisterfullerene – leading to a poset of $>10^{14}$members. And sometimes the poset may be infinite, as with the noted alkane poset. Yet further the same types of progressive reaction posets may occur. For example the posets of ancestors or food webs in biology may often appear in the form of our reaction posets, and thence be susceptible to the same analyses. Thence there is much promise for our presently indicated ideas and techniques.

On a more technical level miss-orderings of associated property values are tolerated, quantitative parameter free predictions are entertained, and consequent standard deviations are describable. Simple quality-of-fit indices (isotonicity score and standard deviation) for the organizational ability of the poset have been designed. Other schemes (i.e., "models") for the prediction of properties might be imagined, and perhaps even useful. More generally parameter dependent fittings as combinations of isotonic functions might be sought. Further it has not escaped our attention that the substitutional progressive reaction networks are very special posets: they are "graded" in the sense that every isotonic path from any one fixed member to a successive fixed member are of the same length, and in fact these posets are even more special. Thence with such additional structure additional theory concerning associated isotonic functions may be relevant. All this presumably should be pursued further. Even with the approaches indicated here, tolerable standard deviations are often obtained, and some predictions beyond the experimentally reported values are made.

Overall notable success is evidenced in the few investigations to date. As such this indicates a potentially wide-range of chemical applications and possibly beyond. Further posetic techniques should be developed and explored.

The authors acknowledge the support (via grant BD-0894) from the Welch Foundation of Houston, Texas.

References

Babić D, Došlić T, Klein DJ, Misra A (2004) Kekulénoid Addition Patterns for Fullerenes and some Lower Homologs. Bull Chem Soc Japan 77:2003-2010

Balaban AT (1994) Reaction graphs. Graph Theoretical Approaches to Chemical Reactivity. Ed. D. Bonchev & O. Mekenyan, Kluwer Academic Publishers Dordrecht pp 137-180

Bertz SH (1984) The Role of Symmetry in Synthetic Analysis. The Concept of Reflexivity. J Chem Soc Chem Commun 218-219

Bertz SH (1986) Synthesis Digraphs and their Vulnerability. The Risk of Failure of Synthesis Plans. J Chem Soc Chem Commun 1627-1628

Brüggemann R, Münzer B, Halfon E (1994) An Algebraic/graphical Tool to Compare Ecosystems with Respect to their Pollution – the German river “Elbe” as an Example – I. Hasse diagrams. Chemosphere 28:863-872

Brüggemann R, Pudenz S, Carlsen L, Sørensen PB, Thomsen M, Mishra RK, (2001) The Use of Hasse Diagrams as a Potential Approach for Inverse QSAR. SAR and QSAR Environ Res 11:473-487

Bytautas L, Klein DJ (2000) Formula Periodic Table for the Isomer Classes of Acyclic Hydrocarbons–Enumerative and Asymptotic Characteristics. Croat Chem Acta 73:331-357

Bytautas L, Klein DJ, Schmalz TG (2000) All Acyclic Hydrocarbons: Formula Periodic Table and Property Overlap Plots via Chemical Combinatorics. New J Chem 24:329-336

Carlsen L, Sørensen PB, Thomsen M (2001) Partial Order Ranking-Based QSAR’s: Estimation of Solubilities and Octanol-Water Partitioning. Chemosphere 43:295-302

Carlsen L., Sørensen PB, Thomsen M, Brüggemann R (2002) QSAR's Based on Partial Order Ranking. SAR and QSAR in Environ Res 13:153-165

Carlsen L (2004) Giving Molecules an Identity. On the Interplay between QSARs and Partial Order Ranking. Molecules 9:1010-1018

Claus A (1866) Theoretische Betrachtungen und deren Anwendung zur Systemik der organischen Chemie, Frieburg

Corey EJ, Choung X-M (1989) The Logic of Chemical Synthesis. Wiley, NY.

Dias JR (1985) A Periodic Table for Polycyclic Aromatic Hydrocarbons. Acc Chem Res 18:241-248

Dias JR (1990) A Formula Periodic Table for Benzenoid Hydrocarbons and the *Aufbau* and Excised Internal Structure Concepts in Benzenoid Enumerations. J Math Chem 4:17-30

Dolfing J, Harrison BK (1993) Redox and Reduction Potentials as Parameters to Predict the Degradation Pathway of Chlorinated Benzenes in Anaerobic Environments. REMS Microbiology Ecology 13:23-30

Došlić T, Klein DJ (2005) Splinoid Interpolation on Finete Posets. J Comput Appl Math 177:175-185

Eigen M (1971) Molekulare Selbstorganisation und Evolution (Self Organization of Matter and the Evolution of Biological Macromolecules). Naturwissenschaften 58:465-523

Eigen M, Schuster P (1977) The Hyper Cycle. A Principle of Natural Self Organization. Part A. Emergence of the Hyper Cycle. Naturwissenschaften 64:541-565

Hendrickson JB (1977) Systematic Synthesis Design. 6. Yield Analysis and Convergency. J Am Chem Soc 99:5439-5450

Hill TL (1977) Free Energy Transduction in Biology. Academic Press NY

Ivanciuc T, Klein DJ (2004) Parameter-Free Structure-Property Correlation via Progressive Reaction Posets for Substituted Benzenes. J Chem Inf Comput Sci 44:610-617

Ivanciuc T, Ivanciuc O, Klein DJ (2005) Posetic Quantitative Superstructure/ Activity Relationships (QSSARs) for Chlorobenzenes. J Chem Inf Model 45: 870-879

Ivanciuc T, Ivanciuc O, Klein DJ (2006) Posetic Quantitative Super-Structure Activity Relationships (QSSAR) for Polychlorinated Biphenyl (PCBs), in Preparation

Ivanciuc T, Klein DJ, Ivanciuc O (2005) Posetic Cluster Expansion for Substitution-Reaction Diagrams and its Application to Cyclobutane, J Math Chem, submitted

Kekulé A (1865) Sur la Constitution des Substances Aromatiques. Bull Soc Chim Fr 3:98-110

Kekulé A (1866) Über die Konstitution und Untersuchung aromatischer Substanzen. Justus Liebigs Ann Chem 137:129

Kekulé A (1872) Über einige Condensationsproducte des Aldehyds. Justus Liebigs Ann Chem 162:77-123

Klein DJ Periodic Tables and the Formula Periodic Table for all Acyclic Hydrocarbons. Proceedings of Wiener Conference on Periodic Tables, in press

Klein DJ, Bytautas L (2000) Directed Reaction Graphs as Posets. MATCH Commun Math & Comput Chem 42:261-289

Klein DJ (1986) Chemical Graph-Theoretic Cluster Expansions. Int J Quantum Chem 20:153-171

Klein DJ, Schmalz TG, Bytautas L (1999) Chemical Sub-Structural Cluster Expansions for Molecular Properties. SAR and QSAR in Environ Res 10:131-156

Körner W (1874) Studj Sull Isomeria Della Cosi Dette Sostanze Aromatiche a sei Atom di Carbonio. Gazz Chim Ital 4:305-446

Körner W (1869) Fatti per Servire Alla Determinazione Del Luogo Chimico Nelle Sostanze Aromatiche. Palermo Giornale di Sciencze Naturali ed Economiche, 5:208-256, as summarized by H. E. Armstrong (1876) J Chem Soc 29:204-241

Ladenburg A (1869) Bemerkungen zur aromatischen Theorie. Chem Ber 2:140-142 & 272-274

LeBel JA (1874) On the Relations which Exist between the Atomic Formulas of Organic Compounds and the Rotatory Power of their Solutions. Bull Soc Chim Fr 22:337

Mendeleev DI, Russ Z (1869) Khim Obshch 1:60 [translated, to German] Berichte 2:553

Randić M (1992) Chemical Structure – What is 'she". J Chem Ed 69:713-718

Randić M, Wilkins CL (1979) Graph Theoretical Ordering of Structures as a Basis for Systematic Searches for Regularities in Molecular Data. J Phys Chem 83:1525-1540

Randić M, Wilkins CL (1979) On a Graph-Theoretical Basis for the Ordering of Structures. Chem Phys Lett 63:332-336

Rocke AJ (1985) Hypothesis and Experiment in the Early Development of Kekulés Benzene Theory. Ann Sci 42:355-381

Rota GC (1964) On the Foundations of Combinatorial Theory. I: Theory of Möbius Functions. Zeit. Wahrscheinlichkeitstheorie und Verw. Gebiete 2:340-368

Shtarev AB, Pinkhassik E, Levin MD, Stibor I, Michl J (2001) Partially Bridge-fluorinated Dimethyl Bicyclo[1.1.1]Pentane-1,3-Dicarboxylates: Preparation and NMR Spectra. J Am Chem Soc 123:3484-3492

Temkin ON, Zeigarnik AV, Bonchev D (1996) Chemical Reaction Networks: A Graph-Theoretical Approach. CRC Press, Boca Raton, Fl

Van't Hoff JH (1874) Sur les Formules de Structure Dans l'Escpace. Archives Nederlandaises des Sciences Exactes et Naturalles 9:445-454

2 Environmental Chemistry and Systems

In this section the multivariate point of view is the central topic. The partial order to be analyzed is a result of a characterisation of objects, e.g., of chemicals by different attributes. Especially the Hasse Diagram Technique (abbreviation: HDT), which plays a central role in this and in the next three sections is devoted to derive and to draw information out of graphs, based on the $\leq$-relation among each single attribute.

In the chapter of Brüggemann and Carlsen some concepts introduced in the chapter of El-Basil are revitalized and explained in the context of the multivariate aspect. Basic concepts, like chain, anti-chain, hierarchies, levels, etc., as well as more sophisticated ones, like sensitivity studies, dimension theory, linear extensions and some basic elements of probability concepts are at the heart of this chapter. The difficult problem of equivalent objects, which lead to the items object sets vs. quotient sets are explained and exemplified.

In the chapter by Brüggemann and Carlsen the degradation of the sediments of Lake Ontario is used as an example, however only playing a minor role. In the chapter of Pudenz the quality of river sediments is the main topic. Sediment quality is assessed by measuring the concentrations of chemicals and by performing biochemical and ecotoxicological tests. The reader will beside others learn how clustering techniques; especially a fuzzy clustering can be combined with partial order.

In the chapter of Carlsen and Walker a hard and real life example is presented. It is a challenge to analyze 50 persistent, bioaccumulating and toxic substances with respect to their hazard toward the environment. Just the Hasse diagram is helpful in getting an overview about these 50 substances.

Introduction to partial order theory exemplified by the Evaluation of Sampling Sites

Rainer Brüggemann[1] and Lars Carlsen[2]

[1] Leibniz-Institute of Freshwater Ecology and Inland Fisheries, Department: Ecohydrology, Germany

[2] Awareness Center, Roskilde, Denmark

Abstract

The first part of this chapter gives a detailed introduction to partial order ranking and Hasse Diagram Technique (HDT). Thus, the construction of Hasse diagrams is elucidated as is the different concepts associated with the diagrams. The analysis of Hasse diagrams is disclosed including structural analysis, dimension analysis and sensitivity analysis. Further the concept of linear extensions is introduced including ranking probability and averaged rank. The evaluation of sampling sites is, in the second part of the chapter, used as an illustrative example of the advantageous use of partial order ranking and Hasse Diagram Technique.

When a ranking of some objects (chemicals, geographical sites, river sections etc.) by a multicriteria analysis is of concern, it is often difficult to find a common scale among the criteria and therefore even the simple sorting process is performed by applying additional constraints, just to get a ranking index. However, such additional constraints, often arising from normative considerations are controversial. The theory of partially ordered sets and its graphical representation (Hasse diagrams) does not need such additional information just to sort the objects.

Here, the approach of using partially ordered sets is described by applying it to a battery of tests on sediments of the Lake Ontario. In our analysis we found: (1) the dimension analysis of partially ordered sets suggests that there is a considerable redundancy with respect to ranking. The partial ranking of the sediment sites can be visualized within a two-dimensional grid. (2) Information, obtained from the structure of the Hasse diagram: For example six classes of sediment sites have high priority, each class exhibits a different pattern of results. (3) The sensitivity analysis identifies one test as most important, namely the test for Fecal Coliforms/*Escherichia coli*. This means that the ranking of samples is heavily influenced by the results of this specific test.

Introduction

Overview

In the present chapter an alternative way of analysing objects, which are characterized by several quantities is presented. Hence, instead of examining the variance (for example leading to principal component analysis) or the distances among objects (for example leading to cluster analysis), we focus on the use of partial ordering in ranking. An important aspect within the concept of partial ordering is the visualization by Hasse diagrams.
More specifically, we study

- the system of comparabilities and incomparabilities between objects which arises, if an order relation between them is defined
- how to set priorities and to detect the pattern which identify objects of high priority
- how to define logical non-contradictory sequences and
- how the selection of criteria influences the ranking of a set of objects.

Beyond this we analyse
- the role of the structure of the Hasse diagram, i.e. levels, hierarchies, articulation points
- the role of order preserving maps among partial orders and especially those order preserving maps whose results are linear orders and
- how we can derive an averaged rank, probability distributions from them and how structural properties of the Hasse diagram can be detected in probability distributions.

Further is discussed, the important concept of
- poset dimension, and
- latent variables are related to partial orders.

Partial order rankings may advantageously be visualized through Hasse diagrams. The program „WHASSE“ allows the construction of Hasse diagrams and provides several tools, which helps to analyse partially ordered sets (Brüggemann et al. 1999 a). A historical and personal view about the development of the programs around partial order is given by Halfon see p. 385. A matrix **W** is introduced, that quantify the importance of the single criteria on the eventual ranking. Additional to the contributions in this book, several textbooks and monographs and journal's publications are recommended to the reader (see reference list: "Introductory references", p. 393).

Partial order

Introductory remarks

Hasse diagrams show the relations of partially ordered sets (posets). In the following is explained why partial order is a useful concept on ranking. Ordering is a logical way to give objects a structure: If for example chemical substances are characterized by their persistence then these substances can be sorted according to the increasing persistence, the sequence of substances corresponds to one characteristic number, namely the persistence. Often however, a single number is not sufficient to characterize objects. For example not only the persistence but also the bioaccumulation of a chemical substance may be important to explain the environmental behaviour of the substance. For further examples see contributions of this book.

Common to these examples is that each object (geographical sites, waste disposal sites, databanks, chemicals, managing options) is characterized by more than one quantity. Objects that are characterized by several quantities (we call them “attributes“ -see later for details-) often cannot be ordered, because there are conflicts between their attributes. Metaphorically we are talking of comparing apples and oranges.

An example may help to understand this. We may have five objects {A, B, C, D, E} characterized by e.g. their environmental persistence „P“ and by their ability to bioaccumulate „B“. As often is the case both attributes do not behave parallel, i.e. it is not automatically given that a persistent substance also is the most bioaccumulating.

We can arrange the five substances according to P or B (Fig. 1):

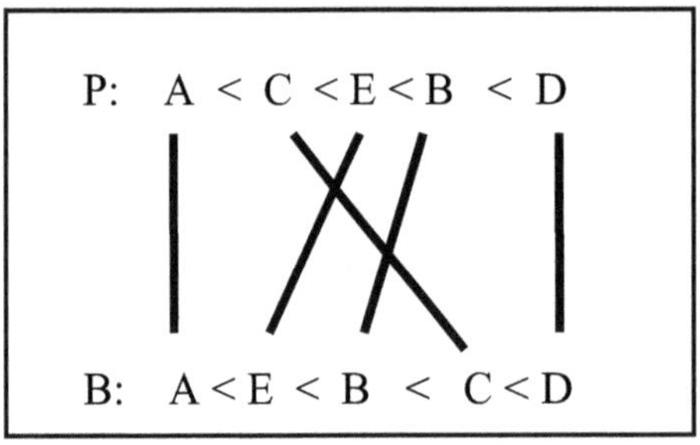

Fig. 1. "Permutation diagram": Two sequences of objects according to two different characteristics.

The type of diagram in Fig. 1 is called a permutation diagram (Urrutia 1989). It shows that there are inversions between the two sequences. Some objects will mutually exchange their positions in dependence which quantity is used to define the sequence (for example C, E). Some other objects do not change their relationships to others, if the sequence defining quantity (here: persistence or bioaccumulation) is changed (For example: A < D or E < B independent whether the persistence or the bioaccumulation is selected; other examples can also be found). Obviously some „rest of order“ remains, if both quantities are considered at the same time and this fact motivate the term „partial order“. Within the given example of five objects partial ordering arises because more than one quantity is used to characterize the single substances. This is often the case, where the complexity of nature prevents the use of a single ranking index (therefore many applications can be found in biology, ecology, ecotoxicology, and chemistry as disclosed through various chapters of this book. Partial order is further a typical tool within operation research, many decision support systems are based -at least implicitly - on partial order. For example in versions of ELECTRE (Roy 1972, 1990) or PROMETHEE (Brans & Vincke 1985, Brans et al. 1986, Heinrich, 2001) a partial order is at least an interim step (see also chapter by Brüggemann et al., p. 237). An access to recent literature may also be found in (Colorni et al. 2001, Lerche et al. 2002).

Obviously, the concept of partially ordered sets appears rather useful in environmental sciences. The „usual“ order, namely the order in which each object can be compared with each other, can be considered as a special case of partial order, i.e., the term "linear" or "total" order is used.
Permutation diagrams become confusing if many objects are included and especially if more than two attributes characterize the objects. In such cases a corresponding number of sequences may arise and for each pair of

sequences a permutation diagram can be drawn. Instead of this troublesome procedure which leads to m·(m-1)/2 pairs of permutation diagrams (m attributes used) the technique of Hasse diagrams provides a useful tool for visualization. From the permutation diagram (Fig. 1) it can be concluded that A < C < D and A < E < B < D, respectively, whereas we cannot say anything concerning the relations between C and E and B, respectively, if both persistence and bioaccumulation are taken into account. Thus, the partially ordered set of the five objects (cf. Fig. 1) is visualized in a "Hasse diagram" (Fig. 2).

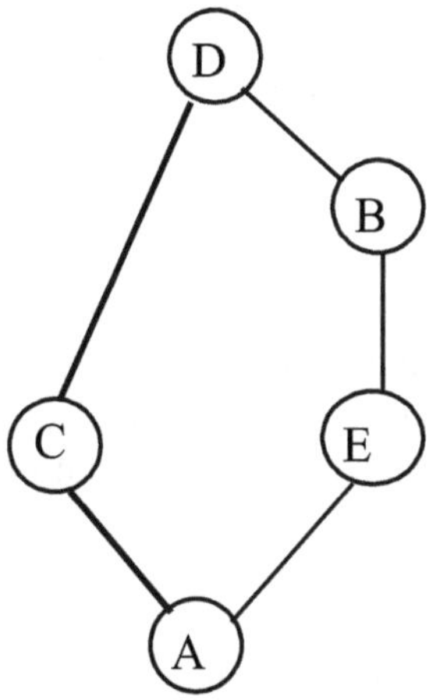

Fig. 2. The Hasse diagram as an alternative to the permutation diagram shown in Fig. 1

The name "Hasse diagram" becomes popular by the German mathematician Helmut Hasse, who worked in Marburg, Berlin and Hamburg, see also chapter by Halfon, p. 385. Often this kind of diagram is simply called "line diagram" or even only "the diagram" (Rival 1985b).

The rationale of using Hasse diagrams

The concept of partial order is described in the chapter of El-Basil, p. 3. Therefore we concentrate ourselves on the specific order relation we are using here, which is known as "Hasse Diagram Technique". In this technique we specifically consider any component of a sequence separately, as it bears its own valuable information with respect to the evaluation. Techniques, motivated by the work of Muirhead 1900, 1906 and Karamata, see

Beckenbach & Bellmann 1971, or by Young diagrams (Ruch 1975, Ruch & Gutman 1979) may be useful too. However, applying these techniques, the components of the sequence, i.e. the attributes would loose their individual meaning, which often is disadvantageous. (with respect to Young diagrams see the chapters by El-Basil, p. 3 and Seitz, p. 367.)

The basis of the Hasse Diagram Technique (HDT) is that we can perform a ranking without the use of a single ordering index (called a "ranking index"), i.e. we rank objects by maintaining all information about them. If an ordering index were used to force the object into a linear order, then information is lost. For example, an object might be ranked higher according to one criterion but lower according to another. Two objects might not be ordered unambiguously because their data are "contradictory" to each other. This ambiguity is not immediately evident when we use a ranking index, still worse: by using a ranking index the two attribute can compensate each other. That means a "bad" value in one attribute can be compensated by a "good" value in another one. Metaphorically speaking you can put one hand in boiling water and one hand in ice water. Discomforting? Yes! However, on an average basis you should feel quite comfortable! Such kinds of potential compensations or conflicts among attributes are immediately evident in a Hasse diagram.

Many problems are governed simply by comparisons, i.e. by the analysis of the order-relation. Typical examples can be found in textbooks of chemistry, when concepts like electronegativity, hardness or softness of compounds, etc is discussed. Many other problems are reducible to an order - relation. Often for example objects may be characterized by a binary bit pattern, representing whether a property is given or not. For example existence or non-existence of chemical functional groups lead to a binary bit pattern, for which a partial order can be defined (see for an example the chapter by Klein & Ivanciuc, p. 35). Partial orders help to analyse Quantitative Structure Activity Relationships (Randić 2002, Brüggemann et al. 2001), see also chapters by Carlsen, p. 163 and Pavan et al., p. 181 and references therein). Other examples are biomarker responses on certain stress factors in ecosystems (see for example, Brüggemann et al. 1995a, 1995b) and the analysis of data sources, see chapter by Voigt and Brüggemann, p. 327 and references therein.

To explain partial order and its visualization by Hasse diagrams, some useful theoretical notations are given in the following section.

Prerequisites

- *Criteria* comprise both quantitative and qualitative properties. Often it is useful to define a criteria hierarchy: Starting with a general criterion, which is hardly quantifiable, one looks for subcriteria to specify the general one, the subcriteria in turn may further specified, until a set of precise criteria is found, which may be quantified by attributes.
- *Attributes* are quantitative, measurable data. We denote these attributes as $q_1, q_2, \ldots, q_m$. It is useful to define the information basis of the evaluation, *IB* to be the set of these attributes: $IB = \{q_1, q_2, \ldots, q_m\}$. Some authors denote attributes as descriptors or possibly parameters. These terms are used synonymously.
- A *case* is a subset of selected attributes, taken from the ground set of attributes, *IB*. The attributes are specific to the problem. Each case corresponds to exactly one Hasse diagram. Thus, a given set of attributes induces a Hasse diagram. More definitions will follow in the text as the need arises.
- An *object* is the item of interest that may be characterized by attributes. Examples of objects can be chemical substances, or geographical sites (see chapter by Myers et al., p. 309), or strategies (see chapter by Simon et al., p. 221) etc. Objects are ranked graphically by Hasse diagrams (see for example Fig. 2). Generally the objects are considered to belong to a set "*E*". Therefore the objects are also often called "elements" and *E* is called a *ground set* or *object set*. The ground set corresponding to the Hasse diagram in Fig. 2 is thus $E = \{A, B, C, D, E\}$ (note: set *E* but element E). We assume that we have n elements of the set *E*.
- *Data* are the numerical values corresponding to each criterion by which a given object is characterized.
- *Equivalent objects* in Hasse diagrams: Different objects that have the same data with respect to a given set of attributes. Equality with respect to a given set of attributes defines an *equivalence relation*, "$\mathfrak{R}$". Objects having the same values of all their attributes form disjoint subsets of *E*, the *equivalence classes*. An equivalence class with only one object is called a *singleton* and is called trivial. The equivalence classes can be considered as elements of a set, the quotient set $E/\mathfrak{R}$. Usually the partial order is based on the quotient set and -if necessary- the equivalent elements are associated with that vertex, where a representative element out of the equivalence

class is drawn. Examples will be given below. Further details, see Patil & Taillie 2004 or Brüggemann & Bartel 1999.

- The *cardinality* of a (finite) set is the number of elements of the set, denoted by card *G* for a set *G*.
- *Numerical representation of objects*: Objects are considered to be elements of the object set *E*. Each object is characterized by attributes. We can create a table where the rows represent the objects and the columns the data of each object corresponding to the column-defining attribute
- Taken an element of *E*, the corresponding row consisting of the data of $q_1,...,q_m$ is often called a *tuple*, and abbreviated by q.
- *Attribute profile or pattern*: If the order of attributes is fixed then the sequence of attribute values for a given object x can be thought of as visualized by a bar diagram. This we have in mind when we are speaking of a profile or pattern.

Further - more specific - terms are explained later.

Graphs and Hasse diagrams

The construction of Hasse diagrams:

A set that has an order relation is called a *partially ordered set (poset)*. An order for a set, for example for the set *E*, is denoted by $(E, \leq)$, the set *E* often being called the ground set (of objects). As the application of partial order, presented here, is based on attributes, just *IB* influences the partial order. Therefore, we often write (*E*, *IB*). If the quotient set is used, then we write $(E/\mathfrak{R}, IB)$.

Partially ordered sets can be visualized through Hasse diagrams, which are quite useful if not too many objects are included. Let a and b be two elements of the object set *E*. Each object is characterized by a set of attributes. The relation '≤' between a and b is valid, if and only if this relation holds for all attributes of a and b. In other words: a ≤ b, if all components of the tuple of a are smaller or equal to the corresponding component of the tuple of b. With help of the notation $q_j(i)$ with i the index for any element of *E*, and j as index for any attribute of *IB* we give a formula:

$$a, b \in E: a \leq b : \text{if and only if } q_j(a) \leq q_j(b), \text{ for all } q_j \in IB \quad (1)$$

We call equation (1) the generality principle, because this equation defines dominance of b over a if all properties of b confirm the ≤-relation.

To illustrate the above, an example may be useful. Consider three objects a, b, c. They are characterized by two attributes, as Table 1 shows.

Table 1. Fictitious example

	q_1	q_2
a	1	1
b	1	2
c	2	1

Obviously $a < b$ and $a < c$, respectively. With respect to the first attribute: $a = b$ and with respect to the second attribute: $a < b$. Therefore $a < b$. A similar argument holds for the a-c relation. Objects for which the $\le$-relation holds are *comparable* to each other. Often it is useful to have a shorthand notation for comparable objects (without specifying the orientation). Thus, if $a < b$, or $b < a$, we write $a \perp b$.

However, the relation $\le$ does not hold for the objects b and c, because with respect to attribute q_1: $b < c$, and with respect to attribute q_2: $b > c$. Hence, objects that cannot be compared with each other, like b, c are called *incomparable*. A shorthand notation to describe two incomparable objects is $b \parallel c$.

Cover relation:

If there is no element „x“ of *E*, for which $a \le x \le b$, $x \ne a, b$, $a \ne b$ holds, then a is *covered* by b, or b *covers* a. Often the cover relation is referred to by its own symbol $<\!\bullet$. Obviously in our example $a <\!\bullet\, b$ and $a <\!\bullet\, c$, the corresponding graphical representation is given in Fig. 3.

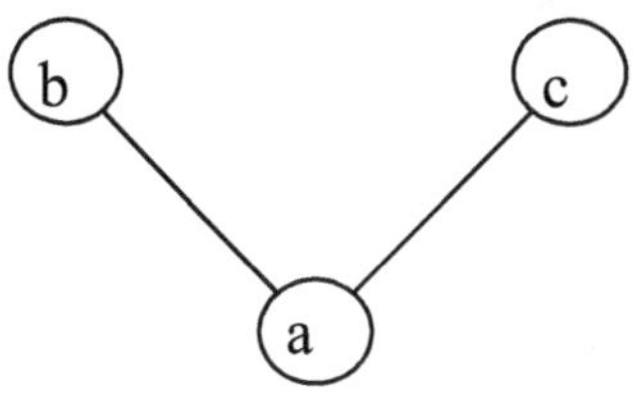

Fig. 3. Visualization of the order relation, induced by the data matrix, shown in Table 1.

Partial orders can be visualized in different ways, see also Chapters written by El-Basil, p. 3 and Seitz, p. 367. An interesting variant can be found in the chapter by Myers et al., p. 309). Other presentations are discussed in Neggers & Kim 1998. In the present chapter, the construction of Hasse diagrams is explained according to the software WHASSE (Brüggemann et al. 1999 a) and is performed with the help of the cover relation as follows:

1. *E* may be represented by a configuration of circles and with an identifier for the objects within and each circle is located in the two-dimensional plane.
2. Note that the program WHASSE only displays a representative within the circle; other objects, having equal data tuples are shown in an extra field of the screen.
3. If a cover-relation holds, then a line between the corresponding object-pair is drawn. The covering pair is oriented corresponding to the ≤-relation.
4. The covered object in the ≤-related pair is located at a lower position on the page. (Alternatively we can, instead of the connecting line segment, draw an oriented arrow, beginning at the covering object and directed towards the covered object; in this case the locations in the two-dimensional plane of the Hasse diagram can be selected arbitrarily. In the practice it is more convenient to select the positions in the plane of the figure, according to the cover-relation.) By this step the lines become an *orientation*, for example "good-bad" or "high-low". See also in chapter by Helm, p. 291.
5. Finally, not all line segments for which the ≤- relation holds are to be drawn. Because of the logical rule of transitivity (which holds by definition for partial orders) lines corresponding to the pair x, z with $x \leq y$ and $y \leq z$ concluding $x \leq z$ are omitted. They do not present a cover-relation.

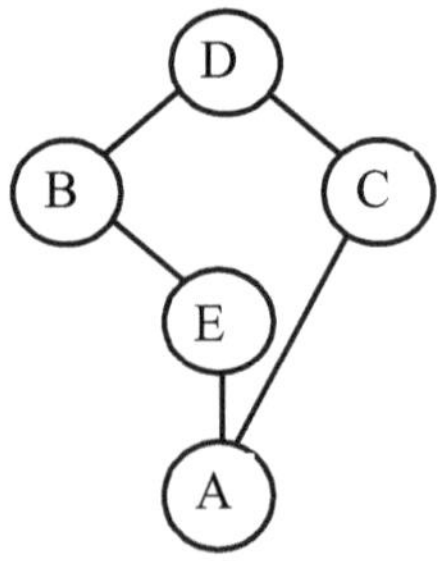

Fig. 4. The Hasse diagram of the example of Fig. 1 and Fig. 2, respectively, drawn by the program WHASSE

In order to introduce further concepts another Hasse diagram is drawn (Fig. 5):

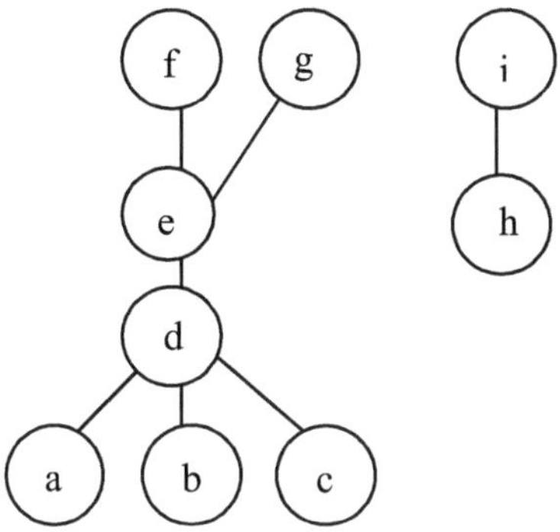

Fig. 5. Arbitrary Hasse diagram

- Elements, which are not covered by other elements, are called *maximal elements*, or -as done for example in the chapter by Carlsen and Walker, p. 153 simply as *maximals*. In Fig. 5 such elements are f, g, i.
- Elements that do not cover any other element are called *minimal element*, or simply *minimals*.
- If there is only one maximal element, then this is also called a greatest element. In Fig. 5, there is no greatest element, however in Fig. 4 element D is a greatest element.
- If there is only one minimal element, then this is also called a least element. In Fig. 5, there is no least element, however in Fig. 4 element A is a least element.
- If in a Hasse diagram there are parts that are not connected then these parts are called *hierarchies*. The suborders ({a, b, c, d, e, f, g}, ≤) and ({h, i}, ≤) are such hierarchies.

Details of the construction of Hasse diagrams „by hand“ are explained by Halfon et al. 1989. There is a useful "four-point-program" how step-by-step Hasse diagrams may be constructed (nevertheless quite tedious, if done by hand). See for a detailed description, (Voigt and Brüggemann, p. 327). There are still many ways to draw a Hasse diagram and some mathematicians are thinking about that point as art, Rival 1989. For example the program WHASSE would draw the Hasse diagram of Fig. 2 as depicted in Fig. 4. In the specific case that a poset can be considered as lattice, i.e. fulfils the axioms of lattices, then Freese 2004 gives an advice how to draw automatically lattices.

According to the scientific background the actual diagram may be constructed such that the results are presented as clear as possible. If there is no such specific background, the Hasse diagram is drawn as symmetric as possible. Incomparable objects are, conservatively located at the same height and as high as possible on the page. For example the object C in Fig. 4 could be located everywhere between objects D and A without hurting the order relations. Because of the above-mentioned convention, incomparable objects are arranged in *levels*. Sometimes a compromise between the symmetry demand and the general clearness of the diagram is to be accepted. The concept of levels is further discussed below.

The concept of order preserving maps plays an important role in applications of Hasse Diagram Technique (HDT). For an introduction, this concept will be exemplified by the so-called level construction (see Fig.6).

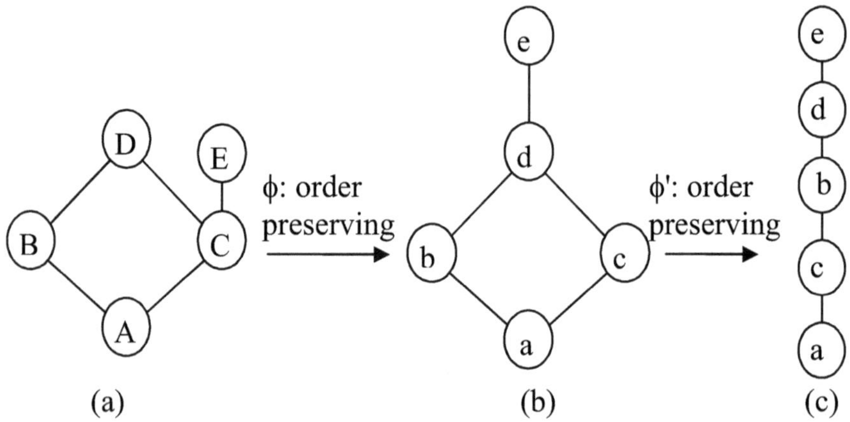

Fig. 6. The Hasse diagram (left side (a)) is mapped onto the Hasse diagram (b). All order relations of the domain set, and order relations (left side) are preserved in the range of the mapping ϕ (right side). Finally an order-preserving map ϕ': $F \rightarrow F$ is applied to obtain a linear order (diagram (c)).

Let E be a set of objects and F another set. Let $x_1, x_2, \ldots$ be the objects of E and $y_1, y_2, \ldots$ the objects of F. An assignment $f(x_i) = y_i$ is order preserving, if any order relation $x_i \leq x_j$ is maintained, i.e. $f(x_i) \leq f(x_j)$ or $y_i \leq y_j$. Thus, if a set {A, B, C, D, E} (Fig. 6a) is assigned to the set {a, b, c, d, e} as follows: f(A) = a, f(B) = b, f(C) = c, f(D) = d, f(E) = e then in order to obtain an order preserving map one has to demand: a < b, b < d, a < c, c < e, c < d as e.g. in Fig. 6b and c. It should be noted that the order C < E is maintained. Thus, f(C) < f(E) or c < e. This is not affected by the creation of a new order d < e. Indeed: Very often an order-preserving map is associated with an enrichment of comparabilities.

Assignments as ϕ are often called mappings, the mapping relate one set (the domain) to another one (the range of a map). Often it is very useful that the order of the image is a linear one. Especially in QSAR applications as shown in chapter by El-Basil, p. 3 the quantity of interest, for example a toxicity of substances, induces a linear order, whereas information on chemicals (say: topological indices or other codes of the chemical structure) leads to a partial order (for example visualized by Young diagrams). Then the art is, to find such topological indices that the partially ordered set can be related to the linear order by an order-preserving map.

There are several possibilities to construct linear orders. Theoretically very important is the concept of linear extensions, which is explained later (vide infra).

Another concept is that of the "levels". Linear orders by a level construction encompasses in HDT the following steps:

1. Set i = 1
2. Consider for the first steps of construction the quotient set $E/\Re$ (not the set of objects E). The set of the maximal elements, MAX, is thus the subset of $E/\Re$.
3. Identify the maximal elements (in $E/\Re$) and label the set MAX_1
4. Reduce the set $E/\Re$ by the maximal elements MAX_1, $E/\Re_{new} = E/\Re_{old} - MAX_1$
5. Draw the elements of MAX_1 in top-position in the drawing plane. All elements of MAX_1 get the same vertical position.
6. Add 1 to i. I.e. $i_{new} = i_{old} + 1$.
7. Identify the new maximal elements of $(E/\Re - MAX_{i-1}, IB)$. Label the new set MAX by i.
8. Reduce the set $E/\Re$ by the maximal elements MAX_i, $E/\Re_{new} = E/\Re_{old} - MAX_i$
9. Draw the maximal elements MAX_i in the same vertical position. Elements of MAX_{i-1} will located below those of MAX_i .
10. Repeat the steps 6-9 till $E/\Re$ is exhausted. The corresponding i is C_{max}, the number of elements in the maximal chain of $(E/\Re, IB)$.
11. Corresponding to the intended application: a) give the top elements the level no C_{max} and the lower levels C_{max} -1 , C_{max}-2, ... ,1 or b) keep the i-labelling as level-label. In that case the bottom elements get the level number C_{max} and the top elements 1.
12. If wanted, the order relations can be added as edges.

This construction is order preserving.

A detailed example may be helpful:

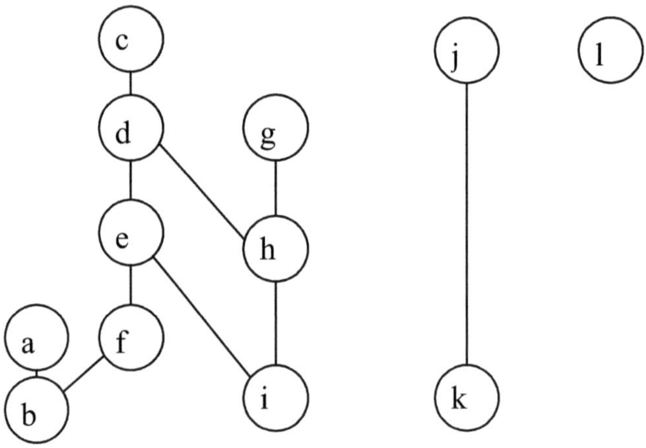

Fig. 7. Hasse diagram of 12 elements. E = {a, b, c, d, e, f, g, h, i, j, k, l}. Note that the Hasse diagram is not drawn following the convention of the program WHASSE in order to clarify the construction

- Step 1: i = 1
- Step 2: E can be identified with $E/\mathfrak{R}$, because there are only trivial equivalence class (i.e. singletons).
- Step 3: MAX_1 = {a, c, g, j, l}
- Step 4: $E/\mathfrak{R}_{\text{new}}$ = {a, b, c, d, e, f, g, h, i, j, k, l} - {a, c, g, j, l} = {b, d, e, f, h, i, k}
- Fig. 8 shows the resulting Hasse diagram:
- Step 5: (see Fig. 9)
- Step 6: i = 2
- Step 7: MAX_2 = {d, k} (see Fig. 8)
- Step 8: $E/\mathfrak{R}_{\text{new}}$ = {b, d, e, f, h, i, k} - {d, k} = {b, e, f, h, i}
- Step 9: (see Fig. 9)
- Step 6: i = 3 (iteration)
- Step 7: MAX_3 = {e, h} (see Fig. 8).
- Step 8: $E/\mathfrak{R}_{\text{new}}$ = {b, e, f, h, i} - {e, h} = {b, f, i}
- Step 9: (see Fig. 9)
- Step 6: i = 4 (iteration)
- Step 7: MAX_4 = {f, i}
- Step 8: $E/\mathfrak{R}_{\text{new}}$ = {b}
- Step 9: (see Fig. 9)
- Step 6: i=5

- Step 7: $MAX_5 = \{b\}$
- Step 8: $E/\Re_{new} = \phi$
- Step 12: $C_{max} = 5$. We follow the labelling of a). See Fig. 9 left side for the level structure and right side for the diagram, supplied with the order relations:

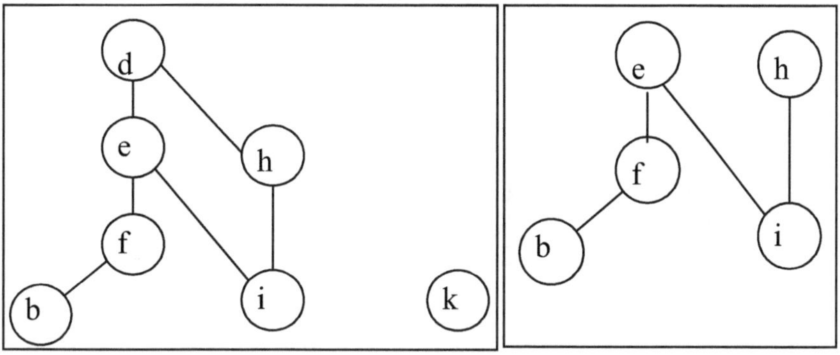

Fig. 8. The resulting poset and its visualization after subtracting the maximal elements of $(E/\Re, IB)$ after the start and the first iteration

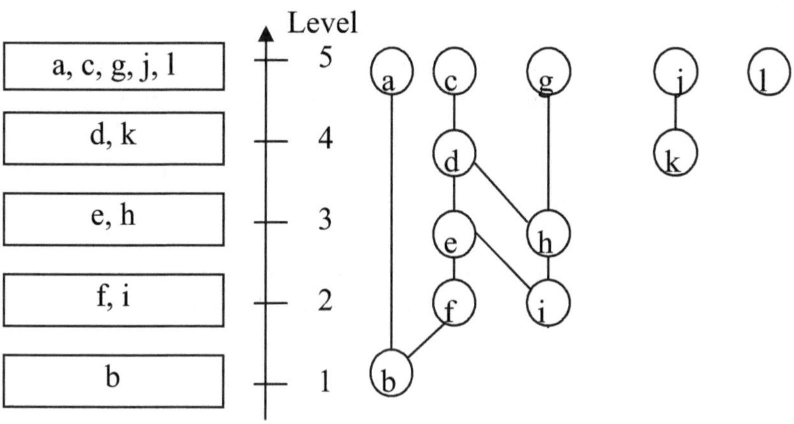

Fig. 9. Example to determine the level structure (Left side: Assignment to levels Right side: the Hasse diagram redrawn)

These steps sound difficult, however they are easily understandable, just by doing! Here some examples (Fig. 10)

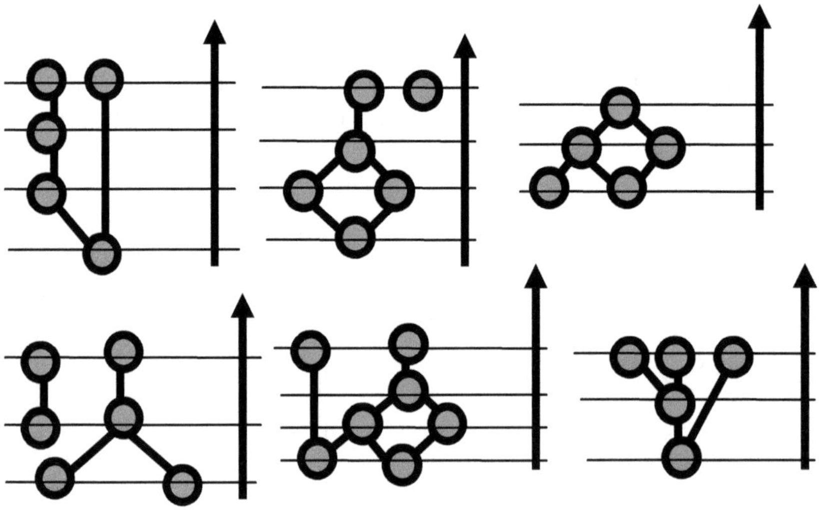

Fig. 10. Example, how to assign the levels. If one vertex contains several equivalent objects, than these objects belong all to the same level. The vertical arrow symbolizes the order induced by the vertical arrangement of the vertices

The levels may be considered as a first very crude evaluation: If a high level is associated with a high hazard, then the sequence of increasing levels coincides with increasing hazard.

In the above advices 1-6, the rule 4 needs additional explanations. In order to do this, we introduce first the concept of graduation and of the rank-function, respectively. If there is a rank function r, then for any element of the ground set the levels are uniquely found. Hence, a poset is graded or possesses a rank function if:

a) $x > y$ implies $r(x) > r(y)$ (order preserving!) and
b) for x covering y a unique function r can be found, such that $r(x) = r(y)+1$.

In the case, shown in Fig. 11 (a) such a rank function exists, whereas in Fig. 11 (b) one cannot find a function r. Obviously, for the Hasse diagram in Fig. 11 (a) all five objects are located at specific levels, whereas the hatched object in the diagram in Fig. 11 (b) may be located either at the level of x or the level of y, respectively. However, corresponding to the level construction the element u belongs to MAX_2. The elements u and z have therefore the same vertical position and are below the top element, which belongs to MAX_1.

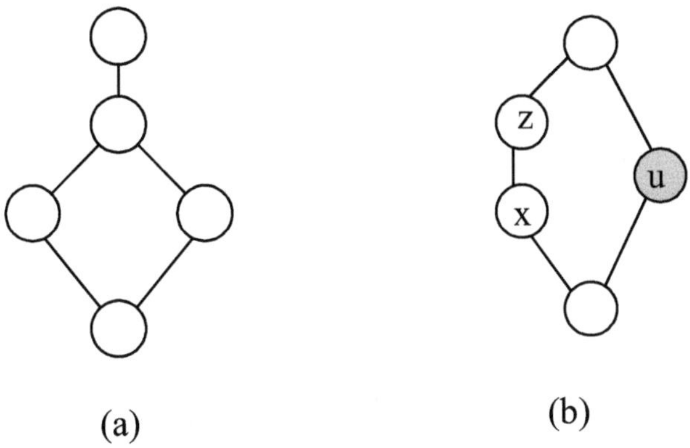

Fig. 11. Graded (a) and non-graded (b) posets (visualized by Hasse diagrams) Grey circle, x and z: (see text)

Posets, which do not have a rank function, give the user of Hasse diagrams the additional freedom, for example to introduce further information. Locating an element as high as possible obviously is a conservative approach. Thus, in, e.g., risk assessment, high values of attributes are associated with high risk. Locating an element of a poset as high as possible has thus a warning function.

A useful theorem to find out whether a rank function r exists, is the so-called Jordan-Dedekind Chain Condition (JDCC) (see also Birkhoff 1984), stating that all maximal chains between the same endpoints have the same finite length. Thus, if a poset satisfies JDCC, a rank function can be found. In Fig. 11 (a) there can be found two maximal chains. Both have the same length. In Fig. 11 (b), once again two maximal chains can be found. However, they differ in their length. Hence the JDCC is hurted in case of the poset, visualized in Fig. 11 (b). A generalization of rank functions for lattices is given in Freese (2004). However, as most empirical posets do not satisfy the axioms of lattices, we will not deepen this concept here.

Hasse diagrams as digraphs

Hasse diagrams can be interpreted as mathematical graphs, i.e. they are called d*igraphs* (directed graph), because of the orientation of the lines. Following the definitions of order the digraphs are acyclic. Interpreted as ordinary graphs, Hasse diagrams are *triangle-free*: Due to the rule of transitivity, line segments corresponding to a < c can be omitted if a < b and at the same time b < c. A digraph consists of a set *E* (or $E/\mathfrak{R}$ if the quotient set is to be partially ordered) of vertices (circles in Hasse diagrams) and a set of oriented edges each connecting two vertices. If the vertices are drawn in the diagram according to the above rules (defining the level-construction) then the arrows can be simply be represented by lines, because then the element x will be arranged below y, if x < y. Therefore the orientation of the line is replaced by the vertical location in the drawing plane. The circles are the objects of *E*, or elements of the set $E/\mathfrak{R}$ to be ranked.

The basic essence is that by the order relation a data matrix is represented by a mathematical graph with objects as vertices and that the structure of this graph tells us somewhat about the data structure. As the data matrix arises from external studies (experimental work, modelling, empirical data) the resulting graph is called an "empirical graph", which may have (hitherto hidden) regularities. A main task in performing partial order as an exploring tool is just to detect (by abstraction, by simplification) regularities or structures in the graph. Helpful, however still not yet fully developed, is that one can establish an algebra among a set of posets, which reveals different kinds of sums, products and exponentiation, see for example Jonsson 1982.

The concepts "hierarchy", "articulation points", "chains" and "antichains" are very basic and simple ones, which direct into the structural analysis of digraphs. These concepts will be explained in the next section.

Simple elements of interpreting a Hasse diagram

Overview

The basics to consider Hasse diagrams are to check

1. the system of comparabilities and incomparabilities
2. the priority elements
3. pattern of attributes and
4. identifying data structures.

Almost all these kinds of analyses of Hasse diagrams can be found in the different chapters of this book.

Example

A simple example is given in the following (Table 2 and Fig. 12).
There are three hierarchies. One of them is a trivial hierarchy as it consists of one element only, i.e., element f that is not comparable to any other elements. Such elements are also called *isolated elements*. If only few isolated elements are found, whereas almost all other are comparable, then the

isolated elements should be examined carefully as very often specific data structures are the reason for their isolation.

Table 2. A more extended example, demonstration of isolated hierarchies

objects\attributes	q_1	q_2	q_3
a	2	2	3
b	1	2	2
c	2	1	2
d	4	3	1
e	4	2	1
f	0	0	5

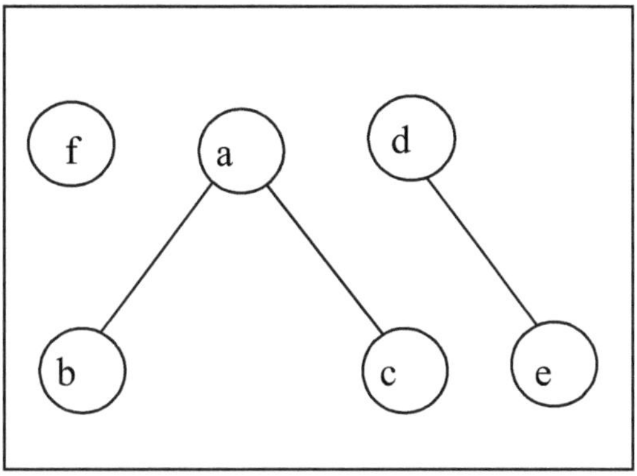

Fig. 12. The partially ordered set of objects of Table 2 has a Hasse diagram with three isolated hierarchies, namely ({a, b, c}, {q_1, q_2, q_3}), ({d, e},{q_1, q_2, q_3}) and ({f}, {q_1, q_2, q_3})

If, on the other hand, all elements of E (or $E/\mathfrak{R}$) are isolated then the attributes should be checked for the degree of anti-correlation (Spearman rank correlation). It depends on the scientific question, whether such a trade-off among attributes (a decreasing sequence of values of one attribute is always accompanied by an increasing sequence of another attribute) should be maintained in the study. There are methods to deal with such cases, see the chapter by Simon et al., p. 221 and by Sørensen et al., p. 259. However, this shall not be further discussed here. The subsets {d, e} as well as {a, b, c} form *nontrivial hierarchies*. Hence, we have three order relations: b ≤ a, c ≤ a, and e ≤ d. The fact that the set E can thus be partitioned into three disjoint subsets is always of great interest with respect to the data structures. Further structural elements, which are of interest in the analysis of Hasse diagrams, are subsequently discussed:

Chain: Subset of the ground set, where all elements are mutually comparable. An example is the chain ({d, e},{q_1, q_2, q_3}) another: ({b, a},{q_1, q_2, q_3}) (Fig. 12). Often it is sufficient, simply to write {d, e} is a chain. Any other element of the ground set added would led to at least one incomparability and thus hurts the definition. Therefore the chains {b, a}, {c, a}, {e, d} are maximal. The identification of chains is of high interest with respect to exploring data structures, because the generality principle demands that for all attributes of objects of a chain it is valid. Thus, if $x < y$, x, y being elements of a chain, then $q_i(x) < q_i(y)$ implies $q_j(x) \le q_j(y)$ for all $j \neq i$. Following the elements of a chain in one direction (from top to bottom or (exclusively) from bottom to top) the attributes are increasing in a weak monotonous manner.

Anti-chain: Subset of the ground set, where all elements are mutually incomparable. An example is the anti-chain ({f, a, d}, {q_1, q_2, q_3}). Any other element of the ground set added to the set {f, a, d} would introduce a comparability. Therefore {f, a, d } is a maximal anti-chain. Attribute profiles being results of monotonous variations as seen in chains are not considered as essentially different. Contrary, attribute profiles through anti-chains are essentially different. Hence the width, Wd(*E*), of the poset is considered as a measure of diversity.

Maximal elements (often also called simply "maximals"): Elements of the ground set $E/\mathfrak{R}$, x_i, for which no $y_i \in E/\mathfrak{R}$ can be found with $x_i \leq y_i$. Maximal elements in the Hasse diagram, shown in Fig. 12 are: f, a, d.

Minimal elements (often also called simply "minimals"): Elements of the ground set $E/\mathfrak{R}$, x_i, for which no $y_i \in E/\mathfrak{R}$ can be found with $x_i \geq y_i$. Minimal elements in the Hasse diagram, shown in Fig. 12 are: f, b, c, e.

Isolated elements: Elements that are both: Minimal and Maximal elements. Maximal/Minimal elements which are not isolated, are often called proper maximal/minimal elements. An isolated element in the Hasse diagram, shown in Fig. 12 is: f.

Hierarchy: Let $E'/\mathfrak{R}$ and $E''/\mathfrak{R}$ be two subsets of $E/\mathfrak{R}$. If for all $x \in E'/\mathfrak{R}$, and all $y \in E''/\mathfrak{R}$: $x \| y$ then $(E'/\mathfrak{R}, IB)$ and $(E''/\mathfrak{R},)$ are hierarchies. In a Hasse diagram they can often be recognized as non connected parts.

Articulation point: If the elimination of one element of $E/\mathfrak{R}$ enhances the number of hierarchies in the residual poset, then this element is called an articulation point. In the Hasse diagram, Fig. 12 the element a is an articulation point.

Long chains, hierarchies and articulation points indicate specific data structures. The role of hierarchies will be explained by a two dimensional scheme (Fig. 13): Several objects may be located as points within the two rectangles H_1 and H_2. Comparing one object of H_1 with one of H_2 will lead to q_1 (of $x \in H_1$) > q_1 (of $y \in H_2$), whereas q_2 (of $x \in H_1$) < q_2 (of $y \in H_2$). Hence no object of H_1 is comparable with that of H_2. In Neggers & Kim 1998 a rather nice wording is found for the objects belonging to the field *F* and *P*: These are the future objects relative to the objects in the field *P*, which are called the objects in the past.

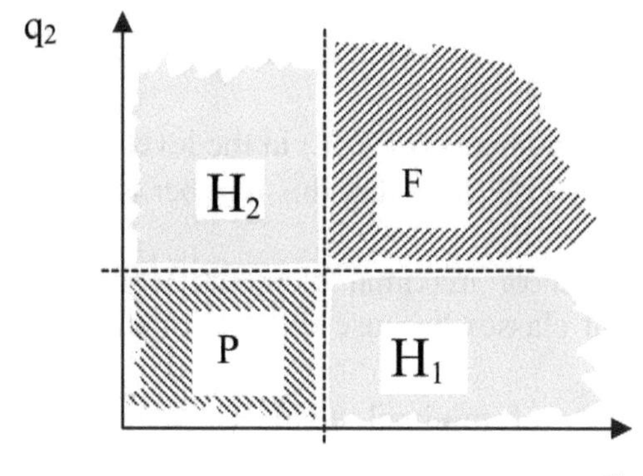

Fig. 13. Role of hierarchies in Hasse diagrams

Note that by construction of levels any level is to be considered as anti-chain. However, this anti-chain may not necessarily be a maximal one. The evaluation of sampling sites for sediment samples of the Lake Ontario is used as a further illustrative example (cf. pp. 94). More details can be found in Brüggemann & Halfon (1997) and Brüggemann et al. (2001 b).

Characterizing a Hasse diagram as a whole

Characteristic Numbers of Posets

In the present section a series of simple characterising numbers is introduced. They are useful to give a general overview and impression of the poset and the corresponding visualizing graph, the Hasse diagram. It is recommended to read the careful discussion by Pavan & Todeschini 2004 and in this book, chapter by Pavan, p. 181. The Hasse diagram of Lake Ontario will exemplify all numbers.

- NECA: Number of equivalence classes with more than one object, i.e., the number of nontrivial equivalence classes.
- Wd(E): The width of a Hasse diagram. It is the maximum number of elements of $E/\Re$, which are found in an anti-chain. In the context of Young diagrams (see Seitz, p. 373) also called a "breadth".
- L(E): The length of a Hasse diagram: The number of line segments in the chain with a maximum number of elements of $E/\Re$.

- H(*E*): The height of a Hasse diagram = C_{max}. H(*E*)=L(*E*)+1. H(*E*) is the number of objects (of *E*/$\mathfrak{R}$) in the maximum chain.
- NL, the number of levels = H(*E*).
- NEL, the number of elements (of *E*/$\mathfrak{R}$) in the level, which contains the most elements of *E*/$\mathfrak{R}$; note that this number is not necessarily the same as Wd(*E*).
- NMAX: The number of maximal elements (called: number of maximal equivalent classes because this information is related to *E*/$\mathfrak{R}$).
- NMIN: The number of minimal elements (notation as for the maximal elements).
- Z: Number of all equivalence classes, including singletons, i.e., Z = card *E*/$\mathfrak{R}$. Note that Z and NECA differ. If NA is the number of elements of *E*, which are contained in nontrivial equivalence classes (NECA) then the following equation holds
  card *E* = NA + Z - NECA (2)
  Some other numbers are also interrelated, for example the relation
  NL = L(*E*)+1 (3)
- P(*IB*): stability of ranking. This quantity is a measure for the effect of extending or reducing the set of attributes on the structure of the Hasse diagram. It is calculated as the quotient of all incomparabilities, U_{total} and Z·(Z-1)/ 2:

$$P(IB) = \frac{2 \cdot U_{total}}{Z \cdot (Z-1)} \quad (4)$$

If P(*IB*) is near 1 or 0, respectively, then extending or reducing, respectively, the set attributes should have a minor effect.

Linear Extensions

The linear extensions are the basis of the dimension theory of posets. Besides the dimension of posets other characterizations may be derived from linear extensions (Carlsen et al. 2002, Lerche et al. 2003, Lerche & Sørensen 2003).

Extensions may be explained by the following: Given a poset (*E*, ≤) then we can assign another poset (*EX(E)*, ≤) which

1. supplies some ||-relations of (*E*, ≤) by < or > -relations
2. maintains all comparabilities of *E* in the correct orientation

Extensions are order-preserving maps from the ground set E into the ground set E; see Davey & Priestley 1990. Linear extensions $(LEX(E), \leq)$ are order-preserving maps from E to E, which assign to $(E, \leq)$ a linear order.
In Fig. 6 (b) an extension is shown (identify A with a, B with b, etc), but not a linear one. An additional preserving map leads to a linear order (Fig. 6 (c)). The diagram in Fig. 6 (c) is a linear extension of that in Fig. 6 (a). Given a poset $(E, \leq)$ then several linear extensions $(LEX(E),\leq)$ are possible. A systematic procedure is described by Atkinson (1989), especially for trees a closed formula can be derived Atkinson (1990). A useful formula to calculate the number of linear extensions is also given by Stanley (1986).

Each relation $x \leq y$, $x, y \in E$ is reproduced in $LEX(E)$. However, the reverse statement is not true. All in all, any linear extension is an image of an order-preserving map. The diagram (Fig. 14) visualizes the concept. All comparabilities $x < y$, $x, y \in E$, of $(E,\leq)$ are reproduced in the first fourteen lines of the table, whereas the last sequence (16th row) in the table illustrates a non order preserving map. The relation $d \leq e$ of $(E, \leq)$ is reversed. This sequence therefore is no linear extension of the poset (shown in the left side of Fig. 14). If the sequences (1) to (14) are considered as partially ordered sets, then they have comparabilities, which are not found in the original poset. For example the elements b and e are comparable in the 14 sequences of Fig. 14, but are incomparable in the original poset. The incomparability of b, e is expressed in the linear extensions by the fact that there are some, where $b > e$, and some where the opposite is true.

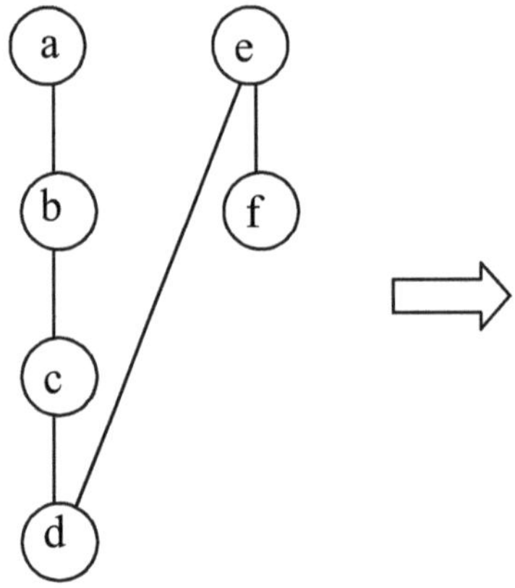

1	a	b	c	e	f	d		2
2	a	b	c	e	d	f		2
3	a	e	f	b	c	d		2
4	a	e	b	f	c	d		4
5	a	e	b	c	f	d		4
6	a	e	b	c	d	f		3
7	e	f	a	b	c	d		1
8	e	a	f	b	c	d		3
9	e	a	b	f	c	d		3
10	e	a	b	c	f	d		3
11	e	a	b	c	d	f		2
12	a	b	e	f	c	d		2
13	a	b	e	c	f	d		4
14	a	b	e	c	d	f		3
	a	b	c	d	e	f		

Fig. 14. Poset (E, ≤) (left side) and its 14 linear extensions (*LEX(E,* ≤)) ((1) to (14).

In Fig. 14 the first column labels the linear extensions, which are represented as sequences in rows 1 - 14. A sequence a b c ... is to be read as a > b > c. Furthermore there is a sequence (last row in the table) which is **not** a linear extension of (E, ≤). Vertical bold lines indicate jumps (see below). The last column indicates the number of jumps of each single linear extension. Consecutive elements in linear extensions (*LEX(E)*, ≤), which have no correspondence in (E, ≤) are called "*jumps*" (see Fig. 14, the vertical bold lines indicating jumps). The jump number, jump ($LEX_i(E, \leq)$), obviously depends on the actual selected linear extension. The jump number of a poset (E, ≤), jump (E, ≤), is just min(jump(LEX_i (E, ≤))), whereby the minimum is to be found by checking all linear extensions. Beside the jump - number there is also a bump - number. Once again the bump number is to be referenced to a specific linear extension. A bump is a consecutive pair of elements in a linear extension, which are comparable in the underlying poset. The bump number of a poset is the maximum about all bump numbers found for the linear extensions. If a linear extension of n elements is formed then n-1 consecutive relations are found in a linear extension. Therefore

$$\text{jump}\ (LEX_i(E, <)) + \text{bump}\ (LEX_i(E, <)) = n - 1 \qquad (5)$$

Linear extensions of a minimal jump number of specific interest: These linear extensions (also called "greedy linear extensions") preserve as much as possible the chain-structure of a Hasse diagram (Rival 1983, Rival & Zaguia 1986). As one can see in Fig. 14 that linear extension with jump number = 1 preserves both chains a > b > c > d and e > f. Consequently, the jump number of a poset may be considered as indicator for "chainyness": Thus, a low jump number indicates that the poset contains subposets, which are long chains. In operation research or queuing plans a jump implies often some cost-intensive rearrangements. Therefore linear extensions with a small number of jumps are preferred in organisation of work. Contrary to that, Patil and Taillie 2004 are discussing in their paper that the jump number may also serve to weight linear extension, where the linear extension with the largest number of jumps gets the highest weight.

If a specific element, say $x \in E$ is selected then its *spectrum* is of interest (Atkinson 1990). It should be noted that other authors (for example Trotter 1991, Schröder 2003) also call the spectrum a projection. However, we favour "spectrum" as the more suitable name for the following construction. Thus, let LT be the number of linear extensions of a poset, then we can find the rank of an element x in the ith linear extension: rank(i, x). Note that this construction should not be confused with the rank function, we discussed above. Conventionally, the bottom element of a linear extension is given the rank 1, thus the top element has the rank n (card E = n). However, if appropriate the top element may be assigned the first priority, such that bottom elements will get numbers > 1 (see for example chapter by Carlsen, p. 163). We call $\lambda_k(x)$ the frequency, how often $x \in E$ gets the rank k. The spectrum spec(x) is a tuple containing n components ($\lambda_1(x)$, $\lambda_2(x)$, ... ,$\lambda_n(x)$). Thus for example the spectrum of element b in Fig. 14 as follows: spec(b) = (0, 0, 3, 6, 5, 0). (i) There is no linear extension, where the rank of b is 1, 2 or 6. (ii) There are 3 linear extensions, where the rank of b is 3. (iii) There are 6 linear extensions, where the rank of b is 4. (iv) There are 5 linear extensions, where the rank of b is 5. Obviously:

$$LT = \sum \lambda_k(..) \qquad k = 1,...,n \tag{6}$$

The set of linear extensions is the basis for probability considerations: Dividing $\lambda_k(x)$ by LT the quantity prob (rk(x) = k) = $\lambda_k(x)$/LT can be interpreted as (ordinal) probability to get the rank k, sometimes also called "absolute rank". Hence, an averaged rank, Rkav can be derived by

$$Rkav(x) = \sum k \cdot \lambda_k(x)/LT \tag{7}$$

and the elements $x \in E$ can be ordered by their Rkav-values. Therefore a total order, however, often including equivalence classes can be derived from a poset, without the numerical combination of attributes to one ranking index. This concept is widely used based on the following arguing: If the attributes are combined, say by weighted sums or any other positively monotonous function, then the result must be (besides ties) one of the linear extensions, as the set of all linear extensions encompasses all results of order preserving maps. However, there are still many open problems due to computational difficulties in handling large object sets, advices can be found in Lerche et al. 2003 or in Patil & Taillie 2004. If n objects are mutually incomparable, then n! linear orders are possible, corresponding to n! permutations. (See also chapter of Sørensen, Lerche, Thomson, p. 259, for a discussion of entropy, related to the number of linear extensions). Hence a crude upper estimation of the number of linear extensions is n! A ground set containing for example 17 elements may have at most ca. $3.5\ 10^{14}$ linear extensions.

Recently an alternative was discussed, in order to use a local model of the partial order, which describes the environment in the directed graph around the element of interest. For further discussions two recent publications should be consulted (Brüggemann et al. 2004, 2005).

A rather good approximation for an element of interest, x, may be obtained, if the successors (all elements "below" x) and predecessors (all elements "above" x), respectively, are organized into a so-called "S-x-P" chain, all remaining elements, i.e. those incomparable to x being considered as isolated. From a combinatorial study follows that the averaged rank of an element x can be expressed as

$$\text{Rkav} = \frac{\sum_{k=0}^{k=U} (S+1+k) \cdot \binom{U}{k} \cdot (S+1)^k \cdot (P+1)^{U-k}}{\sum_{k=0}^{k=U} \binom{U}{k} \cdot (S+1)^k \cdot (P+1)^{U-k}} \tag{8}$$

which can be transformed (Brüggemann et al. 2004) into

$$\text{Rkav} = (S+1) \cdot (S+P+U+2)/(S+P+2) \tag{9}$$

Since N = S+P+U+1 the averaged rank of an element x may be expressed by the following simple relation [1]

$$\mathrm{Rkav}(x) = \frac{(S(x)+1)\cdot(N+1)}{N+1-U(x)} \tag{10}$$

with:

- $S(x) := |\{y \in E : y < x\}|$ is the number of successors of x
- N the total number of elements and
- $U(x) := |\{y \in E : y \parallel x\}|$ is the number elements incomparable to x.
- $P(x) := |\{y \in E : y > x\}|$ is the number of predecessors.

The principle is illustrative demonstrated by determining the averaged rank of element b in the Hasse diagram depicted in Fig. 15.

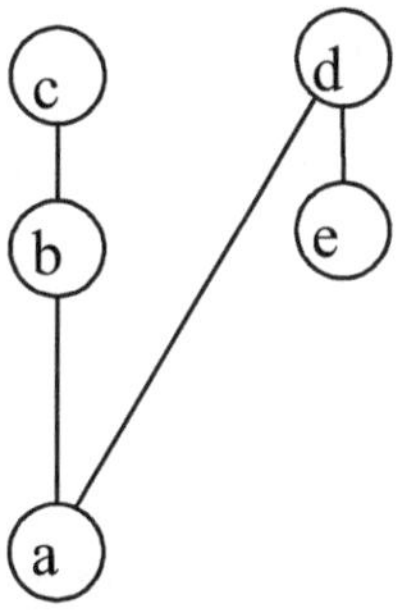

Fig. 15. Example for application of equation 10. The averaged rank of element b is to be estimated

It is immediately seen that N = 5, S(b) = 1 (element c), P(b) = 1 (element a), and U(b) = 2 (elements d and e). Hence, according to equation (10) the averaged rank of element b is estimated to be Rkav(b) (estimated) = (1+1)·(5+1)/(5+1-2) = 3, the exact value - calculated after equation 7 - being Rkav(b) (exact) = 2.889.

Dimension of a poset

The dimension of a poset is based on the set of linear extensions. A linear extension can be considered as a set of ordered pairs. For example the linear extension no 1 in Fig. 14 (right side):

[1] Counting from bottom to top.

{(a, a), (b, a), (c, a), (e, a), (f, a), (d, a), (b, b), (c, b), (e, b), (f, b), (d, b), (c, c), (e, c), (f, c), (d, c), (e, e), (f, e), (d, e), (f, f), (d, f), (d, d)}. Each pair denotes a ≤ -relation. For example (e, b) means that in the linear extension no 1 e ≤ b.

A similar set could be found for any other linear extension, for example no 2:

{(a, a), (b, a), (c, a), (e, a), (d, a), (f, a), (b, b), (c, b), (e, b), (d, b), (f, b), (c, c), (e, c), (d, c), (f, c), (e, e), (d, e), (f, e), (d, d), (f, d), (f, f)}

The intersection of these two sets of pairs leads to:

{(a, a), (b, a), (c, a), (e, a), (d, a), (f, a), (b, b), (c, b), (e, b), (d, b), (f, b), (c, c), (e, c), (d, c), (f, c), (e, e), (d, e), (f, e), (d, d), (f, f)}.

This intersection does not coincide with the set of ordered pairs of the poset itself (Fig. 14 (left side)):

{(a, a), (b, a), (c, a), (d, a), (b, b), (c, b), (d, b), (c, c), (d, c), (e, e), (f, e), (d, e), (f, f), (d, d)}

Thus this kind of troublesome check has to be repeated until the intersection of the set of ordered pairs of the linear extensions coincide with that of the poset. The lowest number of linear extensions -written as ordered pairs as shown above- whose intersection is the actual poset (together with its transitive relations), is its dimension. Following the explanation above one would have to check 14·13/2 intersections, just to verify that the dimension equals 2. If such pair of ordered sets, derived from any two linear extensions is found, one has found a "*realizer*" of the poset (Trotter 1991).

Note, it is not a good policy to derive the dimension by finding explicitly the realizers. Here five useful theorems are taken from the literature (Trotter 1991):

- dim $(E,\leq) \leq \mathrm{Wd}(E)$ (for further on Wd(E), see p. 81) (11)
- Let $(E,\leq)$ a poset and $(C,\leq)$ a chain, $C \subset E$. Then $\dim(E,\leq) \leq 2 + \dim(E\text{-}C,\leq)$ (12)
- Let $(E,\leq)$ a poset, and n:=card $E \geq 4$, then: dim $(E,\leq) \leq n/2$ (13)
- Let $E_A \subset E$ an anti-chain of a poset $(E, \leq)$, then dim $(E, \leq) \leq \max(2, \mathrm{card}(E\text{-}E_A))$ (14)
- If the Hasse diagram, supplied (if necessary) by a greatest and least element can be drawn in the plane without crossing of lines, then the dimension of the poset is 2, (15)
- Let $(E, \leq)$ be a poset and $(E', \leq)$ be a subset of E, then dim $(E, \leq) \geq$ dim $(E', \leq)$ (16)

We apply equation 12 to determine the dimension of the poset shown in Fig. 16:

- Step 1: As the poset is not a linear order we conclude: dim $(E, \leq) > 1$
- Step 2: We select a chain: $C = \{c, b, a\}$
- Step 3: The ground set is now $E\text{-}C = \{d, e\}$. The poset $(\{d, e\}, \leq)$ is a chain.
- Step 4: dim $(\{d, e\}, \leq) = 1$
- Step 5: $1 <$ dim $(E, \leq) \leq 2 + 1$. Thus the dimension of $(E, \leq)$ is either 2 or 3.

Equation 11 would be more useful: As Wd(E) of the poset, shown in Fig. 15 is 2, the dimension must be 2. Generally, for the purposes intended in this chapter equations (15) and (16) are the most interesting theorems.

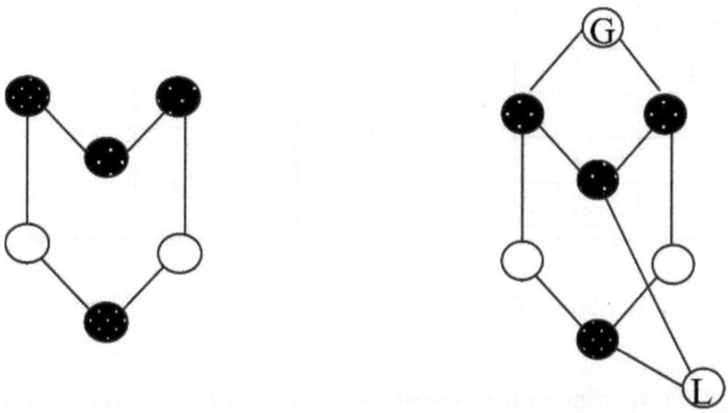

Fig. 16. Hasse diagram of a poset with dimension 3. The Hasse diagram on the left side follows not the convention explained earlier! The Hasse diagram on the right side is supplied by a greatest "G" and least element "L".

The poset, whose Hasse diagram is shown in Fig. 16 (left side) has the dimension 3. A priori, as obviously there is no crossing of lines, the dimension would be expected to be 2. However, this poset must be extended by a greatest, G, and a least element, L. Then a crossing of lines within a plane is not avoidable. Thus posets having such substructure have at least dimension 3. For other examples, compare Trotter 1991. Why is the dimension of posets so interesting? Let us assume we got a Hasse diagram by using 5 attributes. If now, the dimension of the partial order would be 2 then we knew in advance that two linear extensions are sufficient to reproduce the partial order. As each single linear extension can be considered as the linear order induced by an unknown attribute, two attributes are sufficient to obtain the same partial order as by the original five ones. Usually these two attributes cannot be found as a subset of the information base.

They are called latent variables. The original five attributes may be (in a complex manner) mapped onto 2 latent variables.

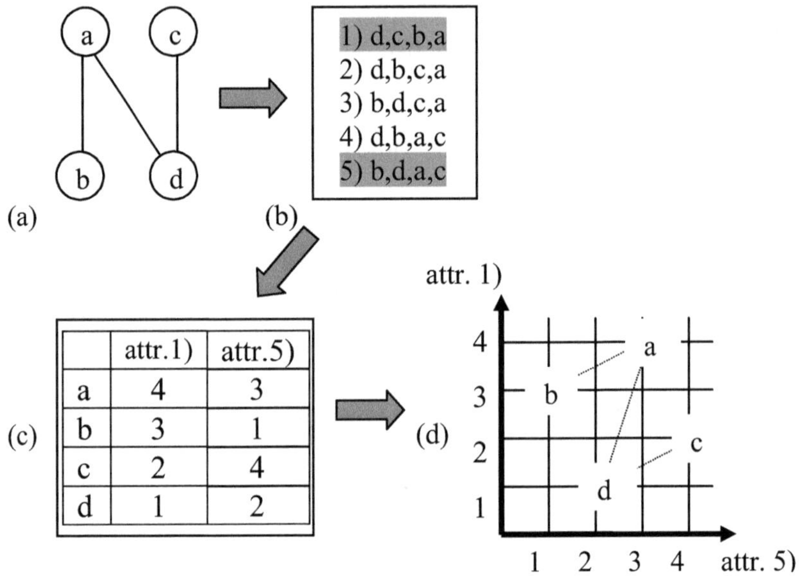

	attr.1)	attr.5)
a	4	3
b	3	1
c	2	4
d	1	2

Fig. 17. The partial order of four elements and the concept of dimension

In Fig. 17 a Hasse diagram (a) and its 5 linear extensions (ordered for increasing values) (b) are shown. Two realizers (grey hatched) are identified and are considered as new attributes ("attr. 1"; "attr.5") (c). The objects are located in a rectangular grid (d) due to values of attr. 1 and attr. 5. A rotation of the coordinate system around ca 45° would reproduce the original Hasse diagram. If the dimension of posets is 2 or 3 then it may be useful, to embed the poset into a two- or three-dimensional grid (see Brüggemann 2001 b). For an example of embedding a poset into a two-dimensional coordinate system, see also the chapter of El-Basil.

On the other hand, any two-dimensional scatter plots can be interpreted as a partial order, if the generality principle is applied to the both coordinates of any point.

Sensitivity study

Mathematical Notation and Background

Preferably a maximal element should be chosen as a starting point for the analysis. This choice, however, is not mandatory. Thus, other elements of E or $E/\Re$ could be chosen too. This selected element is called "key element". We may simultaneously select more than one key element even all elements (no restrictions apply here). For the sake of convenience all key elements are supposed to form a set K ($\subseteq E$).

The analysis of a key element implies a search of all elements located lower than that of the key element, i.e. all elements that can be reached from the key element by a path, a sequence of connecting edges. (Therefore the selection of maximal elements rather than other elements is more meaningful). These elements together with elements equivalent but not identical to the key element are called successors. The set of all successors of the key element "k" is denoted as $G(\text{k},A)$, $A \subseteq IB$. We include the information about the actual set of attributes (i.e. the case) by A. Note the similar concept of "down-sets" in Davey & Priestley (1990): The order ideal (or down set), generated by the key element will be denoted by $O(\text{k},A)$. Then it is valid:

$$G(\text{k},A) = O(\text{k},A) - \{\text{k}\}$$

The operation "-" is the set theoretical subtraction. For example:
$\{a, b, c, d\} - \{a, e\} = \{b, c, d\}$

Those elements of the first set, which also are in the second set, are eliminated. By definition $G(\text{k})$ does not include the key element itself. The successor sets and their cardinalities are the heart of the sensitivity analysis shown here. The successor sets found for two Hasse-diagrams resulting from two attribute-subsets of IB are used to quantify certain differences. The cardinality of successor sets (denoted: card $G(\text{k})$) and of their set theoretical combinations play an important role here.

Residual sets

To assess the influence of each attribute on ranking, we compare Hasse diagrams that arise from subsets *B, C* of *IB*. A straightforward method to perform this task is to choose a key element and quantify the effect of each attribute set on its successor set. For this purpose the residual set, R, i.e. is now introduced.

$$R(\mathrm{k}, B, C) := (G(\mathrm{k}, B) \setminus G(\mathrm{k}, C)) \tag{17}$$

By Venn-Euler diagrams residual sets can easily be understood (see Fig. 18):

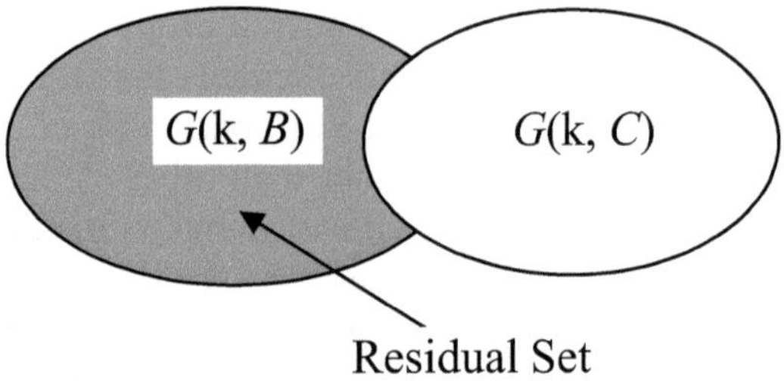

Fig. 18. Venn-Euler diagram of the residual set $R(\mathrm{k}, B, C) = G(\mathrm{k}, B) - G(\mathrm{k}, C)$

In general $R(\mathrm{k}, B, C) \neq R(\mathrm{k}, C, B)$. Therefore the symmetric difference set "$W(\mathrm{k}, B, C)$" of the sets $G(\mathrm{k}, B)$ and $G(\mathrm{k}, C)$ is introduced:

$$W(\mathrm{k}, B, C) := R(\mathrm{k}, B, C) \cup R(\mathrm{k}, C, B) = [G(\mathrm{k}, B) - G(\mathrm{k}, C)] \cup [G(\mathrm{k}, C) - G(\mathrm{k}, B)] \tag{18}$$

If the cardinality of $W(\mathrm{k}, B, C)$ is small, i.e.

$$\mathrm{W(k, B, C)} \ll \min [G(\mathrm{k}, B), G(\mathrm{k}, C)])$$

then subsets *B* and *C* lead to not very different Hasse diagrams. If the difference is large then the two corresponding Hasse diagrams are dissimilar to each other. Those attributes, by which *B* and *C* differ, play a key role in ranking. This finding motivates the introduction of the matrix **W**.

Definition of the matrix W

Calculating the matrix **W**

The matrix **W**(k) assesses the difference of Hasse diagrams induced by the two subsets of attributes with respect to a key element k. This matrix, which is at the heart of the analysis, is called the "dissimilarity-matrix", because the larger the matrix-entries are, the greater is the difference between the successor sets for the element k and hence between the Hasse diagrams (see for more details, below). We define the entry W(k, *B*, *C*) of matrix **W** to be:

$$W(k, B, C) := \text{card}\,[R(k, B, C) \cup R(k, C, B)] \quad (19)$$

For any key-element k the residual sets *R*(k, *B*, *C*) and *R*(k, *C*, *B*) are determined, their elements being counted and summed. The entries of the matrix **W** are subsequently calculated by adding the cardinalities of the *R*-sets. To simplify notation, we now write W(k, i, j) for W(k, *B*, *C*).

Search for the important attributes

Several W(k, i, j)'s, $k \in K$ (*K* is any set of key elements) can be compared to see how a change in attributes affects the partial order with respect to the set of several key elements:

$$W(K,i,j) := \sum_{k \in K \subset E} W(k,i,j) \quad (20)$$

W(*K*) is a symmetrical matrix. **W**(*E*) is the total dissimilarity matrix of the set of *E*. Let be n:=card *E*. Mainly the **W**(k) and the **W**(*E*) matrices are useful. The final steps towards a sensitivity are:

1. If we are interested in comparisons of the full attribute set *IB* with all subsets $A_i \subset IB$, A_i only one row of the matrix **W** is of interest. We can choose the first one without loss of generalization, thus we are left with W(k, 0, 1), W(k, 0, 2), ... , W(k, 0, p), where the index 0 denotes the full attribute set *IB* (i.e. $A_0 \equiv IB$) and $p=2^m-1$.
2. To see the influence of single attributes on a Hasse diagram we compare the Hasse diagrams induced by *IB* with those induced by those attribute sets $A_j \subset IB$ with only m-1 attributes ($A_i = IB - \{q_i\}$). Therefore the effect of dropping exactly one attribute is given by the remaining m entries: W(k, 0, 1), W(k, 0, 2), ..., W(k, 0, m).

3. The m entries W(k, 0, 1), W(k, 0, 2), ..,W(k, 0, m) are put together to form a "sensitivity tuple" of the key element k, s(k) being $[s_1, ..., s_m]$.
4. The larger s_i the larger is the symmetrized difference between $G(k, IB)$ and $G(k, A_i)$ and correspondingly the larger the influence of attribute q_i on the position of key element k within the Hasse diagram under *IB* compared with that under A_i.
5. The matrix **W**(k) depends on the selection of the key element k. If however, more objects are to be analyzed we generalize according to equation (20).
6. **W**(*E*) will be used as a measure of sensitivity. Accordingly we quantify the sensitivity by:
7. $\sigma(i) := W(E, IB, A_i)\ 1 \leq i \leq m$ (21)
with the enumeration scheme of step 3).
8. It can be shown that σ(i) has values between 0 and n·(n-1). Hence a measure of attribute's sensitivity, independent of the number of objects is:
9. $\sigma_{norm}(i) = W(E, IB, A_i)/[n\cdot(n-1)]\ .\ 0 \leq \sigma_{norm}(i) \leq 1$

Evaluation of Sampling Sites

Sediment samples of Lake Ontario as object set and the tests of the battery as information base

A battery of tests developed by Dutka et al. 1986 to test the sediments of near-shore sites of Lake Ontario (Canadian part) is used to exemplify the definitions and some results of HDT. In Lake Ontario 55 sediment samples were tested, thus, the set *E* contains 55 objects. Dutka et al. classified their results and used discrete scores instead of the measured (raw) data. For our analysis we have adopted their classification. Thus, s_i denotes the score of the i-th test of the battery. Five specific tests form the actual battery: (1) Fecal Coliforms „FC", as an indicator designed to control the health state of the sediments, (2) Coprostanol „CP" and (3) Cholesterol „CH" both being indicators of loadings by fecals, (4) Microtox tests „MT" and (5) Genotoxicity tests „GT" disclosing some kind of acute toxicity and the potential for carcinogenicity, respectively (see Table 3).

Table 3. Scores of the 5 test battery results for representatives of the equivalence classes of $E/\mathfrak{R}$

identifier	FC	CP	CH	MT	GT	identifier	FC	CP	CH	MT	GT
1	2	0	0	4	0	17	3	0	0	6	0
2	1	0	0	2	0	18	1	0	0	2	4
3	2	0	0	2	0	23	1	0	0	0	4
4	3	0	0	0	0	25	4	0	0	0	0
5	3	3	2	0	0	27	5	0	0	0	0
7	2	0	0	8	0	31	4	5	4	0	0
9	1	0	0	6	2	32	3	0	0	8	0
11	1	0	0	0	0	91	2	0	0	0	0
12	3	0	0	2	0	92	3	0	0	4	0
14	1	0	0	8	0	95	3	5	2	6	0

By scoring the data many equivalence classes (in fact 20) arise (vide infra). It is convenient to refer only to these classes by specifying a representative for each class Thus, besides the sensitivity study we apply the concept of quotient sets. With the equivalence relation $\mathfrak{R}$ meaning equality in all five scores s_{FC}, s_{CP}, s_{CH}, s_{MT} and s_{GT}, the following sediment samples appeared as equivalent, (Table 4) the quotient set being denoted as $E/\mathfrak{R}$.

Table 4. Nontrivial equivalence classes and their battery of tests pattern. No. of sites in bold letters are later used as representatives for the whole equivalence class ec_i

Equivalence Class (ec)	card (ec)	FC	CP	CH	MT	GT
ec_1={**2**,8}	2	1	0	0	2	0
ec_2={**4**,6,10,13,19,21,22,29,30,48,94}	11	3	0	0	0	0
ec_3={**11**,16,40,41,42,43,44,45}	8	1	0	0	0	0
ec_4={15,**92**}	2	3	0	0	4	0
ec_5={**17**,35}	2	3	0	0	6	0
ec_6={20,24,26,28,34,37,39,49,50,51,**91**,93}	12	2	0	0	0	0
ec_7={**23**,60}	2	1	0	0	0	4
ec_8={**27**,33,46,47}	4	5	0	0	0	0

The sites, referred to as site numbers in bold letters are later used as representatives for the whole equivalence class. The site numbers are used as object identifiers.

The quotient set $E/\mathfrak{R}$ consists of the 8 equivalent classes {ec1, ec2, ec3, ec4, ec5, ec6, ec7, ec8} together with remaining 12 singletons {1}, {3}, {5}, {7}, {9}, {12}, {14}, {18}, {25}, {31}, {32}, {95}.

Now we apply all the characteristic numbers of Hasse diagrams, introduced earlier in this chapter.

n = card E = 55, Z = card $E/\Re$ = 20, NECA = 8, NA = 43

Clearly: card E = NA + SG, SG the number of singletons (here: SG = 12) and SG = Z - NECA.

The information base of the battery of tests is: IB: = $\{s_{FC}, s_{CP}, s_{CH}, s_{MT}, s_{GT}\}$. The partial ordering of the samples arises as explained in sections 2 and 3. The visualization of the partial order by HDT is depicted in Fig. 19.

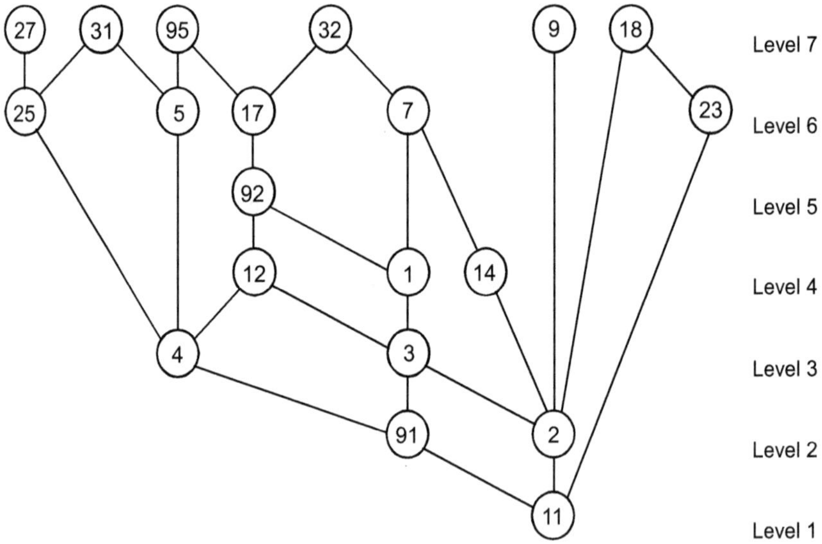

Fig. 19. The comparative evaluation of samples of the Lake Ontario, as generated by the WHASSE software. Hasse diagram of the poset $(E/\Re, \leq)$.

We check now the items discussed in former sections by some illustrative examples. Note, that in the following sections the term sites is used for the single objects/elements covered by the ranking exercise, reflecting the actual nature of the data material.

Comparability:

Taking site 31 as an example it is immediately seen that due to the transitivity (see El-Basil, p. 3) this site is comparable to (and worse than) site 4. Thus, 31 ≥ 4 as we have the sequence 31 ≥ 5 and 5 ≥ 4 from which 31 ≥ 4 follows logically. Likewise, through a longer chain 32 ≥ 17 ≥ 92 ≥ 1 ≥ 3 ≥ 91 ≥ 11 it follows that 32 ≥ 11. We say that 32 are connected to, or comparable to 11, because there is a path, which can be followed without changing the orientation. On the other hand, site 17 is not connected/comparable to site 14, because there is no path, which can be followed from 17 to 14 without changing the orientation: $17 \geq 92 \geq 1 \geq 3 \geq 2$, however: $2 \leq 14$. The relation between the sites 31 and 5, displaying $31 \geq 5$, is a cover relation, whereas the relation between the sites 31 and 4, although $31 \geq 4$ is not a cover relation, as there is an in-between element, i.e., site 5 located between site 31 and site 4. What does comparabilities or chains tell us? By identifying chains we know that the upper object (e.g. site 32) is in all aspects worse than the lower object (e.g. site 3). All attributes increase simultaneously when the path from the lower element, i.e., site 3 to the site 32 is followed. In mathematical terms this can be described as a weak positive monotonous function, i.e. equal or increasing values of all attributes simultaneously following a chain.

Incomparability and Anti-chain:

Site 32 is, e.g., incomparable to site 9 as well as to many others. There is no path (in the digraph) by which we can start from site 32 and stop at site 9 without changing the orientation. It should be remembered that in an ordinary graph there is a path: $32 \rightarrow 7 \rightarrow 14 \rightarrow 2 \leftarrow 9$. However, the arrows recall that in the digraph we have an orientation, whereas in the ordinary graph we only have a line. The set {25, 5, 17, 7, 23} is an example of an anti-chain (cf. Fig. 19). However, this anti-chain is not of maximum length as site 9 could be added without violating the definition of an anti-chain. Large anti-chains indicate a high diversity of attribute profiles. Incomparabilities arise if at least one pair of attributes is antagonistic: i.e. a "walk" from an object x to an object y is accompanied with increasing of at least one attribute and decreasing of at least one other. For an illustration, take the incomparable sites 95 and 32. As the incomparability arises from the fact that CP, CH increase, FC and GT do not change, whereas MT decreases if the path from site 32 to site 95 is followed (cf. Table 3 and Fig. 19).

Priority elements:

As the sampling sites with high responses of the test-battery are of most interest, the maximal elements are taken as priority elements, i.e. the equivalence classes {27, 33, 46, 47}, {31}, {95}, {32}, {9}, {18}. From this we conclude that a) the sites 27, 33, 46, 47, 31, 95, 32, 9, 18 are of specific importance, and b) the set of sites {27, 33, 46, 47} has the same profile of scores, thus, they may be remedied by the same methods, whereas the attribute profiles differ among all other priority objects.

Characterizing numbers:

With the Hasse diagram of Fig. 19 at hand it is easy to derive the remaining characterizing numbers discussed in former sections. Hence, we find L(*E*) = 6, H(*E*) = 7, and NL = H(*E*) =7. These numbers give an impression in which detail the steps from a minimal element to a maximal element may be disclosed. This informs us here about the maximum possible differentiation in the degree of hazards.

In the present case (cf. Fig. 19) a partitioning of *E* (or $E/\mathfrak{R}$) into levels of increasing hazard prevails. Thus, {ec3} < {ec6, ec1} < {ec2, {3}} < {{12},{1},{14}} < {ec4} < {{25}, {5}, ec5, {7}, ec7} < {ec8, {31}, {95}, {32}, {9}, {18}}, the "<" sign reflecting that the sets are ordered corresponding to their level number.

We further find that NEL = 6, which in the present case coincides with NMAX = 6. The number of minimal elements NMIN = 1.

Finally the stability is to be calculated: P(*IB*) = 0.574

This means than on one hand the Hasse diagram will change remarkably, if an attribute is omitted or if an additional attribute is included, leading to new P(*IB*) values of 0.247 and 0.832, respectively. Hence, omitting an attribute changes the Hasse diagram towards a chain, whereas adding a new attribute causes the appearance of several hierarchies, eventually leading to an anti-chain.

See also for another example in chapter by Helm, p. 298.

Linear extensions:

As 20 objects (elements) of $E/\mathfrak{R}$ (= Z) are a rather high number, we would have to expect up to $2\cdot 10^{18}$ linear extensions we restrict our study to the order ideal *O*(95). Its Hasse diagram is shown in Fig. 20.

For the poset, shown in Fig. 20 a total of 66 different linear extensions are possible. In the present context it makes no sense to list them all. For illustration a random selection of 5 linear extensions is listed below:

L1: 11 < 2 < 91 < 3 < 1 < 4 < 5 < 12 < 92 < 17 < 95
L2: 11 < 91 < 2 < 3 < 4 < 1 < 12 < 5 < 92 < 17 < 95
L3: 11 < 2 < 91 < 4 < 3 < 1 < 12 < 5 < 92 < 17 < 95
L4: 11 < 91 < 4 < 2 < 3 < 1 < 12 < 92 < 5 < 17 < 95
L5: 11 < 91 < 2 < 4 < 3 < 1 < 5 < 12 < 92 < 17 < 95

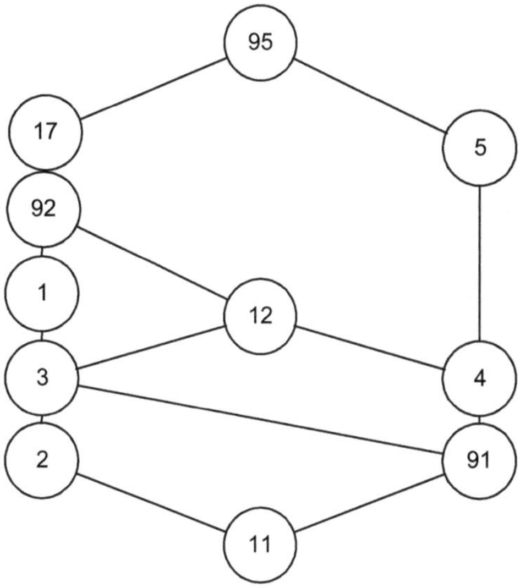

Fig. 20. $O(95)$, the order ideal generated by object (sampling site) 95. All considerations here are based on $E/\mathfrak{R}$

The jump-numbers are 3, 6, 4, 3, 5 for L1 to L5, respectively, the jumps in, e.g., L1 being found between 2 < 91, 1 < 4 and 5 < 12.

What kind of information can be derived from this? If we represent a poset by a set of linear extensions, then those of major interest are those preserving the chains of the poset as far as possible. It is obvious that L2 is a correct representation of the partial order. However, the chains that can be identified (cf. Fig. 20) are separated by many elements, which originally did not belong to chains.

A further use of linear extension is the probability scheme (ranking probabilities) that they provide. Probability plots are depicted for the three sites 1, 17 and 91 (Fig. 21a) and for site 5 (Fig. 21b), respectively. (See also the contributions, chapters by Voigt and Brüggemann, p. 327; Brüggemann et al., p. 237; Carlsen, p. 163.

Remarkable differences can be noted. Thus, in the case of the three sites 1, 17, and 91 rather sharp maxima are developed, indicating that they can safely be assigned to a rank near the maximum of their probability plot. However, the sites differ in their individual ranking position. Thus, site 91 takes a lower rank site 1 a medium rank and site 17 a rather high rank. Therefore a mutual ranking sequence of the sites 1, 17, and 91, i.e., 91 < 1 < 17, can be given since the minimum rank of the one site apparently does not overlap significantly with the maximum rank of a lower positioned site.

The site 5, on the other hand, differs from the above discussed sites as a rather smeared out probability plot is disclosed. Thus, the eventual assignment of a rank for site 5 is uncertain. This can also be seen directly from the visualization in the Hasse diagram (Fig's. 20 and 19). The site 5 is not as strongly connected as the other three elements. In more detail the consequences are discussed in Brüggemann et al. (2001b). We can calculate the local quantity U(x), i.e. the number of incomparabilities of an element x. The larger the values of U(x) the more uncertain the rank of x is. In the case of site 5 it turns out that $U(5) = 6$, whereas the corresponding values for the site 91, 1, 17 are $U(91) = 1$, $U(1) = 2$, and $U(17) = 1$, respectively. Therefore the measure of uncertainty about the ranks is $U(91) = U(17) < U(1) << U(5)$.

As stated earlier, it is possible to calculate averaged ranks; the full list of information is given in Table 5, where the minimum, maximum rank and the local incomparabilities are displayed.

Table 5. Summary of the analysis by linear extensions

Identifier	Min	Rkav	Max	U(x)
1	5	6.67	8	3
2	2	2.85	5	3
3	4	4.70	6	2
4	3	4.33	6	3
5	4	7.67	10	6
11	1	1	1	0
12	6	6.94	8	2
17	9	9.82	10	1
91	2	2.39	3	1
92	8	8.63	9	1
95	11	11	11	0

The analysis by linear extensions is very attractive as it helps to derive a linear ranking, without any subjective preferences. The data lead to a poset, the poset may be analyzed with respect to its structure, this is a combinatorial problem, and finally a ranking probability can be derived. Crucially in this procedure is that very different attribute profiles may lead to the same Hasse diagram and thus to the same set of linear extensions and therefore finally to the same probability characteristics: Thus, the attribute profiles a) (0,0), (1,0), (0,1), (1,1) and b) (0,0), (1,0), (0,5), (4,7) lead to identical Hasse diagrams.

A priori this is fine as the first attribute definitely should compensate the second one. However, the sites, which belong to (1,0),(0,1) on the one side and (1,0), (0,5) on the other side will get the same averaged rank! Thus, the analysis by linear extensions alone should be carried out with appropriate care. We continue the analysis of the poset and discuss the attribute profiles.

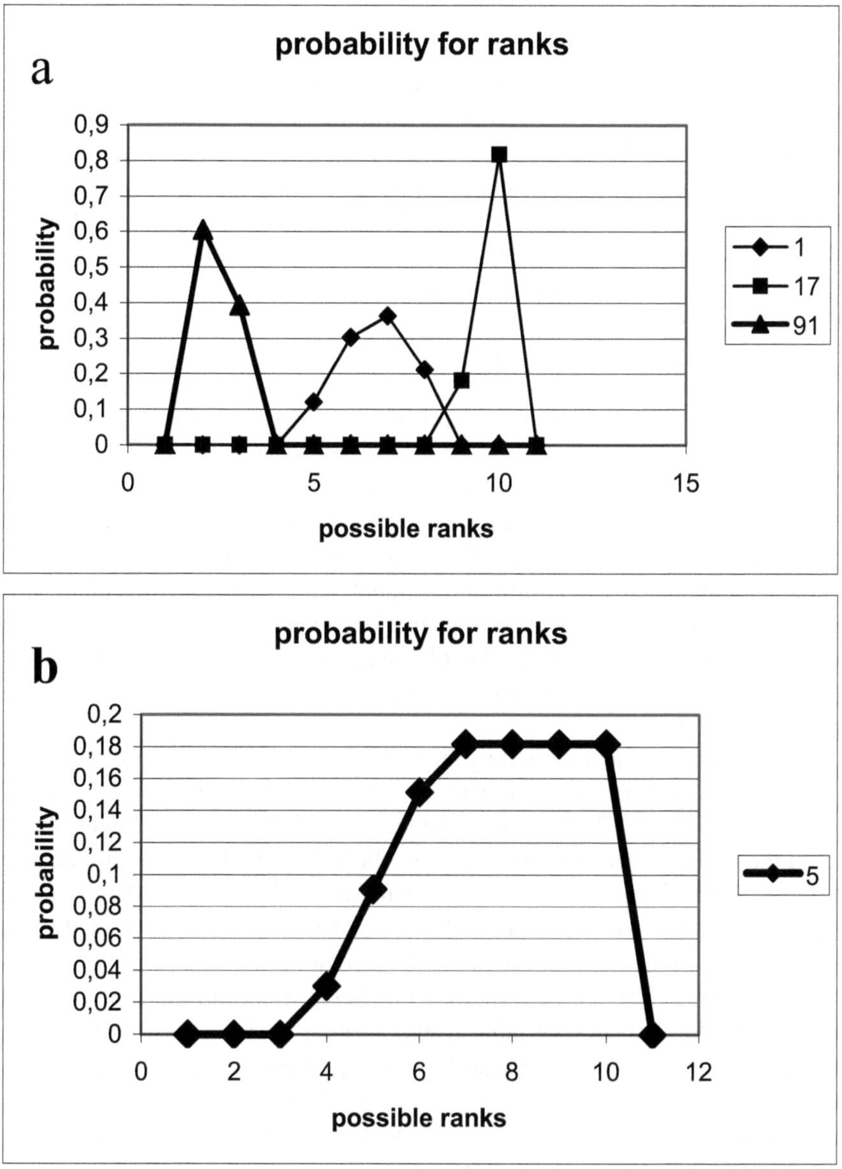

Fig. 21. Probability plots for 4 elements of the poset, shown in Fig. 20

Up to now, we have a quite good overview about the ranking of sites in Lake Ontario. However, does the test battery comprise redundancies? The subsequent dimension analysis will disclose this.

Dimension analysis

We find, applying equation 15 that the poset shown in Figure 19 has dimension 2. Once the dimension d of a poset is found with d < card *IB*, then corresponding many new latent ordering variables l_1, l_2, ... l_d may be used to form the same Hasse diagrams as found by the original attributes. Hence, the same ranking must be possible by a lower number of latent ordering variables and a redundancy within the battery appears possible. However, the numerical relation between the original attributes and the latent ordering variables may be rather difficult to derive and, if even then hard to interpret as it is often the case, e.g., in principal component analysis.

Corresponding to the dimension d = 2, the poset shown in Fig. 19 can alternatively be visualized by a two-dimensional grid as is shown in Fig. 22. Both visualizations have their advantages. Structures within a Hasse diagram, e.g., successor sets, or sets of objects separated from others by incomparabilities, can be more easily disclosed by a representation like that of Fig. 19. In multivariate statistics reduction of data is typically performed by principal components analysis or by multidimensional scaling. These methods minimize the variance or preserve the distance between objects optimally. When order relations are the essential aspect to be preserved in the data analysis, the optimal result is a visualization of the sediment sites within a two-dimensional grid.

Some scores of the test battery are additionally shown. From them the values of the scores of other objects can be estimated or exactly calculated. For example, for site 17, FC must have the value 3, because the lower object 92 and the higher object 95 have $s_{FC} = 3$. The value of CP must be 0 because $s_{CP}(32) = 0$, which is the lowest value. Similarly $s_{CH}(17) = 0$ and $s_{GT}(17)=0$, whereas for $s_{MT}(17)$ only the interval $4 \leq s_{MT}(17) \leq 8$ can be predicted from the knowledge of the neighbours in the Hasse diagram.

The grid (Fig. 22) can be thought of as being a coordinate system, with one axis of a latent order variable l_1 and another by l_2, according to d = 2. By these two latent ordering variables, each element $\in E/\Re$ can be characterized by a pair, which represents correctly the order relations (Compare Figure 17) that are important for ranking but which is clearly not unique with respect to a numerical representation. The interpretation of the latent variables l_1 and l_2 is supported by checking the configurations within the two-dimensional grid in terms of its a priori content (variables FC, CP, CH, MT, GT). A clear correlation can be detected between FC and the latent variable l_1 and also between GT and the latent variable l_2.

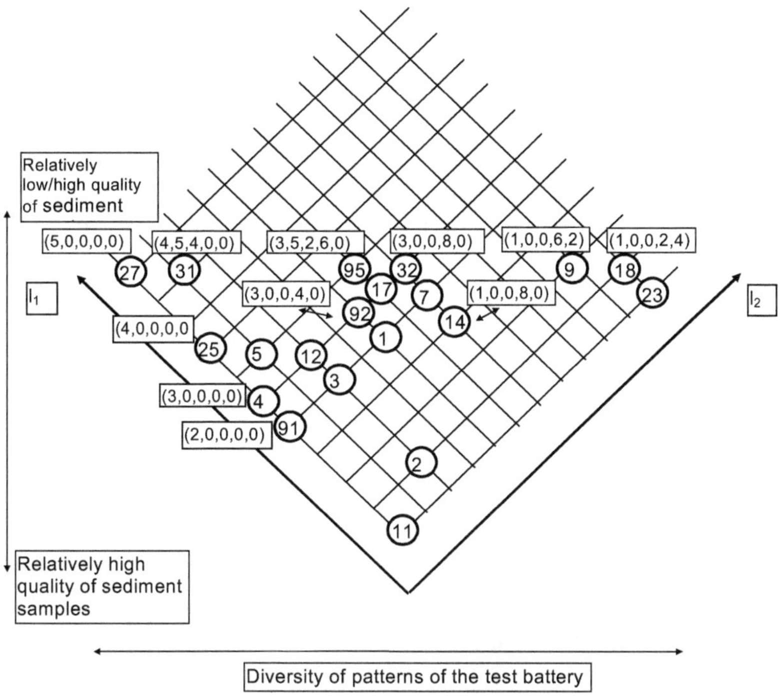

Fig. 22. Visualization of the ranking result of the sediment samples of Lake Ontario after dimension analysis

Sometimes these variables FC and GT with primary meaning are called polar items Shye 1985, Borg & Shye 1995. For further elucidation see also the multivariate technique posac (partial order scalogram analysis with coordinates), which is explained in Brüggemann et al. 2003, Voigt & Welzl, 2002 and for which a tool is provided in the software package Systat (R) 2000.

The other variables accentuate the possibility of discrimination in a nonlinear manner. Therefore, in a qualitative sense, the ranking of the sediment sites of the Lake Ontario seems to be determined by a hygienic and an ecotoxicological component. Some objects could be embedded into the grid on alternative ways. However, the order theoretical information, namely the comparabilities and incomparabilities are maintained. This can be easily proved by verifying that the Hasse diagram induced by five attributes (Fig. 19) is isomorphic to that, induced by the two latent variables (Fig. 22). If the ranking is in mind, then obviously the five tests apparently contain some redundancies, because the decision for "good" or "bad" could also be given on the basis of two coordinate values.

Sensitivity analysis of the ranking

For our example the matrix **W** has the following values (Table 6)

Table 6. Values of the matrix **W** for different combinations of attribute

W	case 0 FC,CP,CH, MT,GT	case 1 CP,CH,M T,GT	case 2 FC,CH,M T,GT	case 3 FC,CP, MT,GT	case 4 FC,CP,C H,GT	case 5 FC,CP,C H,MT
case 0	0	795	0	0	360	124
case 1	-	0	795	795	1155	919
case 2	-	-	0	0	360	124
case 3	-	-	-	0	360	124
case 4	-	-	-	-	0	484
case 5	-	-	-	-	-	0

It is seen that cases 1 to 5 excludes one after another FC, CP, CH, MT and GT, respectively. Thus, comparing these cases to case 0, including all 5 attributes, will disclose the relative importance of the 5 tests comprising the battery. Thus, from this matrix the sensitivities are σ(FC) = 795, σ(CP) = σ(CH) = 0, σ(MT) = 360 and σ(GT) = 124, respectively, unambiguously disclosing the test "FC" as the most important within the attribute set containing the five tests. The tests CP and CH apparently do not have any influence at all on the order theoretical structure of the set of samples, i.e. they do not influence the prioritization of the sites. Their low sensitivities are also found by Dutka et al. 1986, who established a regression model between the two quantities. It is emphasized that this conclusion refers to the classified values of the battery of tests. Hence, the result with respect to FC should be carefully examined as the high sensitivity may be induced by the scoring process.

Fig. 23 shows the Hasse diagram (generated by the computational software, WHASSE (Brüggemann et al. 1999 a) therefore drawn in its standard format: circles, and each object as high as possible in the drawing plane):

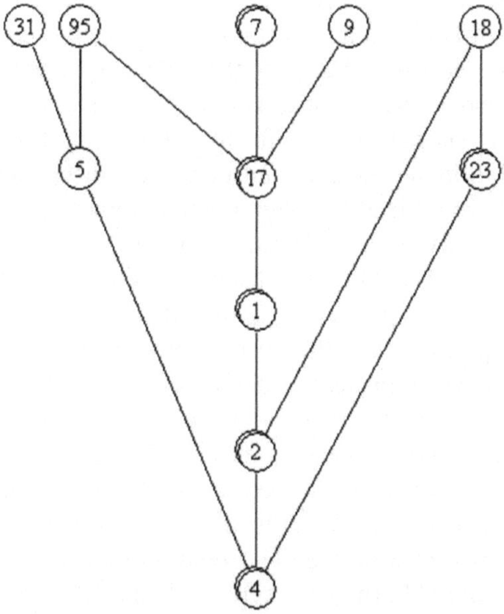

Fig. 23. 55 samples evaluated with the test battery of Dutka, excluding the FC test. Note that many samples are members of non-trivial equivalence classes

The dramatic changes compared to the original Hasse diagram (Fig. 19) are immediately seen.

Discussion and Conclusion

The battery of tests approach helps to evaluate sites using different criteria simultaneously: The decision of which sites are "good" or "bad", i.e. the sorting process is more difficult the larger the number of samples and especially the larger the number of tests, since there is more information that can be used to differentiate among the tested objects. This, in turn, leads to difficulties for ranking, because the complexity of a well-designed battery is being lost, if in order to compare the tested objects, a ranking index like

$$\Gamma = \sum g_i \cdot q_i \quad (22)$$

is constructed. The presentation by a Hasse diagram avoids the arbitrariness in constructing a ranking index. Applying concepts of partially ordered sets must not be performed in isolation. All results depend on the data representation used. The present study aimed at demonstrating the HDT using, and extending the results of Dutka et al. 1986. Therefore we did not need statistical analyses. However, generally, the appropriate data representation is of much concern, Brüggemann & Welzl 2002. The use of cluster analysis and principal component analysis may be helpful to obtain a statistical relevant data representation and to avoid insignificant numerical differences of the attributes, which in turn would lead to insignificant comparabilities and incomparabilities and thus to very complex Hasse diagrams.

A combination of Hasse Diagram Techniques and explorative statistical methods could be a very promising approach to future tasks in environmental sciences. Approaches in this respect were followed on the pollution of regions in Germany with heavy metals, cf. Brüggemann et al. 1999b) and on the contents of environmental databases, cf. Voigt et al. 2004.

The analysis of empirical datasets may lead to empirical partial orders, which do not necessarily fulfill the axioms of lattices. The school around Wille (Wille 1987 and Ganter & Wille 1996) has shown how it is possible nevertheless to construct a lattice. The resulting lattices and the analysis based on them is called "Formal concept analysis". As lattices fulfill more axioms than posets generally, one gets a richer theory of them. Especially it is possible to generate a set of implications. See chapter by Kerber, p. 355 for introductory examples.

The main advantage of a ranking by HDT is that it can be performed without any normative constraints. HDT simply sorts the objects without any additional information. Beyond sorting, many conclusions may be drawn from the Hasse diagrams as they represent a well-defined mathematical structure. Summarizing the following recommendations can be given:

- If the battery of tests is used to test many objects, perform a cluster analysis to get rather numerically robust results. Instead of the measured results for each object use some characteristic values of the cluster (mean values or some other quantities, describing a cluster center).
- Apply HDT to look for priority objects, to identify objects or subsets with characteristic patterns (in mathematical terminology: find "order ideals") or to select sequences (in order theoretical terminology: "chains") of objects.
- Perform a dimension analysis to estimate the redundancy of the test system and a sensitivity study to identify important or less important attributes. The rational for the importance of each attribute cannot be drawn from the HDT; here the scientific background is needed: What are the characteristics for all the tested objects are there any internal correlations among the attributes?

- If an aggregation is done, as, e.g., by eqn. 22 then note that the weights may have an important influence on the ranking results via Γ if objects have an high degree of incomparability, i.e. have a large value for U(x).

References

Atkinson MD (1989) The complexity of Orders. In: Rival, I. (Ed.) Algorithms and Order NATO ASI series. Series C Mathematical and Physical Sciences Vol. 255 Kluwer Academic Publishers, Dordrecht, pp 195-230

Atkinson MD (1990) On the computing the Number of Linear Extensions of a Tree. Order 7:23-25

Beckenbach EF, Bellman R (1971) An Inequality of Karamata. In: Inequalities, 3 edn. Springer-Verlag, Berlin, pp 1-107

Birkhoff G. 1984. Lattice theory. Providence, Rhode Island: American Mathematical Socienty, Vol XXV

Borg I, Shye S (1995) Facet Theory - Form and Content. Sage Publications, Thousand Oaks, California

Brans JP, Vincke PH (1985) A Preference Ranking Organisation Method (The PROMETHEE Method for Multiple Criteria Decision - Making). Management Science 31:647-656

Brans JP, Vincke PH, Mareschal B (1986) How to select and how to rank projects: The PROMETHEE method. European Journ Oper Research 24:228-238

Brüggemann R, Bartel HG (1999) A Theoretical Concept to Rank Environmentally Significant Chemicals. J Chem Inf Comput Sci 39:211-217

Brüggemann R, Bücherl C, Pudenz S, Steinberg C (1999a) Application of the concept of Partial Order on Comparative Evaluation of Environmental Chemicals. Acta hydrochim hydrobiol 27:170-178

Brüggemann R, Halfon E (1997) Comparative Analysis of Nearshore Contaminated Sites in Lake Ontario: Ranking for Environmental Hazard. J Environ Sci Health A32(1):277-292

Brüggemann R, Halfon E, Welzl G, Voigt K, Steinberg C (2001b) Applying the Concept of Partially Ordered Sets on the Ranking of Near-Shore Sediments by a Battery of Tests. J Chem Inf Comp Sc 41:918-925

Brüggemann R, Pudenz S, Carlsen L, Sørensen PB, Thomsen M, Mishra RK (2001a) The Use of Hasse Diagrams as a Potential Approach for Inverse QSAR. SAR and QSAR Environ Res 11:473-487

Brüggemann R, Pudenz S, Voigt K, Kaune A, Kreimes K (1999b) An algebraic/graphical tool to compare ecosystems with respect to their pollution. IV: Comparative regional analysis by Boolean arithmetics. Chemosphere 38:2263-2279

Brüggemann R, Schwaiger J, Negele RD (1995a) Applying Hasse diagram technique for the evaluation of toxicological fish tests. Chemosphere 30(9):1767-1780

Brüggemann R, Simon U, Mey S (2005) Estimation of averaged ranks by extended local partial order models. Math Commun Math Comput Chem 54(3):489-518

Brüggemann R, Sørensen PB, Lerche D, Carlsen L (2004) Estimation of Averaged Ranks by a Local Partial Order Model. J Chem Inf Comp Sc 44:618-625

Brüggemann R, Welzl G (2002) Order Theory Meets Statistics -Hassediagram technique-. In: Voigt K and Welzl G (eds) Order Theoretical Tools in Environmental Sciences - Order Theory (Hasse diagram technique) Meets Multivariate Statistics-. Shaker-Verlag, Aachen, pp 9-39

Brüggemann R, Zelles L, Bai QY, Hartmann A (1995b) Use of Hasse Diagram Technique for Evaluation of Phospholipid Fatty Acids Distribution in Selected Soils. Chemosphere 30(7):1209-1228

Brüggemann R, Welzl G, and Voigt K (2003) Order Theoretical Tools for the Evaluation of Complex Regional Pollution Patterns. J Chem Inf Comp Sc 43:1771-1779

Carlsen L, Lerche DB, Sørensen PB (2002) Improving the Predicting Power of Partial Order Based QSARs through Linear Extensions. J Chem Inf Comp Sc 42:806-811

Carlsen L, Sørensen PB, Thomsen M (2001) Partial Order Ranking - based QSAR's: estimation of solubilities and octanol-water partitioning. Chemosphere 43:295-302

Carlsen L, Sørensen PB, Thomsen M, Brüggemann R (2002) QSAR's based on Partial Order Ranking. SAR and QSAR in Environmental Research 13:153-165

Colorni A, Paruccini M, Roy B (2001) A-MCD-A, Aide Multi Critere a la Decision, Multiple Criteria Decision Aiding. JRC European Commission, Ispra

Davey BA, Priestley HA (1990) Introduction to Lattices and Order. Cambridge University Press, Cambridge

Dutka BJ, Walsh K, Kwan KK, El Shaarawi A, Liu DL, Thompson K (1986) Priority site selection from degraded areas based on microbial and toxicant screening tests. Water Poll Res J Canada 21(2):267-282

Freese R. 2004. Automated Lattice Drawing. In: Eklund P, Ed. ICFCA 2004, Berlin: Springer-Verlag, p 112-127

Ganter B and Wille R (1996) Formale Begriffsanalyse Mathematische Grundlagen, Springer-Verlag, Berlin, pp 1-286

Halfon E, Hodson JA and Miles K (1989) An algorithm to plot Hasse diagrams on microcomputer and Calcomp plotters. Ecol Model 47:189-197

Heinrich R (2001) Leitfaden Wasser - Nachhaltige Wasserwirtschaft; Ein Weg zur Entscheidungsfindung. Wasserforschung e.V., Berlin

Jonsson B (1982) Arithmetic of Ordered Sets. In: Rival I (Ed) Ordered Sets. D. Reidel Publishing Company, Dordrecht, pp 3-41

Lerche D, Brüggemann R, Sørensen PB, Carlsen L, Nielsen OJ (2002) A Comparison of Partial Order Technique with three Methods of Multicriteria Analysis for Ranking of Chemical Substances. J Chem Inf Comp Sc 42:1086-1098

Lerche D, Sørensen PB (2003) Evaluation of the ranking probabilities for partial orders based on random linear extensions. Chemosphere 53:981-992

Lerche D, Sørensen PB, Brüggemann R (2003) Improved Estimation of the Ranking Probabilities in Partial Orders Using Random Linear Extensions by Approximation of the Mutual Probability. J Chem Inf Comp Sc 53:1471-1480
Muirhead RF (1900) Inequalities relating to some Algebraic Means. Proc Edingburgh Math Soc 19:36-45
Muirhead RF (1906) Proofs of an Inequality. Proc Edingburgh Math Soc 24:45-50
Neggers J, Kim HS (1998) Basic Posets. Singapore: World Scientific Publishing Co
Patil GP, Taillie C (2004) Multiple indicators, partially ordered sets, and linear extensions: Multi-criterion ranking and prioritization. Environmental and Ecological Statistics 11:199-228
Pavan,M. and R.Todeschini, 2004. New indices for analysing partial ranking diagrams. Analytica Chimica Acta 515: 167-181.
Randić M (2002) On Use of Partial Ordering in Chemical Applications. In: Voigt K and Welzl G (Ed.) Order Theoretical Tools in Environmental Sciences - Order Theory (Hasse Diagram Technique) Meets Multivariate Statistics. Shaker-Verlag, Aachen, pp 55-64
Rival I (1981) (Ed.) Ordered Sets, NATO ASI Series, Series C. Mathematical and Physical Sciences Vol. 83, Kluwer Academic Publishers, Dordrecht
Rival I (1983) Optimal Linear Extensions by Interchanging Chains. Proc AMS 89:387-394
Rival I (1985a) (Ed.), Graphs and Order, NATO ASI Series, Series C. Mathematical and Physical Sciences Vol. 147, Kluwer Academic Publishers, Dordrecht
Rival I (1985b) The diagram. In: Rival I (1995 a) pp 103-133
Rival I (1989) (Ed.) Algorithms and Order, NATO ASI Series, Series C. Mathematical and Physical Sciences Vol. 255, Kluwer Academic Publishers, Dordrecht
Rival I, Zaguia N (1986) Constructing Greedy Linear Extensions by Interchanging Chains. Order 3:107-121
Roy B (1972) Electre III: Un Algorithme de Classements fonde sur une representation floue des Preferences En Presence de Criteres Multiples. Cahiers du Centre d'Etudes de Recherche Operationelle 20:32-43
Roy B (1990) The outranking approach and the foundations of the ELECTRE methods. In: Bana e Costa (Ed.) Readings in Multiple Criteria Decision Aid. Springer-Verlag, Berlin, pp 155-183
Ruch E (1975) The diagram lattice as Structural Principle. Theor Chim Acta 38:167-183
Ruch E, Gutman I (1979) The Branching Extent of Graphs. J of Combinatorics-Inform System Sciences 4:285-295
Schröder BSW (2003) Ordered Sets An Introduction. Birkhäuser, Boston
Shye S (1985) Multiple Scaling - The theory and application of partial order scalogram analysis. North-Holland, Amsterdam
Stanley P (1986) Enumerative Combinatorics. Volume I, Wadsworth & Brooks/ Cole, Monterey, California
Systat Program http://www.spss.com/software/science/SYSTAT/, SPSS Inc., 2000

Trotter WT (1991) Combinatorics and Partially Ordered Sets Dimension Theory; John Hopkins Series in the Mathematical Science. The J Hopkins University Press: Baltimore

Urrutia J (1989) Partial Orders and Euclidian Geometry, in Rival, I, Ed. 1989, pp 387-434

Voigt K, Welzl G, Brüggemann R (2004) Data analysis of environmental air pollutant monitoring systems in Europe. Environmetrics 15:577-596

Wille R (1987) Bedeutungen von Begriffsverbänden: Beiträge zur Begriffsanalyse (Hrsg.: Ganter B, Wille R, Wolff KE). BI Wissenschaftsverlag Mannheim pp 161-211

Comparative Evaluation and Analysis of Water Sediment Data

Stefan Pudenz[1]*, Peter Heininger[2]

[1] Criterion-Evaluation & Information Management, Mariannenstr. 33, D-10999 Berlin, Germany

*e-mail: stefan.pudenz@criteri-on.de

[2] Federal Institute of Hydrology (BfG), Dept. Qualitative Hydrology, P.O. Box 200253, D-56002 Koblenz

Abstract

With respect to sediment pollution responses of ecotoxicological tests may differ from those of biochemical test systems and moreover both tests are indicating effects instead of simply measuring of chemical concentrations. Because most test results of sediment investigations are commonly given as inhibition values and sediment pollution by chemicals is measured by their concentrations a comparative evaluation of sediments by means of both test results and chemicals at the same time has to consider different scales. Both data transformations on a common scale (standardization) and aggregations lead to loss of information and hamper the interpretation of results. In order to avoid merging of data and to circumvent often-crucial data transformations, partial ordering is used for evaluation of sediment samples from German rivers. The aim here is to compare the evaluation of river sections by different parameter groups, namely biochemical and ecotoxicological tests, as well as concentrations of organic pollutants, heavy metals etc. Fuzzy cluster analysis as a pre-processing step is additionally used to understand the pollution pattern that is given by each test result. It is shown that for most of the river sections, test systems among each other and also compared to chemical concentrations yield different quality pattern and therefore lead to different Hasse diagrams. Sole exception is a bayou where the sediment is undisturbed by shipping traffic and sewage. Moreover, as a consequence of varying pollution pattern during the sampling period (over several years), only for a few river sections it is possible to derive distinct temporal changes: Except for the nematode sediment contact test, where all parameters are significantly correlated, this holds for both ecotoxicological and biochemical tests, and for chemical concentrations. Furthermore, for one river section it could be observed that chemical concentrations indicate a decline of contamination, whereas ecotoxicological parameters point to an increased toxicity. With respect to the development of a classification system for river sediments it is recommended to take care in the selection of parameters and to base it at least at two parameter groups.

Introduction

In order to ensure shipping traffic in rivers and coastal waters fairways have to be dredged continuously. As a consequence thousands of tons of sediments are to be managed yearly. This dredged material can be contaminated with different pollutants. Depending on the degree of contamination dredged material can be relocated within the water or has to be disposed as hazardous waste. However, exactly the question which sediment can be classified as hazardous or not hazardous is a crucial one and a standardized method about how to classify sediments and dredged material respectively, would be a helpful tool not only for administrative purposes but also regarding economic and environmental aspects. Surveying the way of developing such a system several questions arise, which have to be answered a priori:

- What is the state of sediment pollution of all waterways and what kind of contamination is known, currently and in the past?
- How are 'hazardous' to be defined and what parameter should be taken into account respectively, when sediment/dredged material has to be classified?

The German Federal Institute of Hydrology (BfG) holds an extensive database about sediment investigations of Federal Water Ways considering several parameters (Heininger et al. 1998, Heininger et al. 2003). These data can be divided into three groups, namely chemical, ecotoxicological and biochemical parameters. With respect to the questions above and in order to make optimum use of these data the following question arises.

- Is there a difference between a comparative evaluation of sediments when using different parameter groups or is it sufficient to consider one group or certain parameters as representatives for sediment burden?

The questions put here for sediments could be applied to other environmental evaluation problems just as well, for instance to soil or groundwater pollution. However, a common difficulty with these evaluations is that many of the methods mask and aggregate the data, and therefore both valuable information and transparency are lost. An alternative is partial order ranking that avoids the merging of data and thus preserves important elements of the evaluation. Here we will show that partial order ranking has useful qualities in data analysis and it can be applied for preprocessing in the development of classification systems.

For all calculations and graphical presentations of partial ordered sets the ProRank© Software was used (Pudenz 2004).

Database

For many sections of the main waterways in Germany, namely the rivers Rhein, Elbe and Oder, sediment investigations provide results about

- concentrations of priority pollutants like toxic heavy metals (measured in the fine fraction <20 µm) and hazardous organic compounds (detected in the whole sample <2 mm),
- sediment toxicity as revealed in aquatic ecotoxicological tests with Daphnia, Algae and Bacteria using eluates and pore-water as test medium and in an sediment contact test (whole sediment) with Nematodes; in both tests toxicity is expressed in terms of percent inhibition compared to an unpolluted standard
- biochemical tests measuring enzymatic activities (e.g. aminopeptidase activity, glucosidase activity); the test results are given as percent consumption of a specific indicator substance,
- the basic sediment properties like organic carbon concentration, grain size spectrum, water content, for biochemical tests also DNA content; all basic parameters are measured in the whole sample.

A detailed list about parameters and sample sites can be found in Tables 5 and 6 in the appendix.

Partial order ranking requires complete data sets; alternatively data gaps have to be filled or to be cancelled. In case of time series of river sections the missing parameter could be replaced by e.g. the mean of temporal adjoining measurements. The nearer these measurements are the better is the gap filling. However a detailed review of the data set for this study shows that mainly locations, which were investigated only one time per year, had missing values in certain parameters. Alternatively, cancelling of data gaps means loss of information. In order to minimize loss two different procedures are considered:

- exclusion of the parameter with one or more gaps aiming at data sets with maximum sample number (MAX-SN) and
- exclusion of the sample with one or more gaps aiming at a maximum number of parameter (MAX-PN).

Results

Evaluation of Oder sediments using raw data

Chemical pollutants versus aquatic ecotoxicological tests

For the River Oder there are only a few samples with a fully completed data set. Therefore, the largest sample set consisting of chemical and ecotoxicological parameters is used here. The Hasse diagrams (HD) are based on 33 samples (MAX-SN), which were collected along the whole river between 1997 and 2001. Figure 1 shows the result for inhibition values of ecotoxicological tests in eluate and pore-water. All circles are labelled by identifiers for the sampling site and date, for example WD10/99 means Widuchowa at October 1999. Due to many lines the diagram is rather difficult to interpret.

However, compared to the HD based on chemical parameter (see Fig. 2) it shows a distinct level-structure (five levels). That means, for certain sediments a similar pattern concerning ecotoxicological effects in all tests can be observed. Regarding these lines consisting of samples with increasing values in all tests in more detail, we find one maximal chain with five samples: GG6/99 < CB9/98 < KR11/98 < EH10/95 < WD5/98. However, none of these relations is found in the evaluation by chemical concentrations. Moreover, there are only two comparabilities that are common for both Hasse diagrams, namely

- WD10/95 < ZB7/00 and
- WD10/95 < ZB3/00.

More interesting could be an observation about a temporal development of a sample location. However, there are only a few comparabilities indicating a temporal development for a specific site with respect to all tests and inhibition values respectively. For the site Glogau (GG) only the relation GG6/99 < GG5/98 holds, whereas both other samples from there, GG11/97 and GG11/98, are incomparable (see Fig. 1 and Table 1). In addition to Glogau only one more comparability indicates a temporal development with respect to all test results, namely for the site Widuchowa (WD): WD10/95 < WD4/99 (Fig. 1).

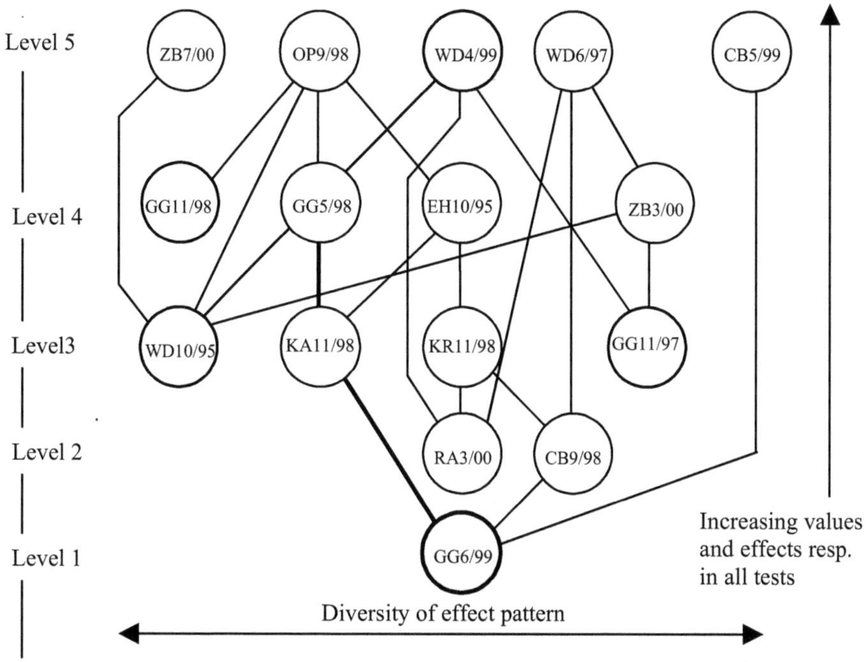

Fig. 1. Hasse diagram of 33 Oder sediment samples concerning six ecotoxicological test results

Table 1. Inhibition (H) of algae (A) and bacteria (B) tests in pore-water (P) and eluate (E); not shown are zero-values for daphnia in all Glogau samples

Sample	HPA	HEA	HPB	HEB
GG11/97	-99,3	-7,4	16,3	3,3
GG5/98	-7,2	-53,1	50,6	19,2
GG11/98	25,7	-7,8	11,4	15,5
GG6/99	-185	-111	15,9	8

In the evaluation by chemical parameters no temporal comparison of a sample location is found. The number of incomparabilities (U=1006) is by far more than the comparabilities (V=25). A high stability value (P(IB)=0,95) c.f. p. 83 indicates that the partial order is very instable against omitting an attribute, where a sensitivity analysis (for details to sensitivity analysis, the reader is referred to e.g. (Brüggemann et al. 2001 and to pp. 91) shows that the evaluation is sensitive against the pollution parameters PCBs, pp'-DDT, PAHs and Sn. However, omitting one of these parameters leads neither to more levels nor to significant more comparabilities as in the diagram based on all parameters.

Summarizing the observations it can be concluded that

- compared to the HD by means of ecotoxicological tests the chemical concentration profile c.f. p. 81 shows by far more diversity (Fig. 2) and yields a different ranking result,
- for both parameter groups temporal developments are hardly to observe.

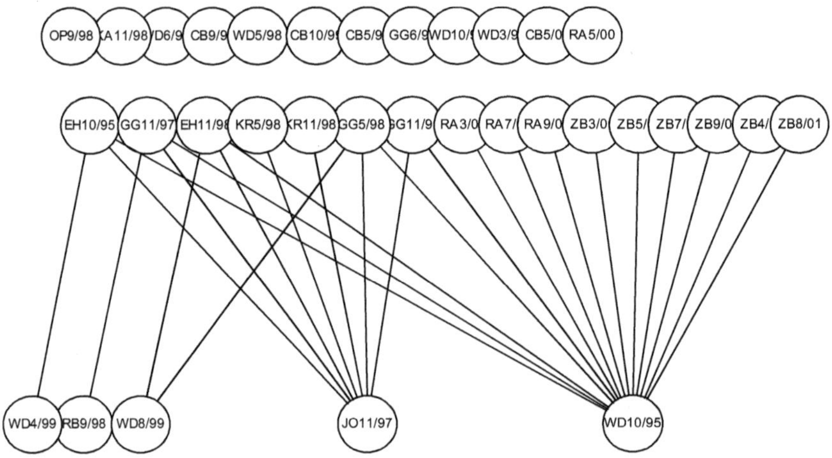

Fig. 2. Hasse diagram of 33 Oder sediment samples concerning concentrations of 22 chemical parameters (except P, B and TBT; see Table 4 in the appendix). Because of shortage of space the upper level consisting of isolated objects is separated

Evaluation of Elbe sediments using raw data

Chemical pollutants vs. aquatic ecotoxicological tests vs. nematode sediment contact test

There are no sediments where all ecotoxicological and biochemical tests and chemical measurements have been carried out together. Therefore we established two sets of samples where the first one contains chemical concentrations and the whole set of ecotoxicological tests (12 samples, see Fig. 3a and 3b) and the other one contains chemical concentrations, biochemical and ecotoxicological tests except nematodes (Fig. 4a and 4b, 28 samples). This data processing procedure leads not only to sets with different samples and size but also to different chemical pollutants that are taken into account. Therefore a comparison between the Hasse diagrams in Fig. 3a-b and Fig. 4a-b is not feasible.

In Fig. 3b it is seen, that the HD for the whole sediment test with nematodes shows most structure compared to the other diagrams in Fig. 3a. It has six levels whereas the diagrams for ecotoxicological tests and chemicals have only three and two levels, respectively. There are several chains with increasing inhibitions in all tests (egg hatch (EH), growth (G), reproduction (R)) simultaneously, for example:

- AE6/00 < FL4/01 < AE10/00 < AK4/01 < FL8/01
- AE6/00 < FL4/01 < FL10/00 < AE4/01 - DE4/01 < FL8/01
- FL6/00 < FL4/01 < FL10/00 < AE4/01 < DE4/01 < FL8/01
- etc.

All four samples of the site FL (Fahlberg List) are comparable (see the bold letters in the sequence shown above), where in year 2000 the toxicity increases from July to October whereas in April 2001 it decreases again and obtains a maximum in August 2001. In contrast to the results of the nematode test the ecotoxicological responses in the other tests indicate a decline of burden from June via October 2000 to August 2001 (as seen in Fig. 3). Moreover, it is noticeable that in the "ecotoxicological HD" the sample FL4/01 (April 2001) is not comparable to all other FL samples. The reason for this antagonism can be easily identified by examining the bar diagram presentation of a HD in Fig. 3: It can be observed that FL4/01 has a relatively high value in the algae test using pore-water (HPA) but a low effect for the eluate (HEA) compared to, for instance FL8/01, which has a lower effect in pore-water and a higher effect in eluate. This may be a hint at different pollution pathways and/or different bioavailability.

Comparing the three HD's in Fig. 3 it is indicated that each of the parameter groups, i.e. ecotoxicological test results with aquatic media, nematodes test results and chemical pollution, lead to different orders and therefore present different effects and responses, respectively.

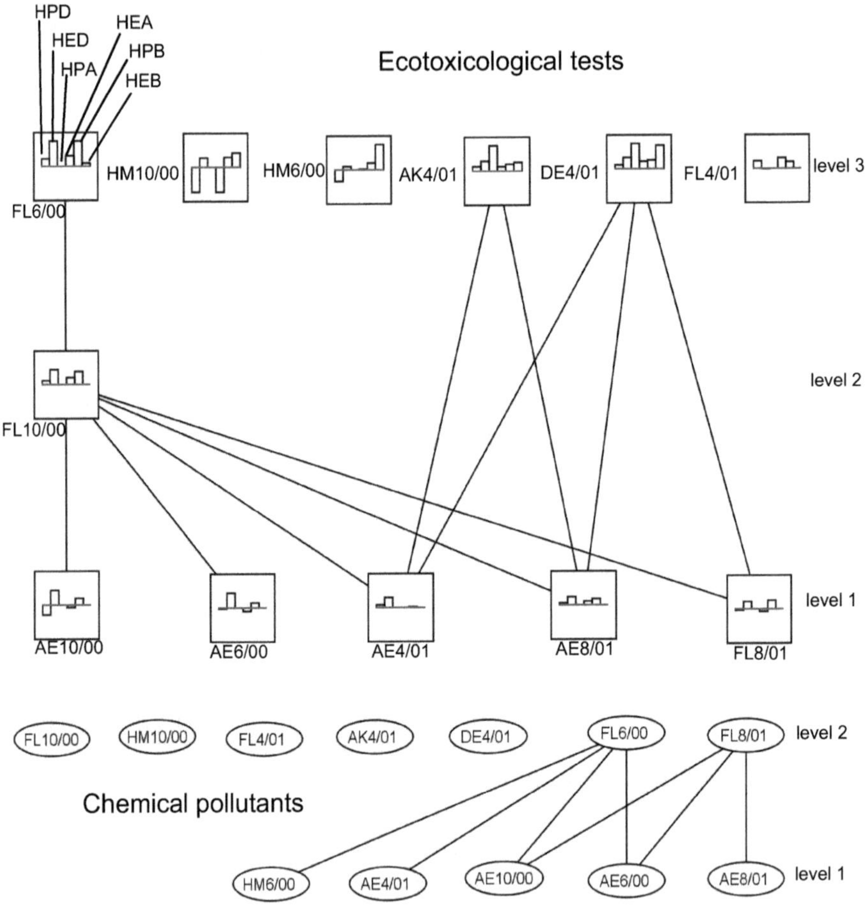

Fig. 3a. Hasse diagrams from evaluation of 12 Elbe sediments for chemical pollutants and ecotoxicological tests. Chemicals without N, S, B, Co, Sn (for abbreviations and speciation of elements, see Table 5 in the appendix). Because of shortage of space evaluation of these samples by nematodes tests is shown in Fig. 3b

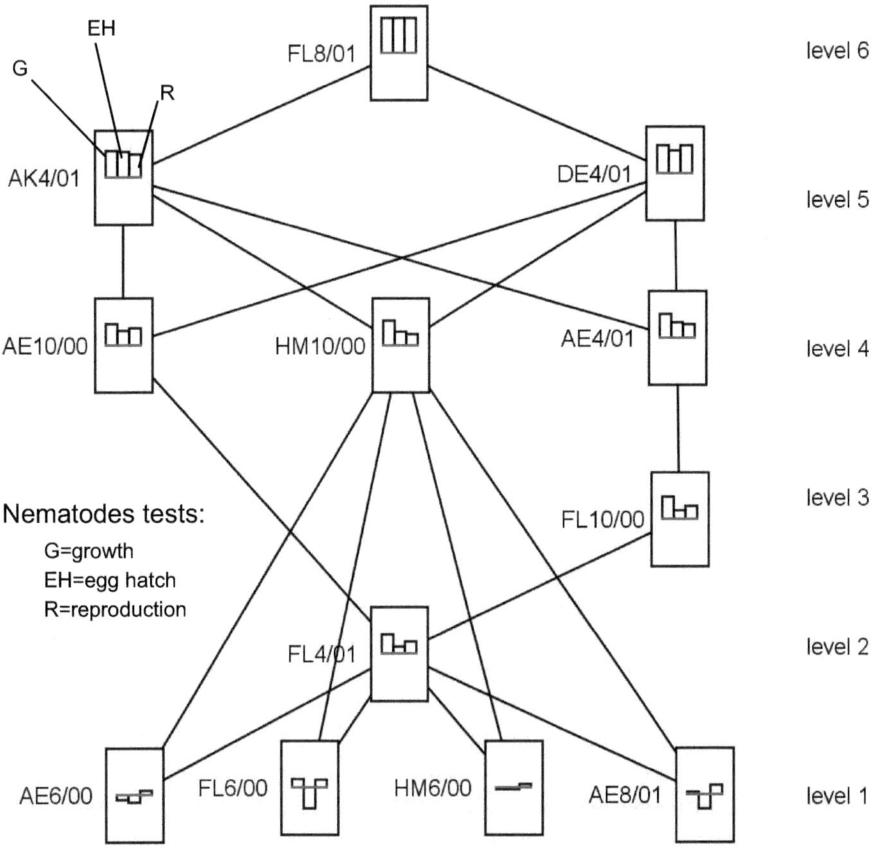

Fig. 3b. Hasse diagram from evaluation of 12 Elbe sediments for nematodes tests corresponding to Fig. 3a

Chemical pollutants versus aquatic ecotoxicological tests versus biochemical tests

Regarding the biochemical tests in comparison to chemical parameters and ecotoxicological tests in Fig. 4 striking differences can be observed too. Instead of a HD consisting of lines and circles, here the so-called level presentation is used. This kind of presentation might be useful when partial ordering results for instance in messy diagrams, as it is the case for ecotoxicological tests indeed. Here, again we want to show that the three parameter groups lead to highly different results, where evaluation by means of chemicals results in solely incomparable samples (a so-called anti-chain) and the biochemical responses are comparable for only three samples. This result may underline the assumption, that the parameter groups yield different responses to sediment quality and therefore to different rankings and classifications respectively. However, it has to be considered that here rough data are used and therefore already small numerical differences may lead to incomparabilities between sediment samples. For example, the HD based on chemical parameters in Fig. 4c consists of only incomparable samples and it is not obvious if data noise is responsible or if it is an effect of different pollution pattern indeed. Therefore, classifying by cluster analysis as pre-processing will be introduced in the following.

HD's after pre-processing by fuzzy clustering

The aggregation of samples is a strategy to get HD's, which is "easier and more robust" to interpret. Here, fuzzy cluster analysis is preferred. In contrast to conventional clustering methods, where each sample will be assigned to a cluster by a "yes/no-decision", fuzzy clustering yields a degree for the assignment of samples to a cluster (membership function with values between 0 and 1). The advantage is that samples, which are located between two clusters because they are outliers or so-called hybrid elements, can be identified (for details, see e.g. Pudenz et al. 2000, Luther et al. 2000). The fuzzy-algorithm used here (k-means fuzzy) requires a default cluster number (FCL) and a threshold value for the membership function (TMF). The TMF determines to which degree a sample belongs to a cluster. Here, preliminary tests have shown that in case of clustering over the whole property space (see below) cluster numbers FCL of six or seven lead to relatively complex diagrams. Therefore a FCL=4 is selected. Correspondingly a high TMF of 0.8 is used, such that hybrid elements and outliers will be identified.

Basically clustering can be distinguished between

- attribute-wise classification, i.e. all samples are clustered for each parameter separately, and
- clustering of all samples by the whole property space, i.e. by means of all parameter at the same time.

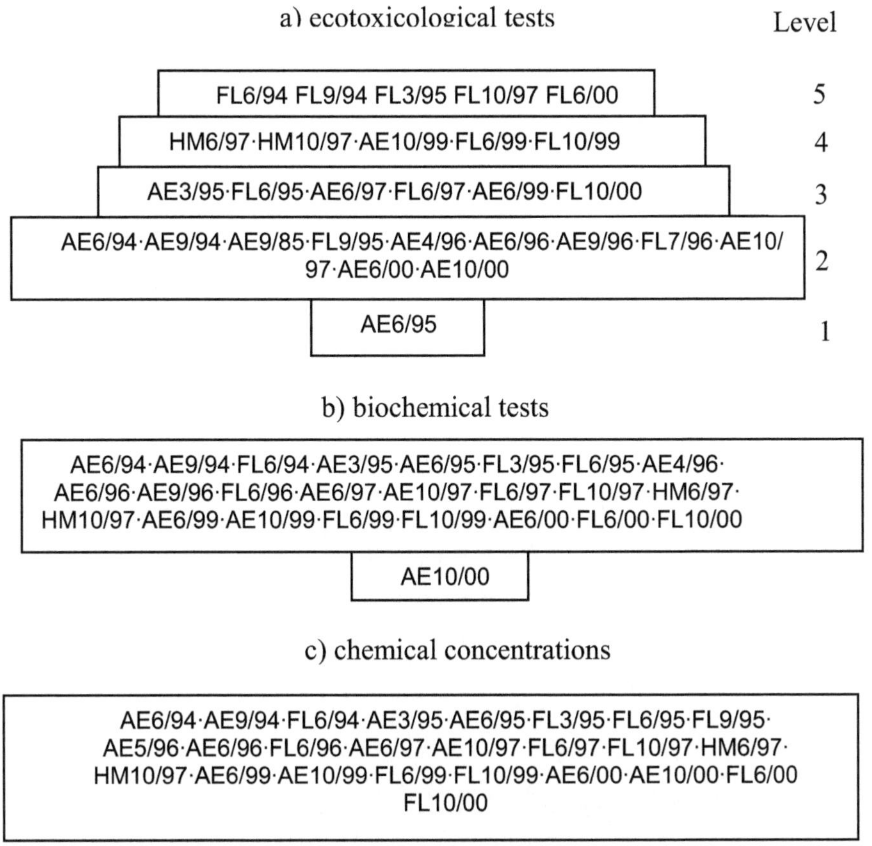

Fig. 4a. HD's as level presentation from evaluation of 28 Elbe samples by means of a) ecotoxicological tests (additionally as Hasse diagram in Fig. 4b), b) biochemical tests and c) chemical concentrations without P, TBT (see Table 5)

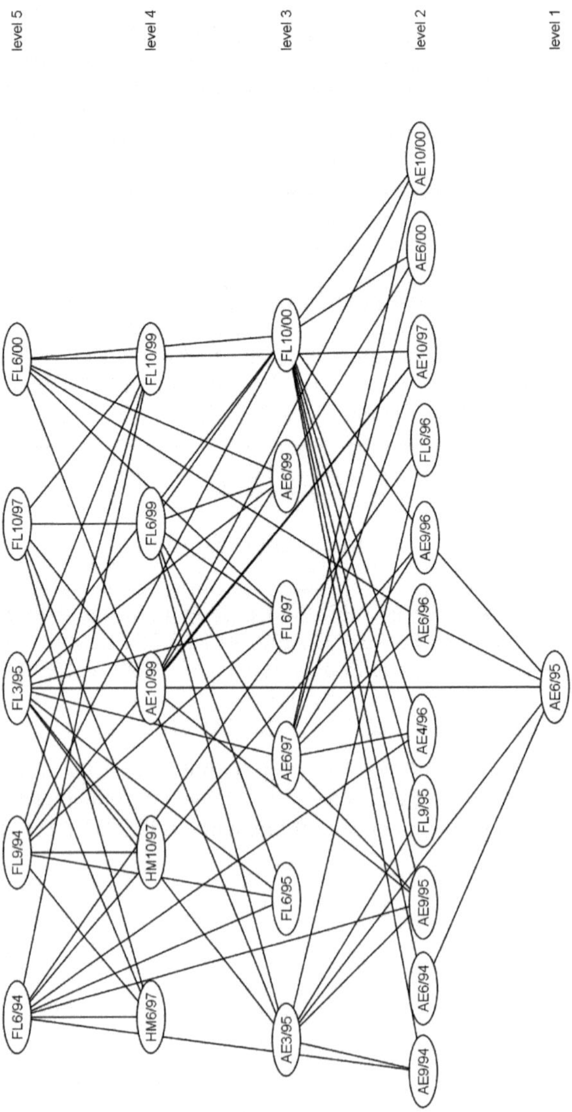

Fig. 4b. Hasse diagram for ecotoxicological tests as shown as level presentation in Fig. 4a

When the whole quality pattern of a location is of interest and samples with similar pattern with regard to all parameter will be identified, then clustering by means of all parameter is convenient. When all samples will be classified by each parameter separately, then efforts are more directed towards neglecting numerical differences between samples. For both methods the aim is, all samples will be ranked by the cluster centre they have been assigned to, instead of their original parameter values. Because of technical software problems, the number of chemical parameter had to be reduced to a maximum number of 20 (only in case of clustering by the whole property space). Here, the following pollutants are seen as the most relevant:

- As, Pb, Cd, Hg, Cu, Cr, Zn
- PAHs, PCBs, pp'-DDT, pp'-DDE, pp'-DDD, AOX, TBT, HCB, α-HCH, γ-HCH

Clustering by the whole property space - River Oder sediments

Chemical pollutants versus aquatic ecotoxicological tests

As mentioned above, if samples are assigned to clusters they are ordered by the coordinates of their cluster centre instead of their original parameter values. In Fig.'s 5a and 5b the clustering provided by four clusters (FCL=4) is represented by equivalence classes *K1*, ..., *K4*. Samples that are not assigned to a cluster are hybrid elements (due to their characteristic burden pattern) and will be denoted as singletons. Clusters that consist of only one sample are denoted as singletons too, whereas the other clusters are called 'nontrivial' ones. Due to very characteristic values these samples have lead to a single cluster and therefore they can be treated like hybrid elements. In Fig. 5a, clustering of river Oder samples by chemical parameters has apparently lead to only three nontrivial clusters (*K1*, *K2*, *K3*). Here, one cluster consists of only one sample in fact, namely RA9/00: As a single sample RA9/00 (Ratzdorf) represents one cluster because of its comparatively high contamination by heavy metals. Because RA9/00 is greater than (above) RA5/00 it can be concluded that concentrations of all pollutants considered have been increased between May and September 2000. However, RA samples from March and July 2000 (RA3/00, RA7/00) are both incomparable to the May and September samples.

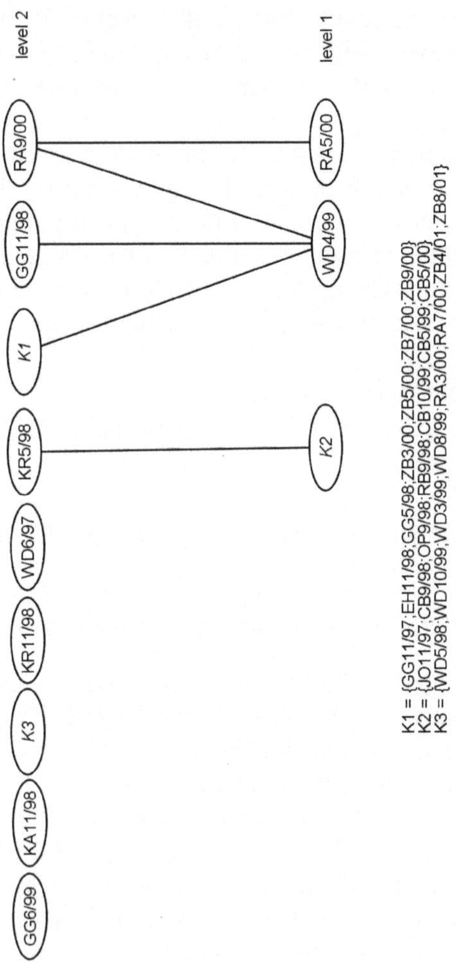

Chemical parameters

Fig. 5a. HD after fuzzy clustering by the whole property space of chemical parameters. River Oder, 31 samples (MAX-PN). Number of hybrid elements = 8 (samples that are not assigned to cluster due to their very characteristic pattern)

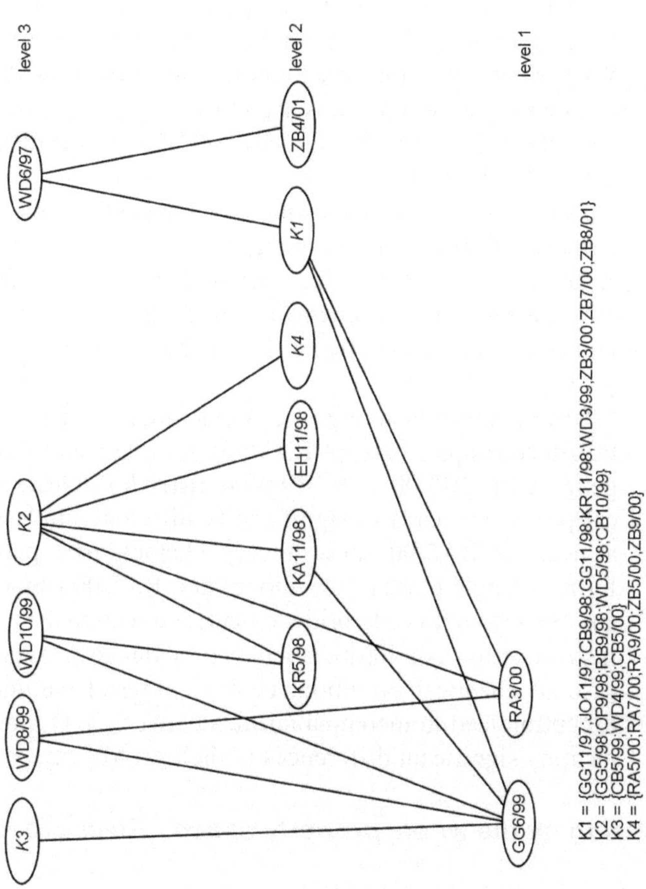

Ecotoxicological parameters

Fig. 5b. HD's after fuzzy clustering by the whole property space of ecotoxicological parameters. River Oder, 31 samples (MAX-PN). Number of hybrid elements = 9 (samples that are not assigned to cluster due to their very characteristic pattern)

Examining the clustering results by chemical parameters in detail, the following can be observed:

- Cluster *K3* and equivalence class *K3,* respectively, contains four out of six samples of Widuchowa (WD). That means, except for June 1997 and April 1999 the pollution pattern by the chemicals considered here is quite similar. The patterns of WD4/99 and WD6/97 are incomparable to these four samples of 1999.
- Comparing *K1* and *K3*, it is seen that the pollution pattern of ZB in year 2000 (ZB3/00, ZB5/00, ZB7/00, ZB9/00) is different to 2001 (ZB4/01, ZB5/01).
- All samples of Cerna Budisovka (CB) have the same pollution pattern over the years of investigation.

A comparison between the diagrams in Fig.'s 5a and 5b once again leads to the assumption that chemical concentrations reproduce another pattern than ecotoxicological effects. This is indicated, for instance, by the following findings:

- Whereas in the clustering of ecotoxicological parameters nearly all Ratzdorf samples (except RA3/00) are assigned to one cluster together with ZB5/00 and ZB9/00 (cluster *K4*), clustering of chemical parameter leads to significantly different similarities: RA9/00 as well as RA5/00 show a very characteristic pattern. RA9/00 forms a single cluster (see above) and RA5/00 cannot assigned to any (therefore it is a hybrid element; see section 3.1).
- Instead of four similarities between Widuchowa samples (WD) in case of chemical pollution, ecotoxicological parameters without exception lead to incomparabilities between WD samples, thus indicating significant differences in their ecotoxicity.

Clustering by the whole property space - River Elbe sediments

Chemical pollutants versus aquatic ecotoxicological tests

Fig.'s 6a and 6b shows the clustering results for 62 Elbe sediment samples. Regarding at first the result of chemical parameters and selecting only the comparable samples with respect to a location, the following relations are found:

- AE9/94=AE6/97=AE10/97=AE10/99=AE6/99=AE10/00=AE6/00 =AE4/01=AE8/01<AE12/92=AE9/93=AE6/94=AE3/95=AE4/96= AE6/96=AE9/96
- FL6/94=FL6/95=FL6/96=FL6/97
- FL12/92=FL9/95=FL10/00=FL6/00=FL4/01
- FL6/94=FL6/95=FL6/96=FL6/97 < FL10/99

- FL6/94=FL6/95=FL6/96=FL6/97 < FL6/99
- FL6/94=FL6/95=FL6/96=FL6/97 < FL9/96
- HM9/95 < HM9/94=HM11/95=HM6/95=HM6/96=HM10/99
- DA6/97=DA9/97
- WB6/96=WB9/96

Except AE6/95 and AE9/95 all samples of Alte Elbe (AE) are assigned to two clusters (*K3*, *K1*) and moreover both are comparable to each other (*K3*<*K1*). Furthermore, since cluster *K3* only contains samples of recent dates, except AE/94, it is indicated that the concentrations of the chemicals considered here are decreased. Moreover, it is striking too that almost only AE samples of prior date are assigned to cluster *K1*. This could be evidence of a higher pollution of the river section AE in this time period and relatively specific burden pattern too. Indeed, since the Alte Elbe is a bayou with limited exchange to the main waterway river Elbe depending on the discharge conditions a stable pollution pattern can be expected over longer periods of time. A similar result is obtained when using aquatic ecotoxicological tests for evaluation: except AE10/99, AE9/93 and AE6/96, all AE samples are assigned to cluster *K1* and have therefore a similar quality pattern.

Considering the other clusters of Fig. 6a it is evident that also *K2* and *K4* consist of samples from almost one river section, namely FL (Fahlberg List) and HM (Meißen harbour). However, in case of FL cluster *K2* is not comparable to the remaining samples of FL: four of overall ten samples from FL are assigned to *K2*. Moreover, most of the samples of FL are incomparable to all other samples (they are isolated elements). Reasons for this specific pattern could be the discontinuous sewage draining from an old contaminated site there.

Cluster *K4* consists of five out of 15 samples from HM. HM9/95 is ≤ *K4* and has therefore lower concentrations. HM9/96, HM6/97 and HM6/00 are assigned to cluster K3 together with many samples from AE and other river sections, while *K3* is not comparable to other HM-samples. Moreover, samples HM10/97, HM9/93, HM10/00, HM6/99, HM6/94 and HM11/92 are isolated (incomparable to all other samples).

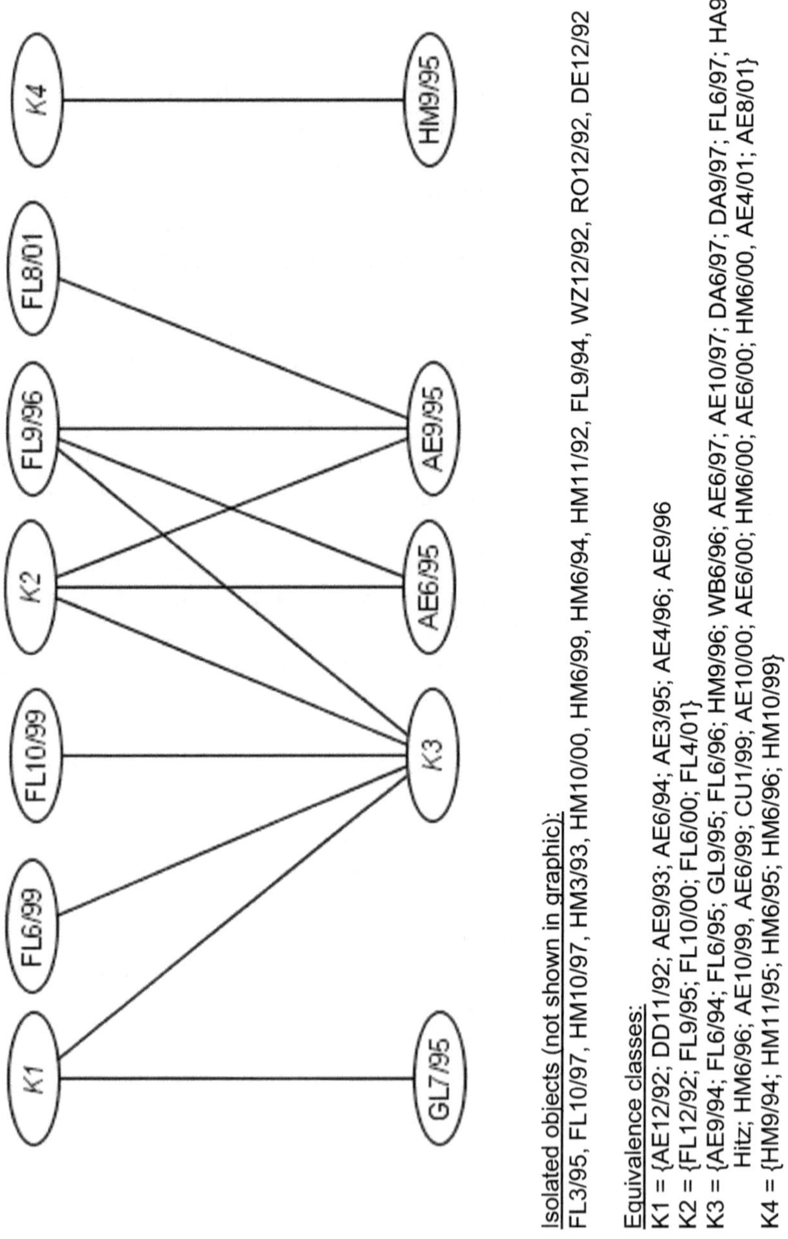

Fig. 6a. HD's after fuzzy clustering by the whole property space of chemical parameters. River Elbe, 62 samples (MAX-PN)

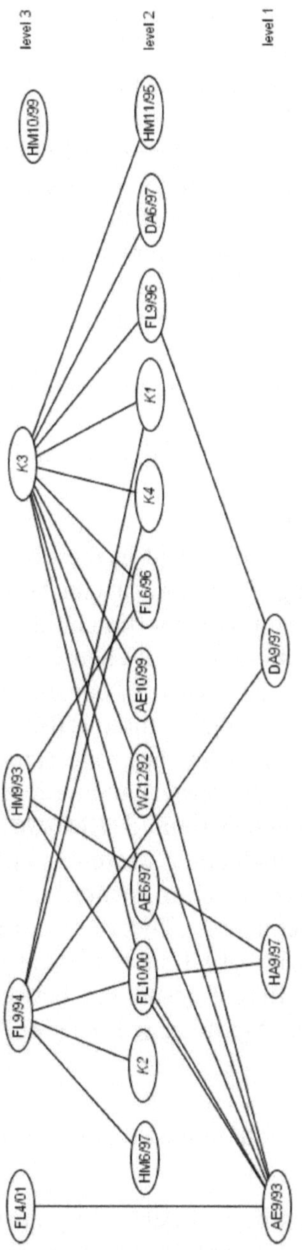

K1 = {AE12/92; DD11/92; HM11/92; RO12/92; AE6/94;AE9/94; AE3/95; AE6/95; AE9/95; FL6/95; FL9/95; GL7/95; GL9/95; HM9/95; AE4/96; AE6/96; AE9/96; HM6/96; WB6/96; WB9/96; AE10(97; FL6/97; Hitz; AE6/99; CU1/99; AE10/00; AE6/00; AE4/01; AE8/01; FL8/01}
K2 = {DE12/92; FL12/92; FL6/94; HM6/94; HM9/94; HM6/95; HM9/956; FL10/99; HM6/00}
K3 = {FL3/95; FL10/95; FL6/00; FL6/00}
K4 = {HM10/97; HM6/99; HM10/00}

Fig. 6b. HD's after fuzzy clustering by the whole property space of ecotoxicological parameters. River Elbe, 62 samples (MAX-PN)

Obviously, within the sampling period the pollution pattern of Meißen harbour varies more than that of other river sections. This may be due to irregular discharges via a creek flowing into the harbour. Using the results of the aquatic ecotoxicological tests only one HM sediment is isolated, i.e. not comparable to any other sediment. Moreover, whereas HM10/00, HM6/99 and HM10/97 have been singletons (isolated) in the chemical approach here they form one cluster (*K4*). However, many of the HM samples are not comparable among each other even though several of them are assigned to one cluster. For example five HM samples belong to cluster *K2*, three to *K4* and three to *K1,* but neither of them is comparable to each other.

Regarding the results of Oder and Elbe sediments it is noticeable that clustering by ecotoxicological parameters

1. leads to more comparabilities in the Hasse diagram and
2. yields other cluster compositions,

compared to chemical parameters.

Ad(1) More comparabilities have been already observed in the evaluation by rough data, i.e. without pre-processing. Therefore it can be expected that also after pre-processing by clustering this proportion holds.

Ad(2) Not many samples can be found in both clusters of chemical and ecotoxicological parameter. For example, in Fig. 6a cluster *K4* consists exclusively of five HM samples. From these five samples, three (HM9/94, HM6/95), HM10/99) are recovered in cluster *K2* (Fig. 6b), one sample (HM11/95) is a hybrid element (not assigned to any cluster) and moreover not comparable to any other HM sample and another sample (HM6/96) is assigned to cluster *K1* in the ecotoxicological evaluation. More examples of different compositions with respect to a certain river section can be found in both clustering of Oder and Elbe sediments. Therefore cluster analysis over the whole property space (by means of all parameters at the same time) strengthens the assumption of different responses between ecotoxicological tests and chemical parameters describing sediment pollution.

Chemical pollutants vs. aquatic ecotoxicological tests vs. nematode sediment contact test

According to the evaluations of Elbe sediments by raw data of a) aquatic ecotoxicological tests, b) chemical measurements, and c) nematode sediment contact tests in Fig. 3, clustering results by means of all parameters simultaneously (i.e. over the whole property space) for each of the groups and additional partial ordering is shown in Fig. 7. A common characteristic for both chemical and ecotoxicological tests is, that

- all AE samples are assigned to one cluster,
- AK4/01 and DE4/01 form a cluster and
- HM10/00 is a singleton and isolated.

The location of AE samples in the lower level of the Hasse diagram is common for all parameter groups. Again, this may result from the special characteristics of undisturbed sediments in this bayou. Another common characteristic is the similarity of samples AK4/01 and DE4/01 (due to cluster analysis) indicating a typical pattern that leads to analogous responses of chemical parameters and test results. However, except the sediments AK4/01, DE4/01 and all AE sediment samples, for FL and HM sediments the parameter groups lead to different compositions of clusters and therefore indicating different responses between the parameter groups.

In contrast to the partial orders from chemical concentrations and aquatic ecotoxicological tests, the results of the sediment contact tests with nematodes lead to a total order. This corresponds to a correlation analysis that shows a significant correlation between growth, reproduction and egg hatch (r=0,7).

Chemical pollutants, biochemical tests and ecotoxicological tests simultaneously

To complete fuzzy clustering over the whole property space and identifying differences in responses between the parameter groups, respectively, clustering results of each biochemical tests, chemical parameters and ecotoxicological tests are combined in a matrix as basis for partial ordering. In addition to the comparisons shown above this presentation may facilitate the identification of similar responses of tests. Fig. 8 shows that partial ordering leads to two equivalence classes[1] containing all AE samples. Once again this fact strengthens the assumption that the bayou Alte Elbe has specific sediment features leading to similar responses of all test systems.

---

1 In contrast to the evaluations above where equivalence classes are a consequence of clustering results (instead of original parameter values samples obtain the values of cluster centres), here equivalence classes are a result of equivalent pattern concerning the three parameter groups (chemicals, ecotox. and biochemical tests).

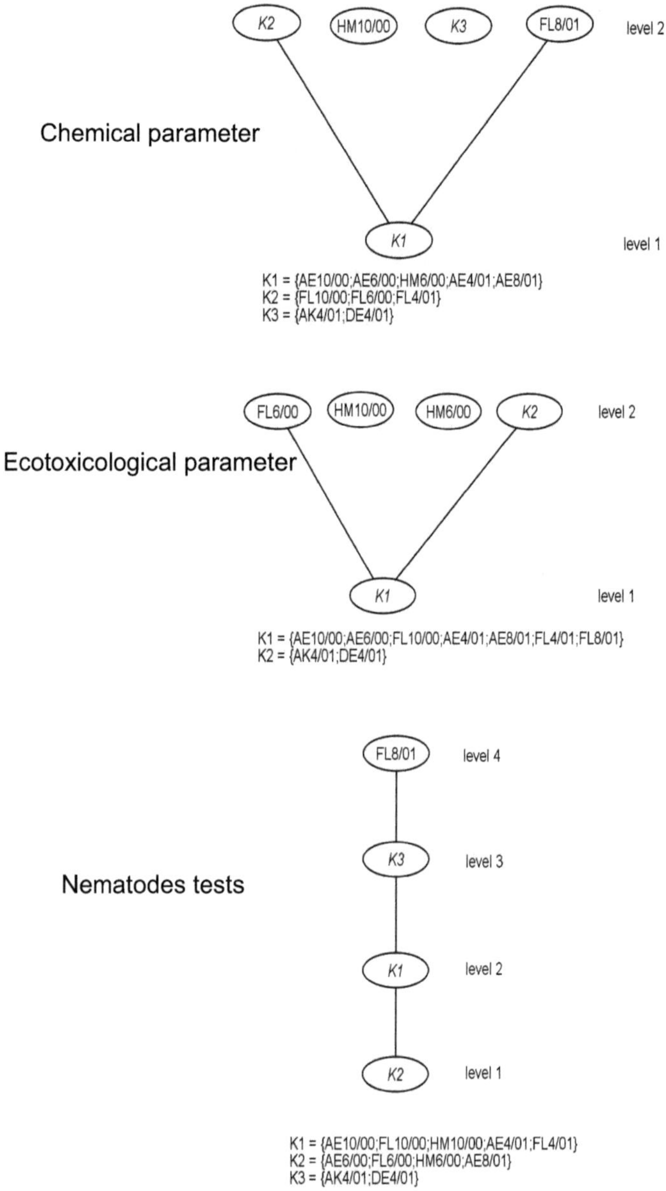

Fig. 7. Clustering results and Hasse diagrams for Elbe sediments

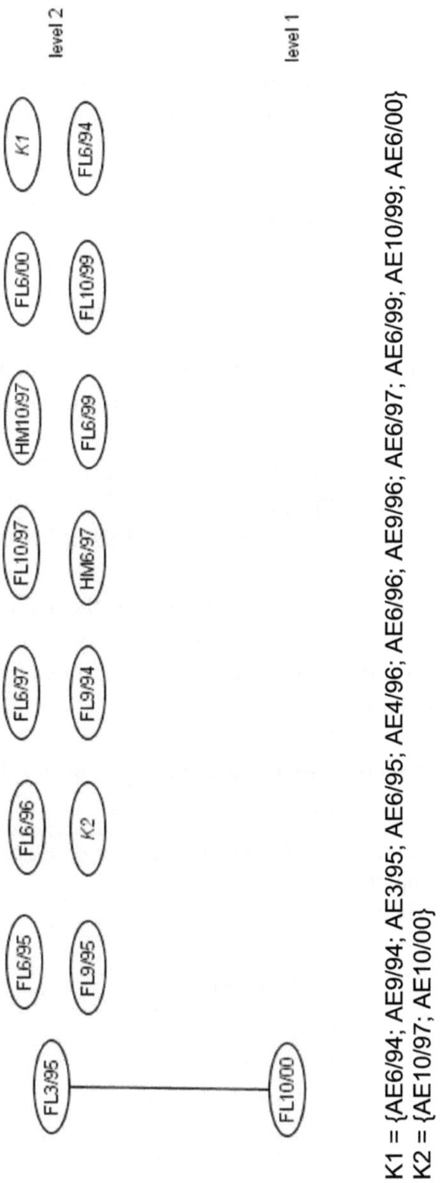

Fig. 8. Hasse diagram of clustering results from three parameter groups: biochemical tests, chemical parameters and ecotoxicological tests (28 Elbe sediments and 36 parameters overall)

The Hasse diagram shows only one comparability, namely between FL10/00 and FL3/95, indicating a decline of concentrations in all chemicals, metals and toxic qualities at the same time. However, all FL samples taken before and in between are not comparable to any other sediment sample, and are therefore expressing different responses of test systems and chemicals.

Attribute-wise clustering - Oder sediments

Aquatic ecotoxicological tests versus chemical pollutants

Fig. 9a and 9b show the evaluations after fuzzy-clustering by means of each parameter separately (attribute-wise clustering). As in the clustering by the whole property space, here for each parameter a cluster number of FCL=4 and a TMF of 0.8 is used. The investigation particularly aims at the discovery of temporal changes in the sediment quality and therefore on the identification of so-called chains[2] with links consisting of samples from a certain river section.

The evaluation by means of chemical concentrations only yields one comparability between samples of a river section, namely ZB4/01 < ZB3/00, whereas the Hasse diagram based on ecotoxicological tests generates several relations as shown in Table 2.

Table 2. Comparabilities of a river section after evaluation by ecotoxicological test

Comparabilities of a river section	Total number of samples of each river section	Incomparable river sections
ZB4/01 < ZB3/00 < ZB7/00 ZB4/01 < ZB8/01	ZB=6	ZB9/00
GG11/97 < GG11/98 GG6/99 < GG11/98 GG6/99 < GG5/98	GG=4	
CB9/98 < CB10/99	CB=4	CB9/96, CB5/00
WD8/99 < WD5/98	WD=6	WD10/99,WD4/99, WD3/99,WD6/97

2 Chains are a sequence of lines in the Hasse diagram indicating that the elements are comparable with each other.

With respect to ecotoxicological responses partial ordering of river section GG (Glogau) indicates an improvement of its pollution status because sample 6/99 is less than 11/98 and 5/98. The incomparability between GG6/99 and GG11/97 is only based on a difference in the bacteria test in eluate (HEB), see Table 2. For river section ZB a conclusion about an improvement is difficult to derive. Though sample ZB4/01 is less than ZB3/00 and ZB7/00, it is not comparable with ZB5/00 and ZB9/00 (Tab. 2).

Both Hasse diagrams in Fig. 9a and 9b indicate neither an increasing nor a decline of sediment burden for a river section, except for GG with respect to ecotoxicological responses.

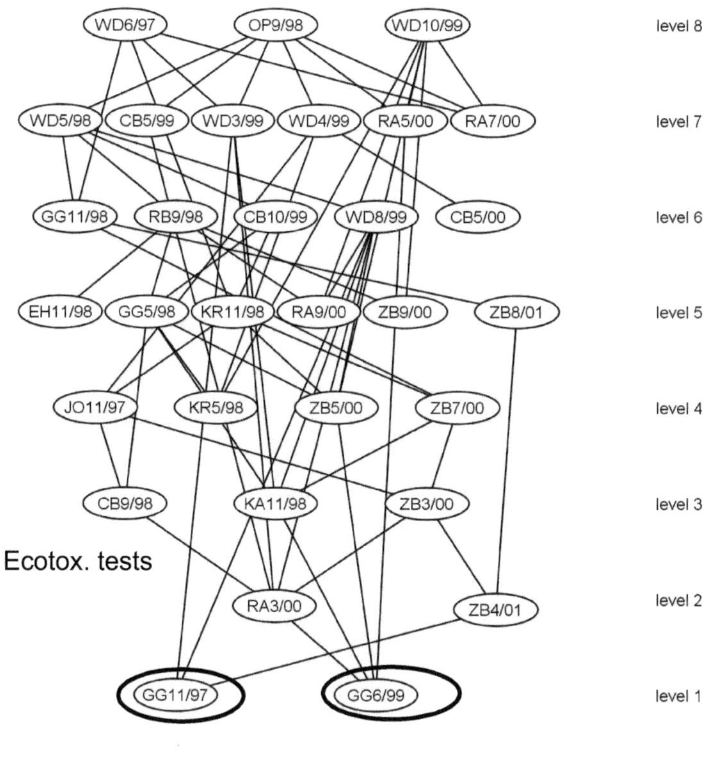

Fig. 9a. Hasse diagram for Oder sediments after attribute-wise clustering of ecotoxicological parameters (FCL=4, TMF=0.8)

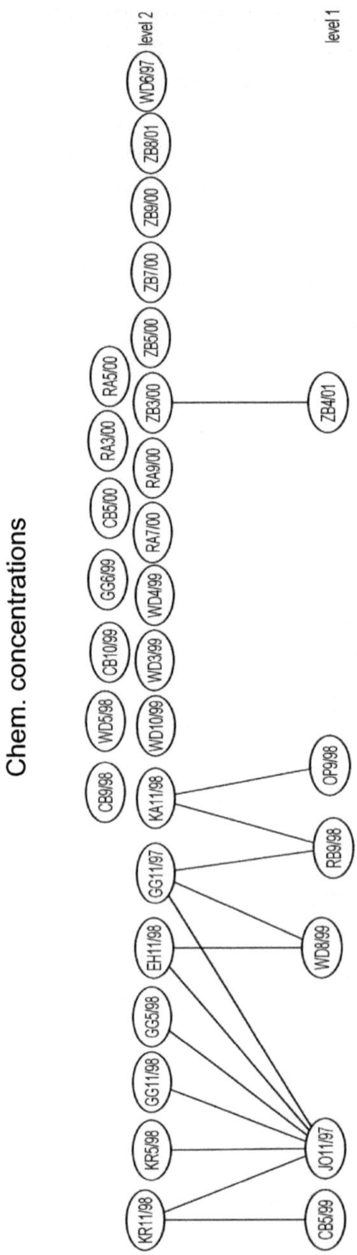

Fig. 9b. Hasse diagrams for Oder sediments after attribute-wise clustering of chem. concentrations (FCL=4, TMF=0.8)

Table 3. Inhibition values of ecotoxicological tests from a selection of samples in Fig. 9. Negative values mean stimulation of test cultures

Sample	HEA	HEB	HED	HPA	HPB	HPD
GG11/97	3,33	1,28	0	-77,33	14,16	0
GG5/98	-44,32	15,87	0	-20,52	47,78	0
GG11/98	3,33	15,87	0	38,63	14,16	0
GG6/99	-96,39	10,11	0	-159,3	14,16	0
ZB3/00	3,33	10,11	0	-20,52	14,16	0
ZB5/00	-96,39	15,87	0	-121,2	32,23	0
ZB7/00	3,33	15,87	0	-20,52	14,16	0
ZB9/00	-198,23	20,4	0	-159,3	47,78	0
ZB4/01	3,33	1,28	0	-20,52	14,16	0
ZB8/01	3,33	5,8	0	7,7	14,16	0

Attribute-wise clustering - Elbe sediments

Aquatic ecotoxicological tests versus chemical pollutants

In contrast to the above made comparisons of quality patterns by chemical concentrations and aquatic ecotoxicological test results for 62 sections (sites) of the River Elbe (see Fig. 6), here we will use attribute-wise clustering to look at temporal changes of only one river section. Afterwards the results of evaluation by a) the nematode test and b) biochemical tests will be compared with the chemical and aquatic ecotoxicological approaches.

Fig.'s 10a and 10b shows the Hasse diagrams after single clustering of each attribute (parameter) for AE sediments using a cluster number of FCL=4 and a TMF=0.8. The equivalence classes *K1* and *K2* in the ecotoxicological evaluation are the only minimal elements, i.e. compared to all samples above their members have the lowest values in all tests. Except *K2* all samples are comparable with *K1*. The fact that except AE9/94, AE6/94 and AE3/95 all samples above *K1* (AE12/92, AE9/93, AE6/95) have been taken at a later date indicates an increasing pollution for AE. Using chemical concentrations for evaluation the temporal trend seems to be contrary. However, both recent samples AE4/01 and AE8/01 are isolated, i.e. not comparable to all other samples. A sensitivity analysis (see e.g. Heininger et al. 2003) shows that the evaluation is most sensitive to the nitrogen content, where omitting this nutrient compensates the isolation of AE4/01 and AE8/01 (see Fig. 11). Moreover, sample AE4/01 is now a minimal element and therefore emphasizes the indication of a decline of the cline of the pollution status with respect to (most of) chemical concentrations.

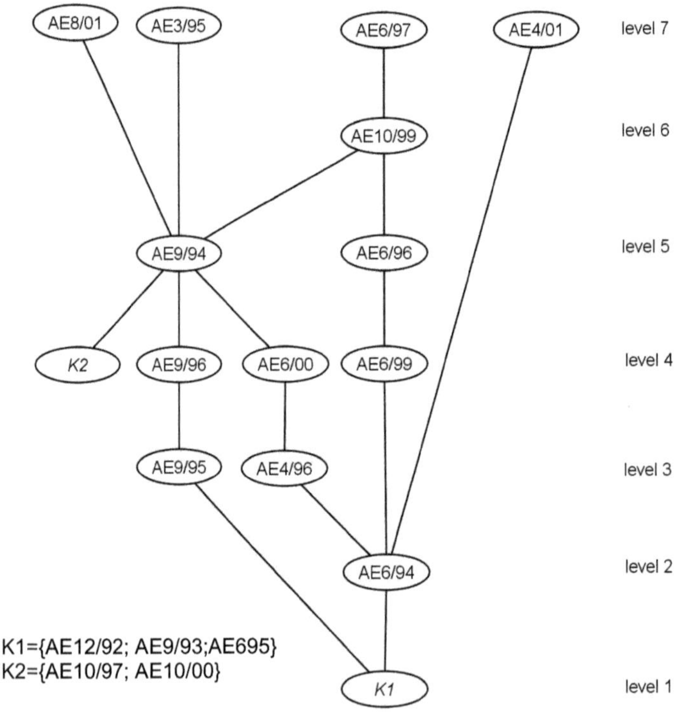

Fig. 10a. Hasse diagrams for AE sediments after clustering of each parameter (FCL=4, TMF=0.8): ecotoxicological tests

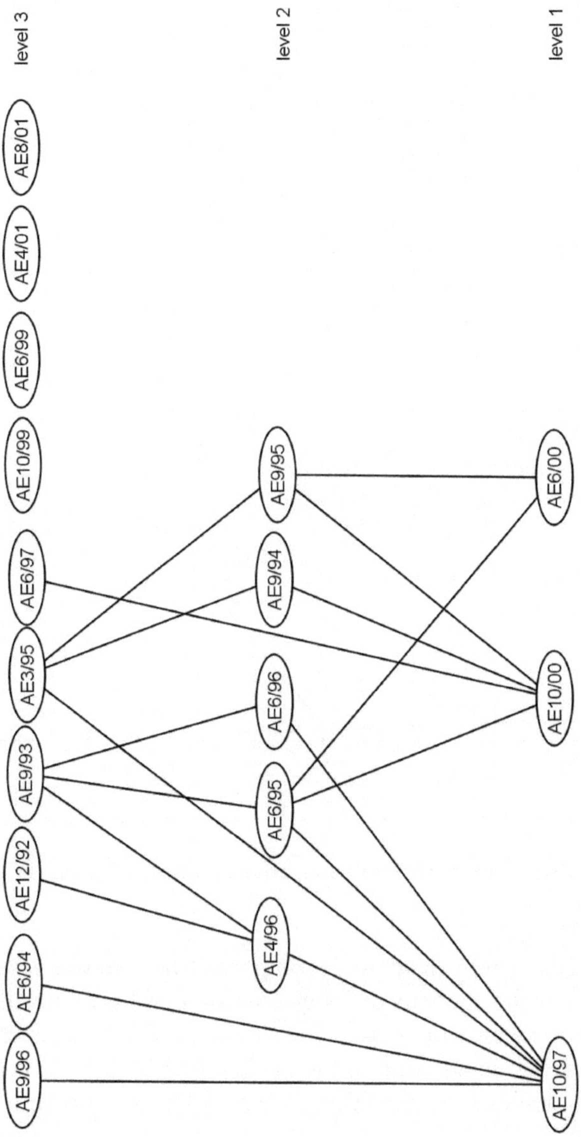

Fig. 10b. Hasse diagrams for AE sediments after clustering of each parameter (FCL=4, TMF=0.8): chemical concentrations

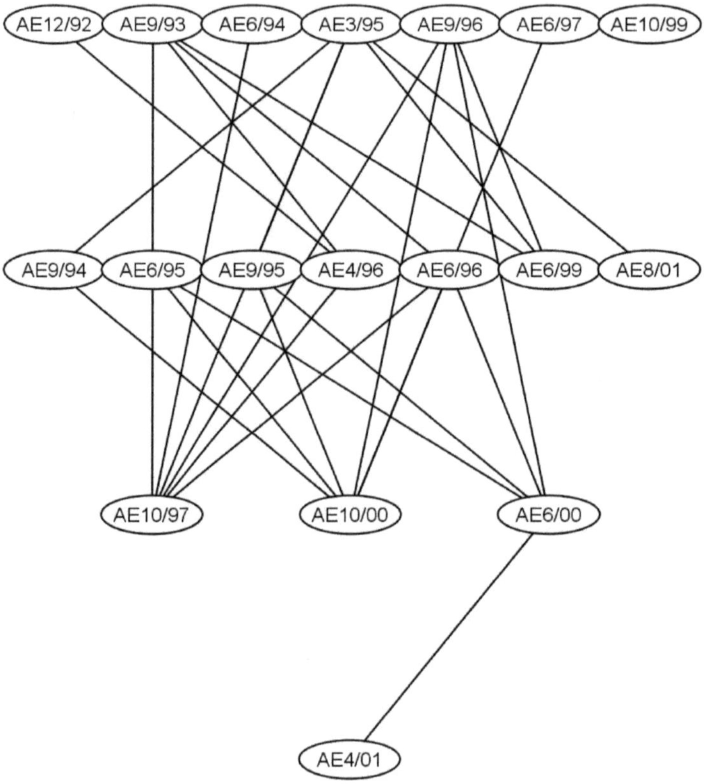

Fig. 11. Hasse diagram for AE sediments after omitting nitrogen (in HD of chem. conc.; Fig. 10)

Seasonal effects may lead to incomparabilities between samples and therefore hamper the analysis of long-term temporal changes. For this reason only annual mean values of each parameter will be clustered as basis for partial ordering. In addition to the evaluation above, here the biochemical test results will also be considered (parameters are shown in Table 5 in the appendix).

Fig. 12 shows that averaging does not facilitate the interpretation of the Hasse diagrams. Different quality patterns between younger and older samples are the reason for incomparabilities. Again, sensitivity analysis may be a method to identify sensitive parameters whose significance may be subject of expert discussion. In case of minor relevance of such a parameter simply omitting may lead to a result that is at least easier to interpret (as shown in Fig. 11). Whereas for biochemical and ecotoxicological tests sensitivity analysis does not yield any striking sensitivity values, the evaluation by chemical parameters is again sensitive to the nitrogen content (c.f. Fig. 12). Now, the results can be discussed as follows:

- Evaluation by means of biochemical tests reveals a positive quality trend of section AE. The pollution patterns of both AE00 and AE99 are less than AE95 and AE97, but incomparable with those of 1996 and 1994. Reasons for that incomparability are higher values (after clustering) in both parameters DHGS and PRV whereas all remaining parameters of AE00 and AE99 have lower values than AE95 and AE97.
- Similar to the trend of biochemical test results the position of AE00 in the Hasse diagram indicates a decline in ecotoxicity. Here, the incomparability of AE00 to the years 1995 and 1994 is due to the bacteria test (eluate) that yields higher values for year 2000 whereas algae (in pore water and eluate) and bacteria in pore water have identical or lower values in 2000 (daphnia tests yield only zero values for all years).
- From the analysis of all AE samples a high sensitivity to nutrient pollution (nitrogen) has been expected (c.f. Fig. 10 and 11). From the water ecological point of view nitrogen is a limiting factor for algae growth and therefore important for evaluation. Considering the management of dredged material (e.g. from AE) this may be important for the relocation in waters, which are sensitive to eutrofication.

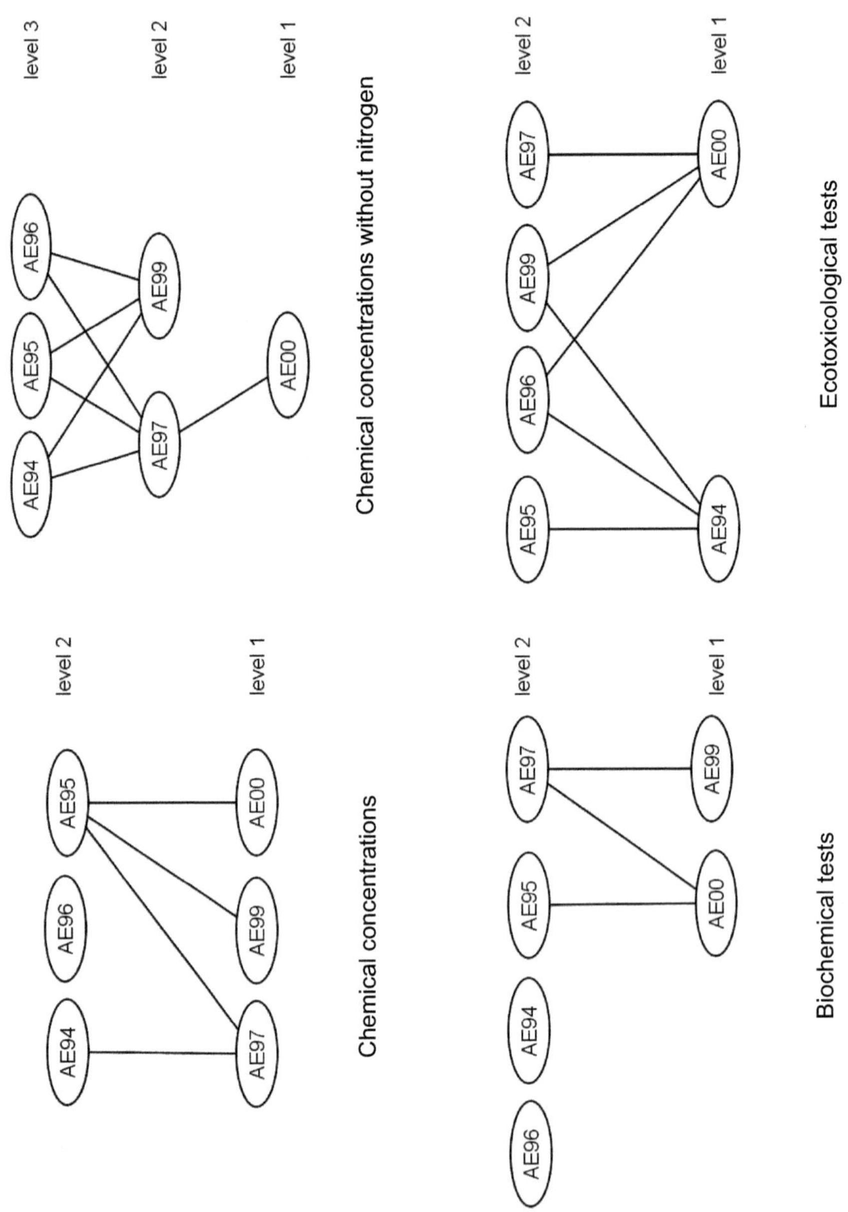

Fig. 12. Clustering and evaluation of annual mean values of river section AE

Attribute-wise clustering – Elbe and Oder sediments

Chemical pollutants vs. aquatic ecotoxicological tests vs. nematode sediment contact test

According to the clustering over the whole property space the evaluation by nematode tests after single clustering will be compared with those of chemicals and aquatic ecotoxicological tests too. However, here the sample set includes additional samples from River Oder due to relatively few samples from River Elbe (nematodes tests started in 2000 only).

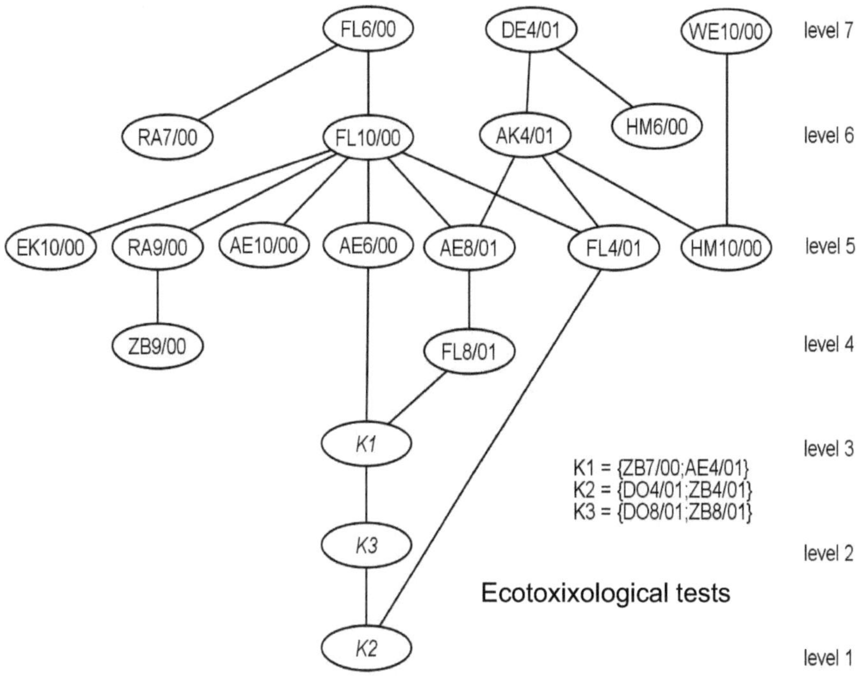

Fig. 13a. Hasse diagram for evaluation of river Elbe and Oder sediment samples after single clustering of parameters (22 samples): ecotoxicological tests

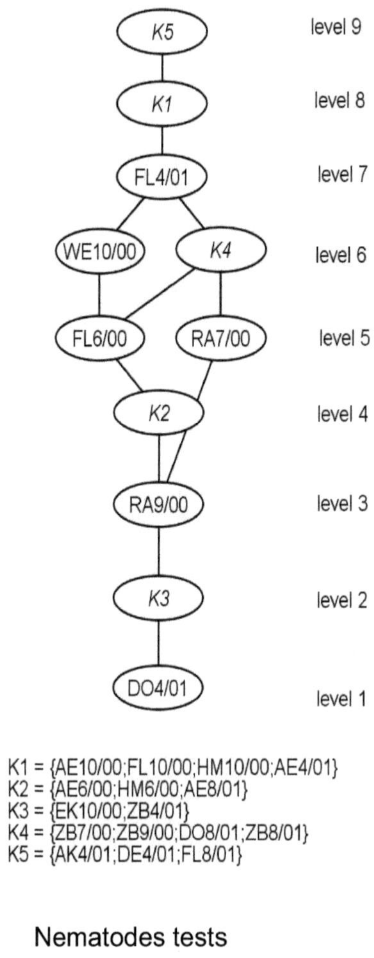

Fig. 13b. Hasse diagram for evaluation of river Elbe and Oder sediment samples after single clustering of parameters (22 samples): nematodes tests

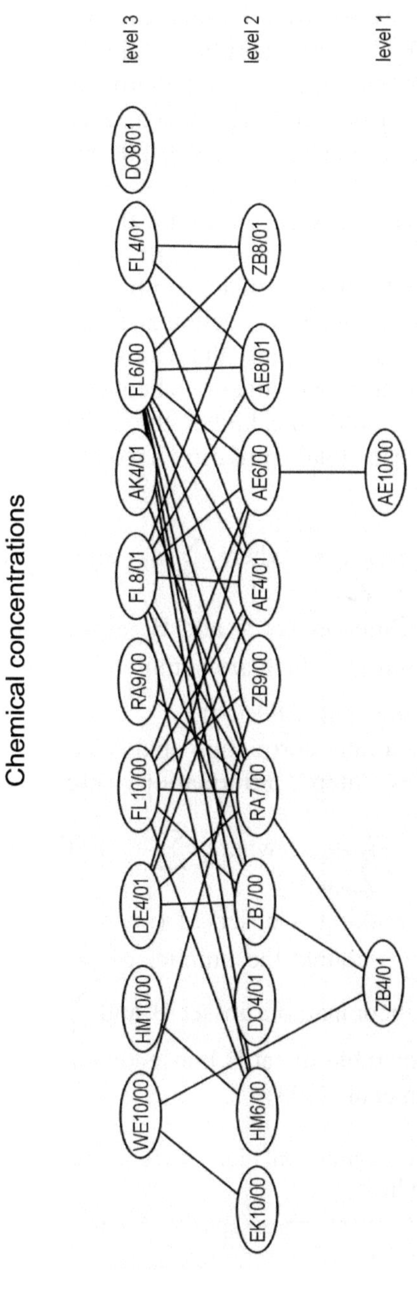

Fig. 13c. Hasse diagram for evaluation of river Elbe and Oder sediment samples after single clustering of parameters (22 samples): chemical concentrations

Compared to the results of nematodes tests, evaluation by chemical concentrations and ecotoxicological tests present a familiar picture (c.f. Fig. 7 and 13a,b,c), which is dominated by incomparabilities (and no equivalence classes) and therefore a higher diversity of pollution pattern. Again, due to high correlation between the three parameters egg hatch, growth and reproduction, evaluation of samples by nematodes tests results in nearly total order.

In contrast to the comparison between biochemical tests, aquatic ecotoxicological tests and chemical concentrations, here averaging is not convenient since nematodes have been investigated only for two years. Moreover, instead of considering only one river section and visual examining of differences between the Hasse diagrams, the Tanimoto index and the W-matrix will be used for similarity investigations between evaluation results. For the Tanimoto index T holds the higher the index the more similarity, where a value of one means total similarity: Two approaches of the Tanimoto index are used here:

1. $T_1 = \frac{A \cap B}{A \cup B} = \frac{\sum_{sr}}{\sum \leq_A + \sum \leq_B - \sum_{sr}}$, where $\sum \leq_A$ and $\sum \leq_B$ are the numbers of comparabilities in set A and B respectively, and $\sum_{sr}$ counts the comparabilities which are common in both sets A and B (for details see Pudenz et al. 1998).
2. T_2 quantifies similarity by a rank correlation analysis and in comparison to T_1 it takes more information into account: $T_2 = \frac{\sum_{sr}}{\sum_{sr} + \sum_{rr} + \sum_{irA} + \sum_{irB}}$, where $\sum_{rr}$ is the sum of pairs for which a reverse ranking is observed (i.e. $x < y$ in set A and $y < x$ in set B), $\sum_{irA}$ counts the number of pairs that are comparable in set A but incomparable in set B and $\sum_{irB}$ is the sum of pairs that are comparable in set B but incomparable in set A (for details see Sørensen et al. 2003).

Table 4. Similarities between Hasse diagrams in Fig. 13 (Elbe and Oder sediments) calculated by two similarity indices

Similarity between A and B:	T_1	T_2
Chemicals – Ecotoxicological tests	0,22	0,22
Chemicals – Nematodes tests	0,15	0,19
Ecotoxicological tests – Nematodes tests	0,26	0,27

For both Tanimoto indices T_1 and T_2 similar results can be observed where the highest value results from the comparison between ranking by nematodes and ecotoxicological tests (Table 4). However, all values show a relative low similarity between the Hasse diagrams in Fig. 13 and therefore the different property pattern of sediments indicated by chemical concentrations, ecotoxicological and nematodes tests should be taken into account for sediment evaluation.

Summary and Conclusions

Partial ordering of sediment data from German rivers Elbe and Oder has shown that Hasse diagram technique is a powerful tool to analyse the sediment status. The diversity of pollution pattern of river sections can be identified and the effects of pollutants on different test systems can be compared without merging of data. However, as shown here a relatively high diversity of the quality pattern may also increase the degree of incomparability between sediment samples and therefore hampers the identification of e.g. temporal changes. But also incomparabilities between samples may give reasons for expert discussion, for instance the consideration of secondary information of river sections (discharge of pollutants) that could be responsible for a specific quality pattern and for incomparabilities, respectively.

In order to reduce incomparabilities and to facilitate the interpretation of Hasse diagrams, pre-processing by two strategies of fuzzy cluster analysis has been applied to sediment data:

- classification of sediments over the whole property space (by means of all parameter at the same time) aiming at the identification of similar quality pattern and
- classification of sediment samples by each parameter separately focusing on disregard of small numerical differences between parameter values on the one hand and in conclusion on ranking with respect to their quality on the other hand.

Cluster analysis over the whole property space often leads to clusters containing time series of samples from one river section. The sediments in particular from the River Elbe section Alte Elbe are characterised by relatively constant pattern over the years of investigation. This is underlined by the fact that for each parameter group similar clustering results can be observed with respect to the site Alte Elbe. An instructive example is the

Hasse diagram in Fig. 8 that considers clustering results of biochemical and aquatic ecotoxicological tests and chemical measurements at the same time, where exclusively all AE samples are assigned to two equivalence classes. It is assumed in this case the specific conditions of a bayou lead to similar responses of different test systems. However, for most of the samples from rivers Oder and Elbe the parameter groups lead to different clustering results and therefore indicate different responses in the sediment contact test with nematodes, in biochemical and aquatic ecotoxicological tests among each other and also compared to chemical concentrations.

This result is also confirmed by partial ordering of sediment samples after an attribute-wise clustering. Here, additional similarity calculations with Tanimoto indices indicate relatively high differences between ranking results.

Whereas the clustering over the whole property space indicates a similar pattern during the sampling period with respect to chemical concentrations and aquatic ecotoxicological tests for river section Alte Elbe (c.f. Fig. 8), attribute-wise clustering enables a more differentiated comparison. Here, the Hasse diagrams indicate contrasting temporal changes for AE. Following the chemical concentrations and additionally omitting nitrogen concentration from the parameter group a decline can be observed whereas the ecotoxicological tests suggest an opposite trend. However, by averaging and therefore eliminating seasonal effects, the opposite trend indicated by ecotoxicological tests is slightly weakened.

In summary, by each parameter group different responses to sediment quality can be expected. Comparing the responses of the different parameter groups, the diversity obtains a maximum when using chemical concentrations for partial ordering, thus hampering a comparative evaluation of sediments.

Regarding the evaluation of dredged material further expert discussions aiming at a detailed selection of parameters should be initiated. Here, sensitivity analysis has shown that for instance omitting nitrogen leads to significant changes in the ranking result.

Furthermore, a basic problem for the evaluation of dredged material seems to be the high diversity of sediment quality as represented by all parameter groups. Though partial ordering is helpful to analyse differences in responses of test systems, due to many incomparabilities it is crucial to derive a decision. Therefore, it has to be investigated whether a linear mapping of the partial order by e.g. linear extensions and an average rank probability (Brüggemann et al. 2004) or other approaches like fuzzy-logic (Ahlf, Heise 2005) are useful for decision purposes.

Appendix

Table 5. Parameter groups and their composition for evaluation of river sediments

Parameter group	Parameters
25 chemical pollutants and basic parameters (concentrations in mg/kg, µg/kg; g/kg)	AOX, pp'-DDT, pp'-DDD, pp'-DDE, HCB, α-HCH, γ-HCH, PAHs (sum of 16 according to EPA 610), PCBs (sum of congeners 28, 52, 101, 138, 153, 180), TBT; N, S, TOC, As*, B, Cd, Cr, Cu, Co, Hg, Ni, P, Pb, Sn, Zn
6 ecotoxicological test results with pore water and eluates (inhibition, %)	Daphnia (HED), algae (HEA), bacteria (HEB) each in sediment eluate and pore-water (HPD, HPA, HPB)
3 nematode test results with the whole sediment (inhibition, %; details in Traunspurger et al. 1997)	egg hatch, reproduction, growth
11 biochemical test results (reduced on the maximum number of tests used here; for details see Heininger, Tippmann 1995).	DHP=Dehydrogenase activity in pore-water DHgS= Dehydrogenase activity in sediment AP=Alanin- Aminopeptidase activity in pore-water AV=Alanin- Aminopeptidase, D1 value of dilution series AS=Alanin- Aminopeptidase activity in sediment βGP=β-Glucosidase activity in pore-water βGV=β-Glucosidase activity, D1 value of dilution series βGS=β-Glucosidase activity in sediment PR=Protease activity PRV=Protease activity, D1 value of dilution series DNAP=DNA- content in pore-water

* Heavy metals, boron (B(V)), arsenic and total phosphorus were determined in the fraction < 20 µm to improve the comparability of the results. This fraction was separated from the freeze-dried and non-milled samples by ultrasonic sieving (Ackermann 1980). Metals were analysed after microwave-assisted digestion with aqua regia at 180 °C in closed vessels by inductively coupled plasma optical emission spectroscopy, atomic fluorescence spectroscopy (mercury) and hydride atomic absorption spectroscopy (arsenic).

Table 6. Sample sites/river sections

Abbr.	Site	Abbr.	Site	Abbr.	Site
AE	Altarm Alte Elbe	HK	Havelkanal	SA	Saale, Buhnenfelder
AK	Hornhafen Aken, Elbe	HL	Hirschsteiner Lache, Elbe	SS	Seddinsee vor Insel
CB	Cermna Budisovka, Oder	HM	Hafen Meißen	TS	Tiefer See
CU	Cumlosen, Elbe	JO	Jocinkou, Odergebiet	UW	Unterwarnow, R6
DA	Damnatz, Elbe	KA	Kaczawa, Oder	VS	Veltener Stichkanal
DD	Dresden, Hafen Pieschen	LA	Lauffen	WA	Warthe, Swierkocin
DE	Dessau, Leopoldhafen	ME	Mescherin	WB	Wittenberge
DÖ	Dömitz, MEW	OD	Oder	WD	Widuchowa
EH	Eisenhüttenstadt, Oder	OK	Oder-Havelkanal	WE	Weiße Elster
EK	Eldenburger Kanal, MEW	OP	Mnichov, Opava, Oder	WO	Westoder
FL	Fahlberg List, Elbe	RA	Ratzdorf	WT	Wettin, Saale
FS	Finowschleuse, Oder	RB	Ramzovskybach, Oder	WU	Wusterwitz, EHK
GG	Glogau, Oder	RG	Rothenburg, Saale	WZ	Wittenberge, Zellwollehafen
HF	Hohensaathe-Friedrichsthaler Wasserstraße	RO	Rodleben, Elbe	ZB	Hohenwutzen, Zollbrücke
				ZD	Zehdenick, OHW

References

Ackermann F (1980) A procedure for correcting the grain size effect in heavy metal analyses of estuarine and coastal sediments. Environ Technol Lett 1:518-527

Ahlf W, Heise S (2005) Sediment Toxicity Assessment – Rationale for Effect Classes. J Soils & Sediments 5(1):16 – 20

Brüggemann R, Halfon E, Welzl G, Voigt K, Steinberg C (2001) Applying the Concept of Partially Ordered Sets on the Ranking of Near-Shore Sediments by a Battery of Tests. J Chem Inf Comput Sci 41:918-925

Brüggemann R, Sørensen P, Lerche D, Carlsen L (2004) Estimation of Averaged Ranks by a Local Partial Order Model. J Chem Inf Comput Sci 41:918-925

Heininger P, Pelzer J, Claus E, Pfitzner S (2003) Results of Long-Term Sediment Quality Studies on the River Elbe. Acta hydrochim hydrobiol 31(4-5):356-367

Heininger P, Pelzer J, Claus E, Tippmann P (1998) Trends in River Sediment Contamination and Toxicity - Case of the Elbe River. Wat Sci Tech 37(6-7):95-102

Heininger P, Tippmann P (1995) Determination of Enzymatic Activities for the Characterization of Sediments. Toxicol Environm Chem 52:25-33

Luther B, Brüggemann R, Pudenz S (2000) An Approach to Combine Cluster Analysis with Order Theoretical Tools in Problems of Environmental Pollution. MATCH - Comm in Math Chem 42:119-143

Pudenz S (2004) ProRank – Software for Partial Order Ranking. MATCH – Comm in Math Chem 54(3):611-622

Pudenz S, Brüggemann R, Komoßa D, Kreimes K (1998) An Algebraic Tool to Compare Ecosystems with respect to their Pollution By PB/CD, III. Comparative Regional Analysis By Applying A Similarity Index. Chemosphere 36(3):441-440

Pudenz S, Brüggemann R, Luther B, Kaune A, Kreimes K (2000) An Algebraic/Graphical Tool to Compare Ecosytems with respect to their Pollution V: Cluster Analysis and Hasse Diagrams. Chemosphere 40(12):1373-1382

Sørensen PB, Brüggemann R, Carlsen L, Mogensen BB, Kreuger J, Pudenz S (2003) Analysis of monitoring data of pesticide residues in surface waters using partial order ranking theory. Environ Tox & Chem 22(3):661-670

Traunspurger W, Haitzer M, Höss S, Beier S, Ahlf W, Steinberg C (1997) Ecotoxicological Assessment of Aquatic Sediments with Caenorhabditis Elegans (Nematoda) - A Method for Testing in Liquid Medium and whole Sediment Samples. Environ Toxicol Chem 16:245-250

Prioritizing PBT Substances

Lars Carlsen[1]*, John D. Walker[2]

[1] Awareness Center
Hyldeholm 4, Veddelev, DK-4000 Roskilde, Denmark

*e-mail: LC@AwarenessCenter.dk

[2] TSCA Interagency Testing Committee (ITC), Office of Pollution Prevention and Toxics (7401), U.S. Environmental Protection Agency, Washington, D.C. 20460, USA

Abstract

The interplay between partial order ranking and Quantitative Structure Activity Relationships (QSARs) constitute a strong decision support tool. By means of partial order ranking it is possible to prioritize and select chemicals for decision-making among a group of substances based on simultaneous evaluation of data related to different endpoints. In the absence of experimental data, QSARs are used to provide estimates. In the present chapter, the identification of chemicals with Persistence and Bioconcentration (PB) potential is used to illustrate the interplay between partial order ranking and QSARs. The endpoints biodegradation and bioconcentration were obtained using the BioWin and BCFWin modules from http://www.epa.gov/oppt/exposure/docs/episuitedl.htm. Partial order theory was used to rank chemicals for PB potential based on QSAR estimates. The proposed approach is suggested as a decision support tool to facilitate pollution prevention activities by regulated and regulatory communities.

Introduction

Persistent, bioaccumulative and toxic (PBT) substances are chemicals that persist in the environment, accumulate in tissues of biological organisms and cause toxic effects. PBT substances are characterized by having persistence characteristics (e.g., an atmospheric half-life of ≥ 2 days, an aquatic half life of ≥ 60 days or a soil or sediment half life of ≥ 6 months), a bioconcentration factor (BCF) ≥ 5,000 and toxicity potential, e.g., an aquatic organism LC_{50} ≤ 1 mg/L (cf. Carlsen and Walker, 2003 and references therein).

It is advantageous to prioritize chemicals for PBT potential by evaluating several criteria. One method for accomplishing this is to include all criteria into a single criterion (for a discussion please see Brüggemann et al., p. 237). As described in this chapter for substances with P and B characteristics, a more effective method for prioritizing chemicals for P and B potential is by simultaneous evaluation of several criteria using partial order ranking.

Materials and Methods

Substances studied

The TSCA Interagency Testing Committee (ITC, http://www.epa.gov/opptintr/itc) screened 8,511 chemicals for PB potential. Walker and Carlsen (2002) described the PB characteristics for 50 of these chemicals (Table 1).

Table 1. Bioconcentration factors (BCF) and Biodegradation potentials (BDP) for the 50 chemicals included in the Walker and Carlsen (2002) study. H and M denotes high and medium estimates for both the bioconcentration (B) and Persistence (P) scores

	CAS RN	Chemical	BCF	BDP	B	P
1	000087-82-1	Benzene, hexabromo-	9417	1.1644	H	H
2	000118-74-1	Benzene, hexachloro-	5153	1.3302	H	H
3	000128-69-8	Perylo[3,4-cd:9,10-c'd']dipyran-1,3,8,10-tetrone	13200	1.5328	H	H
4	000133-14-2	Peroxide, bis(2,4-dichlorobenzoyl)	8478	1.533	H	H
5	000355-42-0	Hexane, tetradecafluoro-	8609	0.5777	H	H
6	000375-81-5	1-Pentanesulfonyl fluoride, 1,1,2,2,3,3,4,4,5,5,5-undecafluoro-	29740	1.0596	H	H
7	000423-50-7	1-Hexanesulfonyl fluoride, 1,1,2,2,3,3,4,4,5,5,6,6,6-tridecafluoro-	7444	0.737	H	H
8	000509-34-2	Spiro[isobenzofuran-1(3H),9'-[9H]xanthen]-3-one, 3',6'-bis(diethylamino)-	25450	1.5815	H	H
9	000596-49-6	Benzenemethanol, 4-(diethylamino)-.alpha.,.alpha.-bis[4-(diethylamino)phenyl]-	4292	1.1758	M	H

10	000678-39-7	1-Decanol, 3,3,4,4,5,5,6,6,7,7,8,8,9,9,10,10,10-heptadecafluoro-	12200	0.3357	H	H
11	001568-80-5	1,1'-Spirobi[1H-indene]-6,6'-diol, 2,2',3,3'-tetrahydro-3,3,3',3'-tetramethyl-	13070	1.994	H	M
12	001770-80-5	Bicyclo[2.2.1]hept-5-ene-2,3-dicarboxylic acid, 1,4,5,6,7,7-hexachloro-, dibutyl ester	29340	1.2935	H	H
13	002379-79-5	Anthra[2,3-d]oxazole-5,10-dione, 2-(1-amino-9,10-dihydro-9,10-dioxo-2-anthracenyl)-	2310	1.9347	M	M
14	002475-31-2	3H-Indol-3-one, 5,7-dibromo-2-(5,7-dibromo-1,3-dihydro-3-oxo-2H-indol-2-ylidene)-1,2-dihydro-	3972	1.0633	M	H
15	002641-34-1	Propanoyl fluoride, 2,3,3,3-tetrafluoro-2-[1,1,2,3,3,3-hexafluoro-2-(heptafluoropropoxy)propoxy]-	1363	-0.5183	M	H
16	003006-86-8	Peroxide, cyclohexylidenebis[(1,1-dimethylethyl)	13560	1.9874	H	M
17	003864-99-1	Phenol, 2-(5-chloro-2H-benzotriazol-2-yl)-4,6-bis(1,1-dimethylethyl)-	14930	1.8338	H	M
18	004051-63-2	[1,1'-Bianthracene]-9,9',10,10'-tetrone, 4,4'-diamino-	5198	1.8572	H	M
19	004162-45-2	Ethanol, 2,2'-[(1-methylethylidene)bis[(2,6-dibromo-4,1-phenylene)oxy]]bis-	7479	1.2501	H	H
20	004378-61-4	Dibenzo[def,mno]chrysene-6,12-dione, 4,10-dibromo-	6110	1.8566	H	M
21	005590-18-1	1H-Isoindol-1-one, 3,3'-(1,4-phenylenediimino)bis[4,5,6,7-tetrachloro-	1916	0.0193	M	H
22	013080-86-9	Benzenamine, 4,4'-[(1-methylethylidene)bis(4,1-phenyleneoxy)]bis-	39730	1.6937	H	H
23	013417-01-1	1-Octanesulfonamide, N-[3-(dimethylamino)propyl]-1,1,2,2,3,3,4,4,5,5,6,6,7,7,8,8,8-heptadecafluoro-	1300	-0.3446	M	H
24	013680-35-8	Benzenamine, 4,4'-methylenebis[2,6-diethyl-	15070	1.8689	H	M
25	014295-43-3	Benzo[b]thiophen-3(2H)-one, 4,7-dichloro-2-(4,7-dichloro-3-oxobenzo[b]thien-2(3H)-ylidene)-	1461	1.3684	M	H
26	015667-10-4	Peroxide, cyclohexylidenebis[(1,1-dimethylpropyl)	28610	1.9254	H	M
27	016090-14-5	Ethanesulfonyl fluoride, 2-[1-[difluoro[(trifluoroethenyl)oxy]methyl]-1,2,2,2-tetrafluoroethoxy]-1,1,2,2-tetrafluoro-	12710	0.8345	H	H
28	017527-29-6	2-Propenoic acid, 3,3,4,4,5,5,6,6,7,7,8,8,8-tridecafluorooctyl ester	45320	0.8418	H	H
29	024108-89-2	Anthra[2,1,9-def:6,5,10-d'e'f']diisoquinoline-1,3,8,10(2H,9H)-tetrone, 2,9-bis(4-ethoxyphenyl)-	14640	0.8899	H	H
30	025637-99-4	Cyclododecane, hexabromo-	6211	1.9548	H	M
31	026628-47-7	Spiro[12H-benzo[a]xanthene-12,1'(3'H)-isobenzofuran]-3'-one, 9-(diethylamino)-	26190	1.8829	H	M
32	029512-49-0	Spiro[isobenzofuran-1(3H),9'-[9H]xanthen]-3-one, 6'-(diethylamino)-3'-methyl-2'-(phenylamino)-	23790	1.5734	H	H
33	031148-95-5	1-Phenanthrenecarbonitrile, 1,2,3,4,4a,9,10,10a-octahydro-1,4a-dimethyl-7-(1-methylethyl)-, [1R-(1.alpha.,4a.beta.,10a.alpha.)]-	13900	1.9209	H	M
34	031506-32-8	1-Octanesulfonamide,1,1,2,2,3,3,4,4,5,5,6,6,7,7,8,8,8-heptadecafluoro-N-methyl-	2355	0.0673	M	H

35	040567-16-6	Butanoyl chloride, 2-[2,4-bis(1,1-dimethylpropyl)phenoxy]-	19450	1.9678	H	M
36	041556-26-7	Decanedioic acid, bis(1,2,2,6,6-pentamethyl-4-piperidinyl) ester	1351	0.9971	M	H
37	050598-28-2	1-Hexanesulfonamide, N-[3-(dimethylamino)propyl]-1,1,2,2,3,3,4,4,5,5,6,6,6-tridecafluoro-	14300	0.3006	H	H
38	051461-11-1	Butanamide, N-(3-amino-4-chlorophenyl)-4-[2,4-bis(1,1-dimethylpropyl)phenoxy]-	4393	1.3375	M	H
39	051772-35-1	1-Naphthalenamine, N-[(1,1,3,3-tetramethylbutyl)phenyl]-	1333	1.8096	M	M
40	054079-53-7	Propanedinitrile, [[4-[[2-(4-cyclohexylphenoxy)ethyl]ethylamino]-2-methylphenyl]methylene]-	3996	1.6579	M	H
41	058798-47-3	3H-Indolium, 2-[[(4-methoxyphenyl)methylhydrazono]methyl]-1,3,3-trimethyl-, acetate	1952	1.9594	M	M
42	064022-61-3	1,2,3,4-Butanetetracarboxylic acid, tetrakis(2,2,6,6-tetramethyl-4-piperidinyl) ester	24930	0.4125	H	H
43	067584-54-7	1-Heptanesulfonamide, N-[3-(dimethylamino)propyl]-1,1,2,2,3,3,4,4,5,5,6,6,7,7,7-pentadecafluoro-	27380	-0.022	H	H
44	067584-57-0	2-Propenoic acid, 2-[methyl[(tridecafluorohexyl)sulfonyl]amino]ethyl ester	29550	0.636	H	H
45	068084-62-8	2-Propenoic acid, 2-[methyl[(pentadecafluoroheptyl)sulfonyl]amino]ethyl ester	7529	0.3134	H	H
46	068259-36-9	1-Naphthalenamine, N-phenyl-ar-(1,1,3,3-tetramethylbutyl)-	1333	1.9294	M	M
47	068555-73-7	1-Heptanesulfonamide, N-ethyl-1,1,2,2,3,3,4,4,5,5,6,6,7,7,7-pentadecafluoro-N-(2-hydroxyethyl)-	35110	0.4216	H	H
48	068555-76-0	1-Heptanesulfonamide, 1,1,2,2,3,3,4,4,5,5,6,6,7,7,7-pentadecafluoro-N-(2-hydroxyethyl)-N-methyl-	14700	0.4526	H	H
49	106246-33-7	Benzenamine, 4,4'-methylenebis[3-chloro-2,6-diethyl-	9015	1.3034	H	H
50	106917-30-0	2,5-Pyrrolidinedione, 3-dodecyl-1-(1,2,2,6,6-pentamethyl-4-piperidinyl)-	1457	1.8888	M	M

QSARs

BCFs were estimated using EPI Suite's BCFWin program (http://www.epa.gov/oppt/exposure/docs/episuitedl.htm). BCFs were estimated from the log octanol-water partition coefficient (log K_{OW}) and a series of structural correction factors (Meylan et al., 1999). The ITC uses BCFs of $\geq$1,000 and $\geq$ 5,000 to screen chemicals for bioconcentration potential. Chemicals with $1000 \leq BCF < 5{,}000$ are assigned a medium (M) bioconcentration potential. Chemicals with $BCF \geq 5{,}000$ are assigned a high (H) bioconcentration potential (cf. Table 1).

Persistence predictions were estimated using EPI Suite's BioWin program (http://www.epa.gov/oppt/exposure/docs/episuitedl.htm). The ultimate aerobic biodegradation probabilities (BDPs) from the ultimate survey model in BioWin were used to predict persistence potential. These predictions were based on expert opinions that different structural groups could be used to estimate a chemical's biodegradation potential (Boethling et al, 1994). The ITC uses BDPs of < 2 and < 1.75 as surrogates for chemicals that are likely to persist for approximately 2 and 6 months, respectively. Chemicals with BDP < 2 were associated with a medium (M) persistence potential. Chemicals with BDP < 1.75 were assigned a high (H) persistence potential (cf. Table 1).

Partial Order Ranking

The theory of partial order ranking has been presented in previous papers (Carlsen et al. 2001, Brüggemann et al. 2001a, Carlsen et al. 2002). In brief, Partial Order Ranking is a simple principle, which a priori includes " " as the only mathematical relation. If a system is considered, which can be described by a series of descriptors p_i, a given compound A, characterized by the descriptors $p_i(A)$ can be compared to another compound B, characterized by the descriptors $p_i(B)$, through comparison of the single descriptors, respectively. Thus, compound A will be ranked higher than compound B, i.e., B   A, if at least one descriptor for A is higher than the corresponding descriptor for B and no descriptor for A is lower than the corresponding descriptor for B. If, on the other hand, $p_i(A) > p_i(B)$ for descriptor i and $p_j(A) < p_j(B)$ for descriptor j, A and B will be denoted incomparable. In mathematical terms this can be expressed as

$$B \leq A \Leftrightarrow p_i(B) \leq p_i(A) \text{ for all } i \quad (1)$$

In partial order ranking – in contrast to standard multidimensional statistical analysis - neither assumptions about linearity nor any assumptions about distribution properties are made. Partial order ranking may be considered as a parameter-free method. Thus, there is no preference among the descriptors. The graphical representation of the partial ordering is typically given in a so-called Hasse diagram (Halfon and Reggiani 1986, Brüggemann et al. 2001b, Brüggemann et al. 1995, Hasse 1952), where comparable elements are connected with lines, whereas incomparable elements appear as unconnected. Substances being ranked identically, i.e. these substances cannot be distinguished by the partial order ranking are located in the same levels in the diagram. Thus, substances that on a cumulative basis are ranked, as the most hazardous, are located in level 1.

Note that the enumeration of levels follows convention. In other chapters of this book the enumeration begins with the bottom level. Patil & Taillie, 2005 introduce in that context the concepts level and co-level. In the present study, the QSAR derived estimates for persistence and bioconcentration were descriptors for the construction of the Hasse diagrams using the WHASSE software (Brüggemann et al., 1995).

Results

Partial order ranking of the substances was made using the 50 BCF and BDP estimates (cf. Table 1) and applying the WHASSE software (Fig. 1).

Fig. 1 consists of 11 levels, 4 maximal elements, i.e., only those connected to lower-ranked elements (15, 28, 43, 47) and 5 minimal elements, i.e., only those connected to higher-ranked elements (11, 23, 41, 39, 46), respectively.

Discussion

Ranking the 4 chemicals in level 1 based on BCF alone (Table 1) would be 28 > 47 > 43 >> 15. However, based only on BDP just the opposite ranking would occur, viz., 15 > 43 > 47 > 28. However, partial order ranking allows both descriptors to be taken into account simultaneously leading to the conclusion that all 4 compounds 15, 28, 43 and 47 apparently are the environmentally more problematic. In the case of compound 15, displaying only a medium level bioaccumulation, the high ranking is associated with a very high environmental persistence.

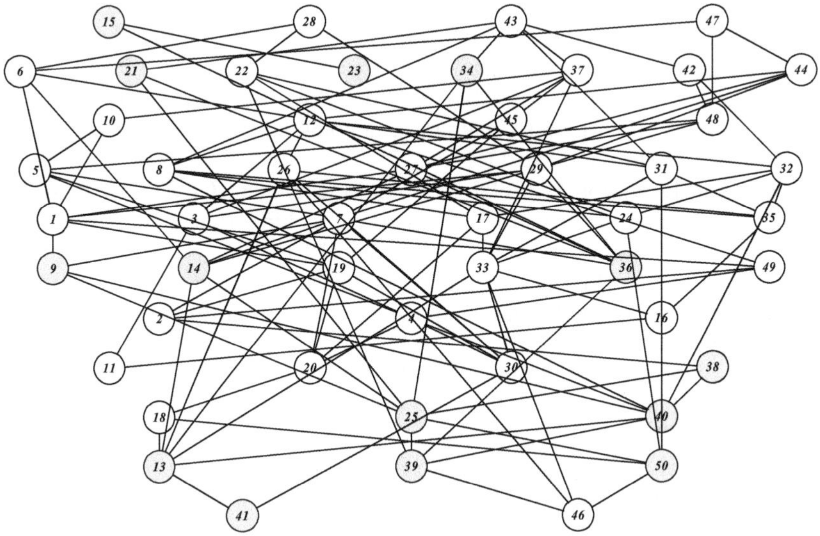

Fig. 1. Hasse Diagram displaying the partial order ranking of the substances studied using the BCF and BDP estimates as descriptors

Conclusions

The present study has demonstrated that substances can be prioritized or ranked using a partial order ranking technique, e.g., based on their PB characteristics. Simple "yes/no" classification or total linear ranking can be obtained based on QSARs alone with reference to selected PB criteria. However, partial order ranking provides more valuable information with regard to which substances are environmentally hazardous because it simultaneously takes into account the persistence and bioaccumulation of the substances under investigation. As such, the combination of QSAR modelling and partial order ranking constitute an effective decision support tool that could be used to facilitate pollution prevention activities by regulated and regulatory communities.

References

Boethling RS, Howard PH, Meylan WM, Stiteler W, Beauman J, Tirado N (1994) Group contribution method for predicting probability and rate of aerobic biodegradation. Environ Sci Technol 28:459–65

Brüggemann R, Halfon E, Bücherl C (1995) Theoretical base of the program "Hasse", GSF-Bericht 20/95, Neuherberg; The software may be obtained by contacting Dr. R. Brüggemann

Brüggemann R, Pudenz S, Carlsen L, Sørensen PB, Thomsen M, Mishra RK (2001a) The use of Hasse diagrams as a potential approach for inverse QSAR. SAR QSAR Environ Res 11:473-487

Brüggemann R, Halfon E, Welzl G, Voigt K, Steinberg CEW (2001b) Applying the concept of partially ordered sets on the ranking of near-shore sediments by a battery of tests. J Chem Inf Comput Sci 41:918-925

Carlsen L, Sørensen PB, Thomsen M (2001) Partial order ranking based QSAR's: Estimation of solubilities and octanol-water partitioning. Chemosphere 43:295-302

Carlsen L, Sørensen PB, Thomsen M, Brüggemann R (2002) QSAR´s Based on Partial Order Ranking. SAR QSAR Environ Res 13:153-165

Carlsen L, Walker JD (2003) QSARs for prioritizing PBT substances to promote pollution prevention. QSAR Comb Sci 22:49-57

Halfon E, Reggiani MG (1986) On the ranking of chemicals for environmental hazard, Environ Sci Technol 20:1173-1179

Hasse H (1952) Über die Klassenzahl abelscher Zahlkörper. Akademie Verlag, Berlin

Lerche DB, Brüggemann R, Sørensen PB, Carlsen L, Nielsen OJ (2002) Comparison of Hasse Diagram Technique with three multicriteria analysis for ranking chemical substances. J Chem Inf Comput Sci 42:1086-1098

Meylan WM, Howard PH, Boethling RS, Aronson D, Printup H, Gouchie S (1999) Improved method for estimating bioconcentration/bioaccumulation factor from octanol/water partition coefficient. Environmental Toxicology and Chemistry 18:664-672

Patil GP, Taillie C (2005) Multiple indicators, partially ordered sets, and linear extensions: Multi-criterion ranking and prioritization. Environmental and Ecological Statistics 11:199-228

Walker JD, Carlsen L (2002) QSARs for identifying and prioritizing substances with persistence and bioconcentration potential. SAR QSAR Environ Res 13:713-725

3 Quantitative Structure Activity Relationships

Chemistry and Environmental Chemistry are confronted with the crucial question how to obtain numerically information about properties of interest. The scientific discipline how to achieve this is associated with the concept: Quantitative Structure Activity Relationships (abbreviation: QSARs). Already in the first section the reader could be introduced to the way of thinking in this field of research. In the chapters in this section the focus is to establish structure - activity relationships by means of order relations. The order relations in turn are derived from sets of properties of the chemicals.

In the first chapter written by Carlsen an important step is the introduction of "noise - deficient" QSARs. The basics of HDT are briefly repeated before the central item how to get information about unknown properties is discussed. In comparison to usual methods, which are well known and widespread in the scientific literature, the approach shown by Carlsen does neither assume linearity in the property-property relationships nor any distribution properties. The method can be considered as parameter free. "Giving the molecules an identity" is the very idea how to rank a chemical whose environmental relevant data are unknown but some structural descriptors. Although the incomparabilities may be considered as "Achilles heel" Carlsen shows in his chapter how to handle such situations by means of linear extensions, probability distributions and the concept of averaged ranks.

In the chapter of Pavan et al. the reader may learn more about the theoretical handling of partial order. Especially a useful description of partial order by matrices can be found here. The main topic, however, is devoted to the QSAR problem. The authors suggest and describe how molecular descriptors can be found in order to find useful partial orders. They describe genetic algorithms to find the best model poset and to derive from these unknown properties. "Experimental ranking and Model ranking" are at the heart of this chapter. The interpolation problem, already discussed in the first section by Klein& Ivanciuc, and later within this section by Carlsen, plays an important role in the chapter of Pavan et al. A discussion of the prediction uncertainty rounds this chapter. As examples phenyl urea herbicides and their toxicity is chosen.

Interpolation Schemes in QSAR

Lars Carlsen

Awareness Center
Hyldeholm 4, Veddelev, DK-4000 Roskilde, Denmark

e-mail: LC@AwarenessCenter.dk

Abstract

The interplay between Quantitative Structure-Activity Relationships (QSARs) and partial order ranking appears as an advantageous method to assess and prioritize chemical substances, e.g., due to their potential environmental hazard taking several parameters simultaneously into account. Especially the application of so-called 'noise-deficient' descriptors is emphasized in order to eliminate the natural fluctuation of experimental as well as simple QSAR derived data. Further partial order ranking appears as an attractive alternative to conventional QSAR methods that typically rely on the application of stochastic methods. The latter use of partial order ranking may be applicable both to direct QSARs as well to solving inverse QSAR problems. The present chapter summarizes the various types of interplay between of partial order ranking and QSAR modelling.

Introduction

The number of chemical substances that are in use and constitute a potential risk to the environment exceeds today 100.000 (EEA 1998). Even with the proposed new system for registration, evaluation and authorisation of chemicals, REACH, the number of chemicals that will be included in this scheme will be approx. 30.000 (COM 2001; COM 2003). It is obvious that it is not practically possible experimentally to generate all necessary input for the risk assessment of these compounds. Information concerning the fate and effects of these substances in the environment is needed and may be obtained through modelling, e.g., by comparison with structurally related, well-investigated compounds. Thus, within the REACH scheme a widespread use of QSAR modelling to retrieve physico-chemical and toxicological data are foreseen.

A priori the evaluation and prioritization of chemical substances can be based on experimental or QSAR generated data alone. This would give rise to a classification of the substances based on fulfilment of single criteria only. However, typically it is desirable to include a series of criteria simultaneously in the assessment. Thus, to further qualify the assessment the substances may be ranked by a simultaneous inclusion of a series of criteria such as, e.g., biodegradation, bioaccumulation and toxicity hereby disclosing those substances that on a cumulative basis appear to be the environmentally more problematic. In this respect partial order ranking appears as a highly attractive tool (Brüggemann et al. 2001b; Carlsen et al. 2001; Carlsen et al. 2002; Davey and Priestley 1990).

To further elucidate the mutual ranking of the substances linear extensions may be brought into play, leading to the most probably linear (absolute) rank of the substances under investigation (Brüggemann et al. 2001a; Davey and Priestley 1990; Fishburn 1974; Graham 1982). Further the concept of average rank (Brüggemann et al. 2004) can be applied.

Partial order techniques may also be applied directly as QSAR method as illustrated by the use of the QSAR descriptors as input to the ranking (Carlsen et al. 2001; Carlsen et al. 2002). On the other hand partial order ranking based on QSAR descriptors may also be applied as "inverse" QSARs, i.e. to disclose specific characteristics of new substances to be synthesized, e.g., as substitutes for environmentally harmful counterparts (Brüggemann et al. 2001b) or simply to give a given chemical compound an identity by comparing specific characteristics to those of other, possibly less harmful substances (Carlsen 2004).

Methods

Obviously the successful interplay between QSARs and partial order ranking depends on the single techniques. Thus, in the following QSARs and partial order ranking, including linear extensions will be shortly presented.

QSARs

The basic concept of QSARs can in its simplest form be expressed as the development of correlations between a given physico-chemical property or biological activity (endpoint), *P*, and a set of parameters (descriptors), D_i, that are inherent characteristics for the compounds under investigation

$$P = f(D_i) \tag{1}$$

The properties (endpoints), *P* that has been subjected to QSAR modelling comprises physico-chemical properties as well as biological activities.

In general models that describe/calculate key properties of chemical compounds are composed of three types of inherent characteristics of the molecule, *i.e.* structural, electronic and hydrophobic characteristics. Depending on the actual model few or many of these descriptors may be taken into account. Thus, eqn. 1 can be rewritten as

$$P = f(D_{structural}, D_{electronic}, D_{hydrophobic}, D_x) + e \tag{2}$$

The descriptors reflecting structural characteristics may e.g. be element of the actual composition and 3-dimensional configuration of the molecule, whereas descriptors reflecting the electronic characteristics may e.g. be charge densities, dipole moment etc. The descriptors reflecting the hydrophobic characteristics are related to the distribution of the compound between a biological, hydrophobic phase, and an aqueous phase. The fourth type of characteristics, D_x, accounts for possible underlying characteristics that may be known or unknown, such as environmental or experimental parameters as, e.g., temperature, salt content etc. The data may often be associated with a certain amount of systematic and non-quantifiable variability in combination with uncertainties. These unknown variations are expressed as "noise". Thus, the parameter, *e*, account for possible noise in the system, i.e., the variation in the property that cannot be explained by the model.

In principle all types of QSAR models can be used to generate descriptors for subsequent use in partial order ranking, i.e. commercially available generally applicable QSARs as well as more specialized custom made QSARs. However, as partial order ranking due to its inherent nature only focusing on the relation "≤" (vide infra) may be hampered by random fluctuations in the descriptors, the so-called 'noise-deficient' QSARs (Carlsen 2004, Carlsen 2005a; Carlsen 2005b) advantageously can be applied.

Thus, recent studies on organophosphates appear as an illustrative example on the application of 'noise-deficient' QSAR-derived endpoints as input for a subsequent partial order ranking. The descriptors are generated through QSAR modelling, the EPI Suite being the primary tool (Carlsen 2005a, Carlsen 2005b; Carlsen 2004)[1].

Based on the EPI generated values for solubility (log *Sol*), octanol-water partitioning (log K_{OW}), vapour pressure (log *VP*) and Henry's Law constants (log *HLC*) new linear QSAR models are build by estimating the relationships between the EPI generated data and available experimental data for up to 65 organophosphor insecticides, the general formula for the descriptors, D_i, to be used being

$$D_i = a_i \cdot D_{EPI} + b_i \tag{3}$$

D_{EPI} being the EPI generated descriptor value and a_i and b_i being constants. The log K_{OW} values generated in this way are subsequently used to generate log *BCF* values according to the Connell formula (Connell and Hawker 1988)

$$\log BCF = 6.9 \cdot 10^{-3} \cdot (\log K_{ow})^4 - 1.85 \cdot 10^{-1} \cdot (\log K_{ow})^3 + 1.55 \cdot (\log K_{ow})^2 - 4.18 \cdot \log K_{ow} + 4.72 \tag{4}$$

The model was somewhat modified (Carlsen 2005a, Carlsen 2005b; Carlsen 2004). Thus, a linear decrease of log *BCF* with log K_{OW} was assumed in the range $1 < \log K_{OW} < 2.33$, the log $BCF = 0.5$ for log $K_{OW} \leq 1$, the latter value being in accordance with BCFWin (EPI 2000).

Subsequently, these QSAR generated endpoints may be applied for a partial order ranking of the substances using two or more of the endpoints as descriptors for the ranking exercise.

[1] The EPI Suite is a collection of QSAR models for physical chemical and toxicity endpoint developed by the EPA's office of Pollution Prevention Toxics and Syracuse Research Corporation (EPI 2000).

Partial Order Ranking

The theory of partial order ranking is presented elsewhere (Davey and Priestley 1990) and application in relation to QSAR is presented in previous papers (Carlsen et al. 2001; Brüggemann et al. 2001b; Carlsen et al. 2002; Carlsen and Walker 2003). In brief, Partial Order Ranking is a simple principle, which a priori includes "≤" as the only mathematical relation. If a system is considered, which can be described by a series of descriptors p_i, a given compound A, characterized by the descriptors $p_i(A)$ can be compared to another compound B, characterized by the descriptors $p_i(B)$, through comparison of the single descriptors, respectively. Thus, compound A will be ranked higher than compound B, i.e., $B \le A$, if at least one descriptor for A is higher than the corresponding descriptor for B and no descriptor for A is lower than the corresponding descriptor for B. If, on the other hand, $p_i(A) > p_i(B)$ for descriptor i and $p_j(A) < p_j(B)$ for descriptor j, A and B will be denoted incomparable. In mathematical terms this can be expressed as

$$B \le A \Leftrightarrow p_i(B) \le p_i(A) \text{ for all i} \qquad (5)$$

Obviously, if all descriptors for A are equal to the corresponding descriptors for B, i.e., $p_i(B) = p_i(A)$ for all i, the two compounds will have identical rank and will be considered as equivalent. It further follows that if $A \le B$ and $B \le C$ then $A \le C$. If no rank can be established between A and B these compounds are denoted as incomparable, *i.e.*, they cannot be assigned a mutual order.

In partial order ranking – in contrast to standard multidimensional statistical analysis - neither assumptions about linearity nor any assumptions about distribution properties are made. In this way the partial order ranking can be considered as a non-parametric method. Thus, there is no preference among the descriptors. However, due to the simple mathematics outlined above, it is obvious that the method a priori is rather sensitive to noise, since even minor fluctuations in the descriptor values may lead to non-comparability or reversed ordering. An approach how to handle loss of information by using an ordinal in stead of a matrix can also be found in the chapter by Pavan et al., see p. 181).

In partial order ranking – in contrast to standard multidimensional statistical analysis - neither assumptions about linearity nor any assumptions about distribution properties are made. Partial order ranking may be considered as a parameter-free method. Thus, there is no preference among the descriptors. A main point is that all descriptors have to the same designations, i.e., “high” and “low” (cf. p. 70). This means that some descriptors may be multiplied by –1 in order to achieve identical designations. As an example bioaccumulation and toxicity can be mentioned. In the case of bioaccumulation, the higher the number the more problematic the substance, whereas in the case of toxicity, the lower the figure the more toxic the substance. Thus, in order to secure identical directions of the two descriptors, one of them, e.g., the toxicity figures, has to be multiplied by –1. Consequently, both in the case of bioaccumulation and in the case of toxicity higher figures will now correspond to more hazardous compounds.

The graphical representation of the partial ordering is often given in a so-called Hasse diagram (Hasse 1952; Halfon and Reggiani 1986; Brüggemann et al. 2001a; Brüggemann et al. 1995). In practice the partial order rankings are done using the WHASSE software (Brüggemann et al. 1995).

Linear Extensions

The number of incomparable elements in the partial ordering may obviously constitute a limitation in the attempt to rank e.g. a series of chemical substances based on their potential environmental or human health hazard. To a certain extent this problem can be remedied through the application of the so-called linear extensions of the partial order ranking (Fishburn, 1974; Graham 1982). A linear extension is a total order, where all comparabilities of the partial order are reproduced (Davey and Priestley 1990; Brüggemann et al. 2001a). Due to the incomparisons in the partial order ranking, a number of possible linear extensions corresponds to one partial order. If all possible linear extensions are found, a ranking probability (cf. p. 99) can be calculated, i.e., based on the linear extensions the probability that a certain compound have a certain absolute rank can be derived. If all possible linear extensions are found it is possible to calculate the averaged ranks (cf. p. 86) of the single elements in a partially ordered set (Winkler 1982; Winkler 1983).

Averaged Ranks

The average rank is simply the average of the ranks in all the linear extensions. On this basis the most probably rank for each element can be obtained leading to the most probably linear rank of the substances studied.

The generation of the averaged rank of the single compounds in the Hasse diagram is obtained applying the simple relation recently reported by Brüggemann et al. (2004) (see also p. 86). The averaged rank of a specific compound, c_i, can be obtained by the simple relation

$$Rk_{av} = (N+1) - (S+1)\cdot(N+1)/(N+1-U) \tag{6}$$

where N is the number of elements in the diagram, S the number of successors to c_i and U the number of elements being incomparable to c_i (Brüggemann et al. 2004), counting from top to bottom.

Partial Order based QSARs

QSAR - Quantitative Structure Activity Relationships - in general terms denotes models, which, based on the variation in structural and/or electronic features in series of selected, molecules, describe variation in a given end-point of these molecules. These end-points may be, e.g., biological effects or physical-chemical parameters, which experimentally can be verified. Based on the developed QSAR model end-points of new, structurally related compounds, hitherto not being experimentally studied, may be predicted.

Since the variation in, e.g., biological effects or physical-chemical parameters typically cannot be described by one single descriptor QSAR modelling relies heavily on statistical methods. Further, since QSAR modelling may often involve seeking unknown relations between several descriptors and a given end-point, traditional statistical approaches such as simple multiple linear regression (MLR) may not be the ideal choice although widely used. Thus, development of QSAR models are often successfully based on multivariate projection methods, such as principal component analysis (PCA) followed by MLR using the principal components as descriptors or, more common, partial least square (PLS) projection, as the modelling in many cases can be described by linearization of complex unknown relations.

Partial Order Ranking (Brüggemann et al. 1995), which from a mathematical point of view constitute extremely simple, appears as an attractive and operationally simple alternative to the above rather demanding statistical method.

The partial order ranking method allows ranking of series of well investigated compounds, e.g., octanol-water distribution coefficients based on structural and/or electronic parameters of the compounds. The mutual ranking of the compounds can then be compared to the ranking based on the experimentally derived values for octanol-water distribution coefficients. If the ranking model resembles the experimental ranking of the parameters under investigation, the model is validated and other compounds not being experimentally investigated, can be assigned a rank in the model and hereby obtain an identity based on the known compounds, see however chapter Klein and Ivanciuc, p. 35.

Direct QSARs

An example of the possible applicability of partial order ranking as a tool for QSAR modelling has been reported by Carlsen et al. (2002). Thus, a series of non-hydrogen bond donor molecules, which have previously been studied using statistically based QSAR's in order to verify the applicability of the partial order ranking method to a well-known system were selected. Thus, octanol-water distribution coefficients (Kamlet et al. 1988) and solubilities (Kamlet et al. 1987) were retrieved for a group of approx. 40 compounds exhibiting rather different structural and electronic characteristics. The experimental data was closely mimicked through a Linear Solvation Energy Relationship (LSER) approach (Carlsen 1999; Kamlet et al. 1977; Kamlet et al. 1988), the corresponding statistical approach being MLR. Carlsen et al. (2002) successfully applied the same molecular descriptors as the LSER studies, i.e., the molecular volume ($V_i/100$), the polarity (π^*) and the hydrogen bond basicity (ß) (Kamlet et al. 1987; Kamlet et al., 1988) as demonstrated using the same basis set of compounds.

Contrary to the method reported by Pavan et al. (see p. 181) giving the results as intervals, the approach by Carlsen et al. (2002) suggested specific values. Thus, the model derived values for a given compound X (*ValueX*) was obtained by simple arithmetic means between the lowest value of the comparable compounds ranked above X (*minAbove*) and the highest value of the comparable compounds ranked below X (*maxBelow*).

$$ValueX = (minAbove + maxBelow)/2 \qquad (7)$$

The predicted values are compared to the corresponding experimentally derived values as depicted in Fig. 1.

It is immediately noted (Fig. 1) that in the partial order ranking based models solubilities reasonably well reproduce the experimentally derived values. However, it should be noted that the actual distance between the *minAbove* and *maxBelow* elements is crucial. Thus, the larger the distance between these two values the larger the potential uncertainty in the prediction (Carlsen et al. 2001).

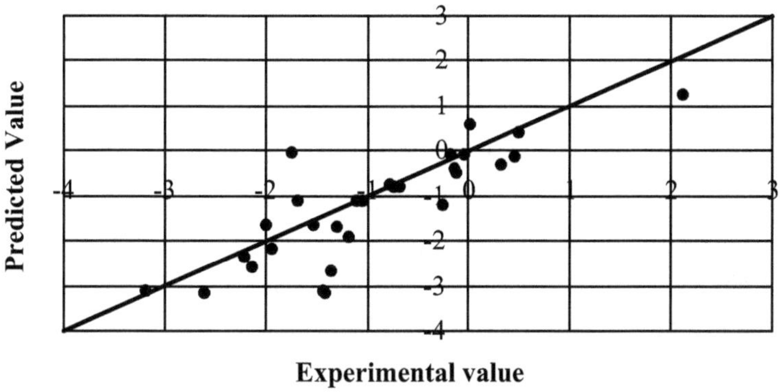

Fig. 1. Experimental vs. predicted solubilities

Inverse QSARs

Quantitative structure-activity relationships are often based on standard multidimensional statistical analyses and applying sophisticated local and global molecular descriptors, assuming linearity as well as implying normal distribution behaviour of the latter. Thus, the aim is to develop a tool helpful to define a molecule or a class of molecules that fulfils predescribed properties, i.e. an inverse QSAR approach. However, if QSARs based on highly sophisticated descriptors are used for this purpose, the structure of potential candidates and thus the actual synthetic pathways may be hard to derive. On the other hand, descriptors, from which the synthesis recipe can be easily derived, seem appropriate to be included in such exercises. Unfortunately, if descriptors simple enough to be useful for defining syntheses recipes of chemicals are used, the accuracy of an arithmetic expression may fail. Brüggemann et al. (2001b) suggested a method, based on the very simple elements of the theory of partially ordered sets, to find a qualitative basis for the relationship between such fairly descriptors on the one side and a series of ecotoxicological properties, on the other side. The obvious advantage of the partial order ranking method has to be sought for in the fact that this method does not assume neither linearity nor normal distribution of the descriptors.

In the study of Brüggemann et al. (2001b) a series of synthesis specific descriptors, i.e. simple structural descriptors such as the number of specific atoms and the number of specific bonds were included in the analyses along with graph theoretical and quantum chemical descriptors. On this basis a 6-step procedure was developed to solve inverse QSAR problems.

Although the approach a priori appears as an attractive alternative more chemicals have to be considered in order further to develop the technique. Assuming this lead to more comparabilities and more neighbouring objects for a specific chemical, then the property space stretched by the order theoretical environment is smaller, which may lead to higher accuracy for estimation of toxicity data for a "new" chemical.

Giving molecules an identity

The basic idea of using partial order ranking for giving molecules an identity is illustrated in Fig. 2. Thus, let us assume that a suite of 10 compounds has to be evaluated and that the evaluation should be based on 3 pre-selected criteria, e.g., persistence, bioaccumulation and toxicity. Let the resulting Hasse diagram be the one depicted in Fig. 2A. If we apply the 3 descriptors representing biodegradation, bioaccumulation and toxicity, respectively, so the more persistent, the more bioaccumulating and the more toxic a substance would be the higher in the diagram it would be found, Fig. 2A discloses that the compounds in the top level, i.e., compounds 1, 3, 4, 7 and 8 on a cumulative basis can be classified as the environmentally more problematic of the 10 compounds studied with respect to their PBT characteristics, whereas compound 10 that a found in the bottom of the diagram is the less hazardous.

Subsequently we can introduce compounds solely characterized by QSAR derived data in order to give this new compound, X, an identity, e.g., in an attempt to elucidate the environmental impact of X. Adopting the above discussed 10 compounds and the corresponding Hasse diagram (Fig. 2A) we introduced the compound X. The revised Hasse diagram, now including 11 compounds is visualized in Fig. 2B. It is immediately disclosed that compound X has now obtained an identity in comparison to the originally well-characterized compounds, as it is evaluated as less environmentally harmful than compounds 4 and 7, but more harmful than compound 10. Thus, through the partial order ranking the compound, X, has obtained an identity in the scenario with regard to its potential environmental impact.

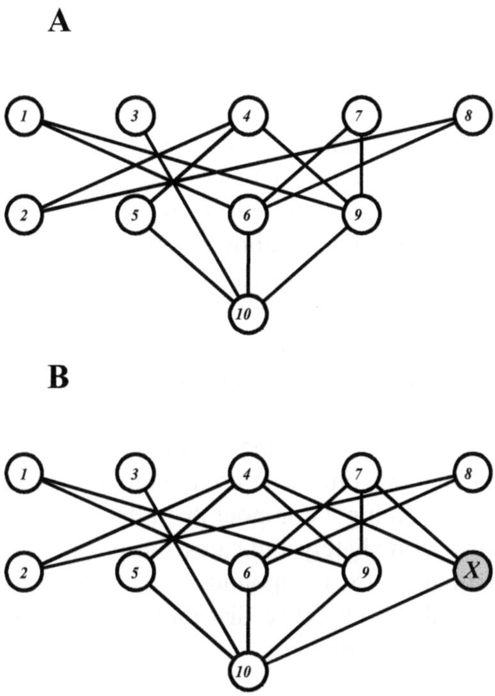

Fig. 2. Illustrative Hasse diagram of A: 10 compounds using 3 descriptors and B: the same 10 compounds plus 1 new compound X

Hasse diagrams are characterized to the presence of a number of comparisons. The actual number of incomparisons is roughly speaking a result of interplay between the number of compounds and the number of descriptors (Sørensen et al. 2000). Thus, increasing the number of descriptors will, for the same number of compounds, increase the number of incomparisons.

A priori the incomparisons may turn out as an Achilles' heel of the partial order ranking method. However, the adoption of the linear extension approach apparently remedies this, at least to a certain extent.

Turning back to the model diagram (Fig. 2B) it can be noted that e.g. the compounds 4 and 7 are incomparable, i.e. looking just for these two compounds it cannot from the Hasse diagram be concluded which of them are the more hazardous. However, bringing the linear extensions into play gives us the probability for these two compounds to have a certain absolute rank. In Fig. 3A the probability distribution for the compounds 4 and 7 for the possible absolute ranks is visualized. It is easily seen that the probability for finding compound 4 at rank 1 or 2 are higher than for compound 7 (Rank 1 is equal to top rank). On the other hand, compound 7 is more probable to be found at rank 4-7 than compound 4. On this basis we can conclude that comparing compounds 4 and 7, the most probable absolute ranking will place compound 4 above compound 7. In Fig. 3B the probability distribution for compound 10 is shown. The probabilities of finding compound 10 at rank 11 are approx. 70 % and at rank 10 approx. 30 %. The incomparability between compounds 10 and 2 accounts for this since compound 2 has an approx. 30 % probability to be occupy rank 11.

The 'new' compound, X, introduced in the diagram displayed in Fig. 2B apparently is comparable only with compound 4, 7 and 10 and thus incomparable with the remaining 7 compounds in the scenario. The high number of incomparisons immediately indicates the presence of a relative broad probability distribution for compound X. This is nicely demonstrated in Fig. 4 displaying the probability distribution of compound X for being found at specific absolute ranks.

The probability distribution of compound X in relation to compounds 4, 7, 10 and X is visualized in Fig. 5. It must in this connection be remembered that although the probability distribution of compound X overlaps those of compounds 4, 7 and 10, compound X must be located between compounds 4 and 7 and compound 10 (cf. Fig. 2B).

To further elucidate how the single compounds under investigation can be assumed to behave on a combined basis, e.g. taking all descriptors simultaneously into account the concept of average rank (Carlsen 2005a; Carlsen 2005b; Carlsen 2004; Brüggemann et al. 2004; Lerche et al. 2002) can be adopted. In Table 1 the averaged rank calculated according to eqn. 6 is given.

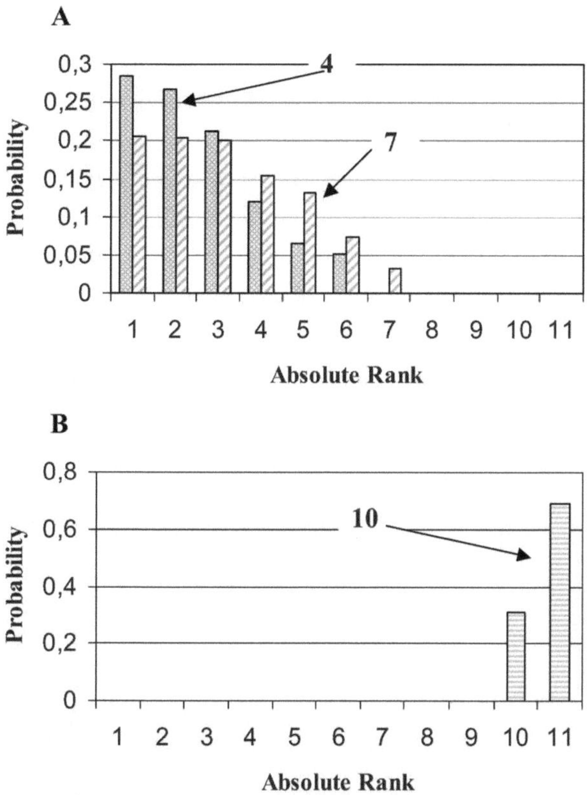

Fig. 3. Probability distribution of A: compounds 4 and 7 and B: compound 10 to occupy specific absolute ranks (rank 1 and 11 is top and bottom rank respectively)

Thus, from the above discussion on the probabilities for specific ranks, it was concluded that the new compound X must be located between the compounds 4 and 7 and compound 10 (cf. Fig. 2B), which is further substantiated by the figures in Table 1. Assuming that if the averaged ranks, Rk_{av}, of two compounds are close, the two compounds will on an average basis display similar characteristics as being determined by the set of descriptors applied, the analysis of average rank discloses, cf. Table 1, that compounds X most closely resembles compounds 6 and 9. Consequently, compound X has now obtained an identity compared to the basis set of compounds, i.e., compounds 1-10.

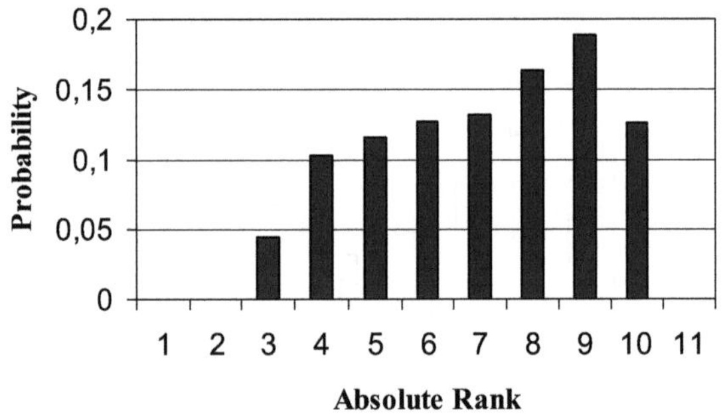

Fig. 4. Probability distribution of compound X to occupy specific absolute ranks

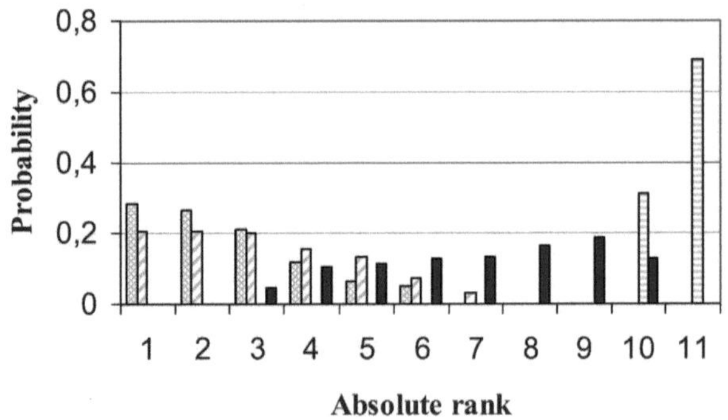

Fig. 5. Probability distribution of compound X in relation to compounds 4, 7 and 10 to occupy specific absolute ranks

Table 1. Averaged ranks of the 11 compounds included in the Hasse diagram displayed in Figure 2B

Compound No.	Averaged Rank (Rk_{av})
1	2.4
2	9.0
3	4.0
4	1.7
5	6.0
6	8.0
7	2.0
8	2.4
9	8.0
10	10.9
X	7.2

Conclusions and Outlook

Partial order ranking and QSAR modelling supplement each other and constitute an effective tool in various areas of chemical sciences. Thus, the interplay between QSAR and partial order ranking constitute an effective decision support tool to assess the chemical substances, e.g. in relation to their potential environmental hazard. Thus, the combined application of QSAR modelling and partial order ranking offers the possibility to assess a large number of chemicals based on several parameters, such as, e.g., persistence, bioaccumulation and toxicity simultaneously and through this effectively disclose the environmentally more problematic substances that requires immediate attention. Thus, this decision support tool may well find extended application in connection with the new proposed chemical legislation, i.e., REACH, within the European Union. It is in this connection worthwhile to note that also economic parameters may be included in the partial order ranking analyses.

The QSAR-partial order ranking system further appears as an appropriate tool to give specific molecules an identity in relation to others and thus constitute as a support tool in the development of less hazardous substitutes to acknowledged harmful substances. In this connection partial order ranking potentially also constitute a rather strong tool to solve inverse QSAR problems, e.g., to develop suitable synthetic pathways for new substances.

The direct application of partial order ranking as QSAR modelling tool provides an attractive alternative to conventional methods, as partial order ranking is a parameter free method. The predicting ability of the partial order models is acceptable and the technique may accommodate otherwise non-comparable descriptors. However, further improvement of the precision of the models is desirable (cf. also Pavan et al., p. 181).

References

Brüggemann R, Halfon E, Bücherl C (1995) Theoretical base of the program "Hasse", GSF-Bericht 20/95, Neuherberg, The software may be obtained by contacting Dr. R. Brüggemann

Brüggemann R, Halfon E, Welzl G, Voigt K, Steinberg CEW (2001a) Applying the concept of partially ordered sets on the ranking of near-shore sediments by a battery of tests, J Chem Inf Comput Sci 41:918-925

Brüggemann R, Lerche D, Sørensen PB, Carlsen L (2004) Estimation of averaged ranks by a local partial order model, J Chem Inf Comput Sci 44:618-625

Brüggemann R, Pudenz S, Carlsen L, Sørensen PB, Thomsen M, Mishra RK (2001b) The use of Hasse diagrams as a potential approach for inverse QSAR. SAR QSAR Environ Res 11:473-487

Carlsen L (1999) Linear Solvation Energy Relationships (LSER). A method to assess the solubility of inorganic species, Research Notes from NERI, No. 107, National Environmental Research Institute, Roskilde, 38 pages

Carlsen L, Sørensen PB, Thomsen M (2001) Partial order ranking based QSAR's: Estimation of solubilities and octanol-water partitioning. Chemosphere 43:295-302

Carlsen L (2005a) A QSAR Approach to Physico-Chemical Data for Organophosphates with Special Focus on Known and Potential Nerve Agents. Internet Electron J Mol Des 4:355-366

Carlsen L (2005b) Partial Order Ranking of Organophosphates with Special Emphasis on Nerve Agents. MATCH-Commun Math Comput Chem 54(3):519-534

Carlsen L (2004) Giving molecules an identity. On the interplay between QSARs and Partial Order Ranking, Molecules 4:1010-1018

Carlsen L, Sørensen PB, Thomsen M Brüggemann R (2002) QSAR´s Based on Partial Order Ranking. SAR and QSAR Environ Res 13:153-165

Carlsen L Walker JD (2003) QSARs for Prioritizing PBT Substances to Promote Pollution Prevention. QSAR Comb Sci 22:49-57
Carlsen L, Sørensen PB, Thomsen M (2001) Partial order ranking based QSAR's: Estimation of solubilities and octanol-water partitioning. Chemosphere 43:295-302
COM (2001) White Paper, Strategy for a future Chemicals Policy, COM (2001) 88 final, http://europa.eu.int/comm/environment/chemicals/0188_en.pdf

COM (2003), Proposal for a REGULATION OF THE EUROPEAN PARLIAMENT AND OF THE COUNCIL concerning the Registration, Evaluation, Authorisation and Restriction of Chemicals (REACH), establishing a European Chemicals Agency and amending Directive 1999/45/EC and Regulation (EC) {on Persistent Organic Pollutants} Proposal for a DIRECTIVE OF THE EUROPEAN PARLIAMENT AND OF THE COUNCIL amending Council Directive 67/548/EEC in order to adapt it to Regulation (EC) of the European Parliament and of the Council concerning the registration, evaluation, authorisation and restriction of chemicals, COM 2003 0644 (03) http://europa.eu.int/eur-lex/en/com/pdf/2003/com2003_0644en.html

Connell DW, Hawker DW (1988) Use of polynomial expressions to describe the bioconcentration of hydrophobic chemicals in fish. Ecotox Environ Safety 16:242-257

Davey BA, Priestley HA (1990) Introduction to Lattices and Order. Cambridge University Press, Cambridge

EEA (1998) Chemicals in the European environment: Low doses, high stakes ? European Environment Agency, Copenhagen, 33 pages

EPI (2000) Pollution Prevention (P2) Framework, EPA-758-B-00-001, http://www.epa.gov/pbt/framwork.htm

Fishburn PC (1974) On the family of linear extensions of a partial order, J Combinat Theory 17:240-243

Graham RL (1982) Linear Extensions of Partial Orders and the FKG Inequality. in: Ordered Sets, I Rival (ed) pp 213-236

Halfon E, Reggiani MG (1986) On the ranking of chemicals for environmental hazard. Environ Sci Technol 20:1173-1179

Hasse H (1952) Über die Klassenzahl abelscher Zahlkörper Akademie Verlag, Berlin

Kamlet MJ, Doherty RM, Abraham MH, Carr PW, Doherty RF, Taft RW (1987) Important differences between aqueous solubility relationships for aliphatic and aromatic solutes. J Phys Chem 91:1996-2004

Kamlet MJ, Doherty RM, Abraham MH, Marcus Y, Taft RW (1988) An improved equation for correlation and prediction of octanol/water partition coefficients of organic nonelectrolytes (including strong hydrogen bond donor solutes). J Phys Chem 92:5244-5255

Lerche D, Brüggemann R, Sørensen P, Carlsen L, Nielsen OJ (2002) A comparison of partial order technique with three methods of multi-criteria analysis for ranking of chemical substances. J Chem Inf Comput Sci 42:1086-1098

Sørensen PB, Mogensen BB, Carlsen L, Thomsen M (2000) The influence of partial order ranking from input parameter uncertainty. Definition of a robustness parameter. Chemosphere 41:595-601

Winkler PM (1982) Average height in a partially ordered set. Discrete Mathematic 39:337-341

Winkler PM (1983) Correlation among partial orders. Siam J Alg Disc Meth 4:1-7

New QSAR Modelling Approach Based on Ranking Models by Genetic Algorithms - Variable Subset Selection (GA-VSS)

Manuela Pavan[1]*, Viviana Consonni[1], Paola Gramatica[2] and Roberto Todeschini[1]

[1] Milano Chemometrics and QSAR Research Group - Dept. of Environmental Sciences, University of Milano-Bicocca, P.za della Scienza, 1 - 20126 Milano (Italy)

[2] QSAR and Environmental Chemistry Research Unit – Dept. of Structural and Functional Biology, University of Insubria, via Dunant 3 - 21100 Varese (Italy)

*e-mail: manuela.pavan@jrc.it

Abstract

Partial and total order ranking strategies, which from a mathematical point of view are based on elementary methods of Discrete Mathematics, appear as an attractive and simple tool to perform data analysis. Moreover order ranking strategies seem to be a very useful tool not only to perform data exploration but also to develop order-ranking models, being a possible alternative to conventional QSAR methods. In fact, when data material is characterised by uncertainties, order methods can be used as alternative to statistical methods such as multiple linear regression (MLR), since they do not require specific functional relationship between the independent variables and the dependent variables (responses).

A ranking model is a relationship between a set of dependent attributes, experimentally investigated, and a set of independent attributes, i.e. model variables. As in regression and classification models the variable selection is one of the main step to find predictive models. In the present work, the Genetic Algorithm (GA-VSS) approach is proposed as the variable selection method to search for the best ranking models within a wide set of predictor variables. The ranking based on the selected subsets of variables is compared with the experimental ranking and evaluated both in partial and total ranking by a set of similarity indices and the Spearman's rank index, respectively. A case study application is presented on a partial order ranking model developed for 12 congeneric phenylureas selected as similarly acting mixture components and analysed according to their toxicity on *Scenedesmus vacuolatus*.

Introduction

The increasing complexity of the systems analysed in scientific research together with the significant increase of available data require availability of suitable methodologies for multivariate statistics analysis and motivate the endless development of new methods. Moreover, the increasing of problem complexity leads to the decision processes becoming more complex, requiring the support of new tools able to set priorities and define rank order of the available options. The huge number of chemicals used and released in the environment is one of the complex problems the scientific community has to deal with. Since it is not possible to generate experimentally all necessary input for the risk assessment of these chemicals, information on the environmental fate and effects of the chemicals is usually performed by Quantitative Structure - Activity Relationships (QSAR) regression modelling.

In QSAR models structural, steric and/or electronic features in series of selected chemicals are associated with modification in a given biological or physical-chemical end-point of the chemicals. QSAR modelling usually looks for unknown relations between several descriptors and the endpoints; however, when a relationship between a toxic activity and molecular descriptors is searched for, it should be kept in mind that toxicity data are typically multiple response endpoints, i.e. the chemical toxicity is analysed at different concentrations to detect both acute and chronic effects. Furthermore, toxicity data often include large uncertainties and measurements errors. Thus, if the aim is to point out the more toxic and thus hazardous chemicals and to set priorities before final decisions are taken and data material is characterised by uncertainties, order models can be an attractive complement to statistical methods such as multiple linear regression (MLR). Despite conventional QSAR methods, order ranking strategies do not require a priori knowledge of specific functional relationships. Moreover, they are suitable in all those environmental problems whose aim is to define order relations among several chemicals, to point out the more hazardous chemicals and to set priorities before final decision are taken. (Halfon et al. 1986; Halfon 1989; Halfon et al. 1998). For these purposes order-ranking models, which allow finding out not a quantitative response for each chemical but the inter-relationships among different chemicals, seem a promising approach in supporting environmental decision-making processes. By partial and total ranking method, compounds can be ranked on independent variables (model ranking) and the resulting ranking can be compared to the ranking based on the experimentally derived values for given end-points (experimental ranking). If the model ranking agrees with the experimental ranking of the end-points under investigation, predictions of the experimental ranking of other compounds, not being experimentally investigated, can be performed using the ranking model. (Brüggemann et al. 2001a; Carlsen et al. 2001; Carlsen et al. 2002a; Carlsen et al. 2002b; Sørensen et al. 2003).

The Genetic Algorithm (GA-VSS) approach is here used as the variable selection method to search for the best ranking models within a wide set of candidate variables. The models based on the selected subsets of variables are compared with the experimental ranking and evaluated both in partial and total ranking by a set of similarity indices and the Spearman's rank index, respectively. Only the best quality models are retained in the population undergoing the evolution procedure. After a few iterations, the evolving population is usually composed of different combinations of variables that correlate well with the experimental ranking.

In the present study a partial ranking model for 12 phenylurea herbicides selected as similarly acting mixture components (Gramatica et al. 2001; Backaus et al. 2004) is illustrated: the aim to compare their concentration-response curves by the partial ranking model (Hasse diagram) to provide a priority list of these chemicals for the aquatic system according to their overall toxicity on the freshwater algae *Scenedesmus vacuolatus*, contemporary accounting for their toxicity at the complete range of effect, and finally to model the ranked toxicity profiles (cf. attribute profiles, p. 68) by structural molecular descriptors.

Theory

Partial ranking method: Hasse diagram technique

The Hasse diagram technique is a very useful tool to perform partial order ranking (POR). It has been introduced in environmental sciences by Halfon. (Halfon et al. 1986; Halfon 1989; Halfon et al. 1998) and refined by

Brüggemann (Brüggemann et al. 1999; Brüggemann et al. 2001b). In this approach the basis for ranking is the information collected in the full set of attributes, which is called the "information basis" of the comparative evaluation of elements.

The typical data matrix contains n elements (rows) and R attributes (columns). The entry y_{ir} of the matrix is the numerical value of the *r-th* attribute of the *i-th* element. Let IB be the information basis of evaluation, i.e. the set of R attributes, and E the set of n elements: the two elements s and t are comparable if for all $y_r \in$ IB either $y_r(s) \leq y_r(t)$ or $y_r(s) \geq y_r(t)$. If $y_r(s) \leq y_r(t)$ for all $y_r \in$ IB then $s \leq t$, while if $y_r(s) \geq y_r(t)$ for all $y_r \in$ IB then $s \geq t$.

The request "for all" is very important and is called the generality principle:

$$\begin{aligned} &s, t \in \mathrm{E};\ s \leq t \Leftrightarrow y(s) \leq y(t) \\ &y(s) \leq y(t) \Leftrightarrow y_r(s) \leq y_r(t) \text{ for all } y_r \in \mathrm{IB} \end{aligned} \quad (1)$$

If there are some y_r, for which $y_r(s) < y_r(t)$ and some others for which $y_r(s) > y_r(t)$ then s and t are incomparable, and the common notation is $s \| t$. A partial order ranking is easily developed by the Hasse diagram technique comparing each pair of elements and storing this information in the Hasse matrix which is a (n x n) antisymmetric matrix: for each pair of elements s and t the entry h_{st} of this matrix is:

$$h_{st} \begin{cases} +1 & \text{if } y_r(s) \geq y_r(t) \text{ for all } y_r \in IB \\ -1 & \text{if } y_r(s) < y_r(t) \text{ for all } y_r \in IB \\ 0 & \text{otherwise} \end{cases} \quad (2)$$

The results of the partial order ranking are visualized in a diagram, named Hasse diagram, where each element is represented by a small circle, comparable elements which belong to an order relation are linked, while incomparable elements are not connected by a sequence of lines. Conventionally the elements are located in the drawing plane at the same geometrical height and as high as possible in the diagram, thus the diagram exhibits a level structure (see also Brüggemann and Carlsen). The elements at the top of the diagram are called maximal elements and they have none element above; the elements which have none element below are called minimal elements. In environmental field the main assumption is that the lower the numerical value of the criteria the lower the hazard. Therefore, the maximal elements are the most hazardous and are selected to form the set of priority elements.

Total ranking methods

Total order ranking methods (TOR) are multicriteria decision making techniques used for the ranking of various alternatives on the basis of more than one criterion. The criteria, which are the standards by which the elements of the system are judged are not always in agreement, they can be conflicting, motivating the need to find an overall optimum that can deviate from the optima of one or more of the single criteria.

Desirability and utility functions are well-known multicriteria decision-making methods (Massart et al. 1997; Keller et al. 1991; Hendriks et al. 1992; Lewi et al. 1992; Harrington 1965), based on the definition of a desirability/utility function for each attribute in order to transform values of the attributes to the same scale. Each attribute is independently transformed into a desirability d_{ir} by an arbitrary function, which transforms the actual value of each element into a value between 0 and 1:

$$d_{ir} = f_r(y_{ir}) \qquad 0 \le d_{ir} \le 1 \tag{3}$$

r being the selected criterion, f the function chosen and y_{ir} the actual value of the *i-th* element for the *r-th* criterion.

Once the kind of function and its trend for each criterion is defined, the global desirability D and the global utility U of each *i-th* element can be evaluated as follows:

$$D_i = \sqrt[R]{d_{i1} \cdot d_{i2} \cdot \ldots \cdot d_{iR}} \qquad 0 \le D_i \le 1 \tag{4}$$

$$U_i = \frac{\sum_{r=1}^{R} d_{ir}}{R} \qquad 0 \le U_i \le 1 \tag{5}$$

In addition each criterion can be weighted in order to take into account criterion importance in the decision rule. In the case of weighted functions the overall desirability and utility of the *i-th* element are defined as follows:

$$D_i = (d_{i1}^{w_1} \cdot d_{i2}^{w_2} \cdot \ldots \cdot d_{iR}^{w_R}) \qquad 0 \le D_i \le 1 \tag{6}$$

$$U_i = \sum_{r=1}^{R} w_r \cdot d_{ir} \qquad 0 \leq U_i \leq 1 \tag{7}$$

w_r being the weight of the *r-th* criterion and $\sum_{r=1}^{R} w_r = 1$.

The use of a weighting scheme introduces arbitrariness into the analysis and thus it can be useful whenever additional information is available and the decision maker opinion can to be taken into account; see chapters by Brüggemann et al., pp. 237 and Voigt, Brüggemann, pp. 327 for a more detailed discussion of this point.

Once D or U for each element has been calculated, all the elements are ranked according to their D (or U) value. The overall desirability is calculated combining all the desirabilities through a geometrical mean. It must be highlighted that the desirability product is very strict: if any desirability d_{ir} is equal to 0 the overall desirability D_i will be zero, whereas the D_i will be equal to one only if all the desirabilities have the maximum value of one. The overall utility is calculated less severely: in fact the overall quality of an element can be high even if a single desirability/utility function is zero.

Order ranking models

Partial and total ranking methods have been widely used to perform data exploration, investigate the inter-relationships of objects and/or variables and set priorities. However it appears a very useful tool even for modelling purposes. Mathematical models have become an extremely useful tool in several scientific fields like environmental monitoring, risk assessment, QSAR and QSPR, i.e. in the search for quantitative relationships between the molecular structure and the biological activity/ chemical properties of chemicals.

A ranking model is defined as a relationship between one or more dependent attributes, investigated experimentally, and a set of independent attributes, also called model attributes, which are usually theoretical calculated variables such as molecular descriptors:

$$rank_i(y_{i1}, y_{i2}, \ldots, y_{iR}) = f(x_{i1}, x_{i2}, \ldots, x_{ip}) \tag{8}$$

where f is a ranking function applied on the training set elements (TS), R the number of dependent attributes and p the number of independent attributes. A model ranking development is based on the following steps:

1. Experimental ranking: a total or partial ranking method is applied to experimental attributes (dependent attributes).
2. Model ranking: the total or partial ranking method is applied to a subset of selected model attributes (independent attributes).
3. Experimental and model ranking comparison: evaluation of the degree of agreement between two rankings, i.e. analysis of model ranking reliability.
4. Model ranking evaluation: for each element the interval of each experimental attribute is compared with the interval derived from the model ranking.

Thus, the ranking model is given by the chosen ranking function and the ordered training set.

In the first phase, elements are ranked according to the experimental attributes describing them. Thus, a partial or total ranking method is selected and applied to the experimental attributes providing a diagram of partially ordered elements or a totally ordered element sequence, respectively. In the second phase the same ranking method previously applied to the experimental attributes, is now applied to a selected subset of model attributes, and the elements are ranked according to the selected model attributes.

Then, the two rankings are compared to evaluate the model ranking capability to reproduce the element ranking based on the experimental attributes. In this way the similarity between two partially or totally ordered sequences, is measured. Finally, if the agreement between the model ranking and the experimental ranking is considered satisfactory, the model ranking can perform predictions of the ranking of other elements, not being investigated experimentally. As in multiple linear regression (MLR) methods, the selection of variables (attributes) is crucial to developing an acceptable ranking model. The aim of variable subset selection is to reach optimal model complexity in predicting response variables by a reduced set of independent variables (Hocking 1976; Miller 1990). Ranking models based on the optimal subsets of a few predictor attributes have the great advantage of being more statistically stable, interpretable and showing higher predictive power. One of the simplest techniques for variable selection, -"sentimental selection"-, is based on the a priori selection of a few variables, by experience, tradition, availability, opportunity or previous knowledge. Another more mathematically based, but common, method of performing variable selection is the one based on an exhaustive examination of all the possible k variables models (the model size) obtained by a set of p variables. However, when many variables are available, an exhaustive examination of all possible models is not feasible as, given the extremely high number of possible variable combinations, it requires extensive computational resources and is time consuming. In such cases a variable selection technique is needed. The Genetic Algorithm (GA-VSS) approach is used here as the variable selection method to search for the best ranking models within a wide set of variables.

GA-VSS applied to partial ranking models

Genetic algorithms (GA) are an evolutionary method widely used for complex optimisation problems in several fields such as robotics, chemistry and QSAR (Goldberg 1989; Wehrens et al. 1998). Since complex systems are described by several variables, a major goal in system analysis is the extraction of relevant information, together with the exclusion of redundant and noisy information. A specific application of GA is variable subset selection (GA-VSS) (Leardi et al. 1992; Leardi 1994; Luke 1994; Leardi 1996; Todeschini et al. 2004). Variable selection is performed by GAs by considering populations of models generated through a reproduction process and optimised according to a defined objective function related to model quality. The procedure is illustrated in Figure 1.

It consists in the evolution of a population of models, i.e. a set of ranked models according to some objective function, based on the crossover and mutation processes, which are alternatively repeated until a stop condition is encountered (e.g., a user-defined maximum number of iterations) or the process is ended arbitrarily.

It is to be highlighted that the GA-VSS method provides not a single model but a population of acceptable models; this characteristic allows the evaluation of variable relationships with response from different points of view. Moreover, when variable subset selection is applied to a huge number of variables, the genetic strategy can be extended to more than one population, each based on different variable subsets, evolving from each other independently. In this case, after a number of iterations, these populations can be combined according to different criteria, obtaining a new population with different evolutionary capabilities (Todeschini et al. 2004).

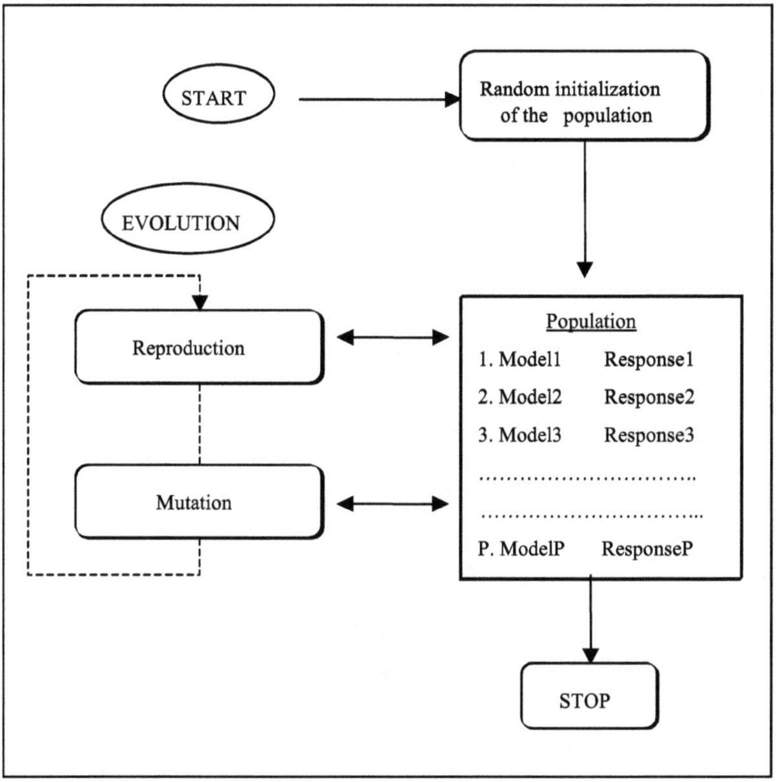

Fig. 1. Genetic algorithm procedure

Partial ranking optimisation parameters

Variable subset selection is performed by GAs, optimising populations of models according to a defined objective function related to model quality. In partial ranking models objective function is an expression of the degree of agreement between the element ranking resulting from experimental attributes and that provided by the selected subset of model attributes.

For the same *n* elements the correlation between the experimental partial ranking and the model ranking (denoted as E and M, respectively) can be evaluated by a set of similarity measures, called Tanimoto indices (Rogers et al. 1960; Brüggemann et al. 1995; Bath et al. 1993; Moock et al. 1998; Sørensen et al. 2003). Each Tanimoto index can be used as the measure of "goodness of fit" (degree of agreement) as it is the ratio of the number of agreements over the number of disagreements, i.e. contradictions in the ranking of two elements in the model and experimental ranking, weighting differently model and experimental incomparabilities.

Another similarity index is here proposed as a measure of the agreement between two partial rankings. It is calculated comparing the experimental and model Hasse matrices, denoted **E** and **M** respectively, according to the following expression:

$$S(\mathbf{E},\mathbf{M}) = 1 - \frac{\sum_{st} \left| h_{st}^{E} - h_{st}^{M} \right|}{2n \cdot (n-1)} \qquad 0 \leq S(\mathbf{E},\mathbf{M}) \leq 1 \tag{9}$$

where h_{st} is the entry of the Hasse matrix for each pair of elements s and t and S(**E**,**M**), being a similarity index, ranges from 0 (no similarity) to 1 (complete similarity) and expresses the differences between the two compared matrices; if two elements (s and t) have the same mutual rank in both rankings, their contribution is 0. Thus it can be forecast that if two elements (s and t) have different ranks, but not opposite ones, in the two rankings ($\boldsymbol{h}_{st}^{E} = \pm 1$ and $\boldsymbol{h}_{st}^{M} = 0$, or $\boldsymbol{h}_{st}^{E} = 0$ and $\boldsymbol{h}_{st}^{M} = \pm 1$), then their contribution is 1, while if the mutual ranks are opposite ($\boldsymbol{h}_{st}^{E} = +1$ and $\boldsymbol{h}_{st}^{M} = -1$, or $\boldsymbol{h}_{st}^{E} = -1$ and $\boldsymbol{h}_{st}^{M} = +1$), their contribution is 2. In this way the discrepancies due to opposite mutual rankings are evaluated more deeply than those due to comparable element pairs that have become incomparable, and *vice versa*.

Fig. 2 shows the procedure used to compare the partial experimental ranking and the partial model ranking.

Total ranking optimisation parameters

In total ranking models the degree of agreement between the element ranking resulting from experimental attributes and that provided by the selected subset of model attributes is measured by the Spearman's rank index (Kendall 1948).

Applying a total order ranking method, like desirability or utility functions, to the experimental attributes $y_1, \ldots, y_R$, an experimental ranking, Γ_{exp}, is calculated. According to the experimental ranking, a specific experimental rank is associated to each *i-th* element:

$$\Gamma_i^{exp} \equiv f(y_{i1}, y_{i2}, \ldots, y_{iR}) \rightarrow rank_i^{exp} \tag{10}$$

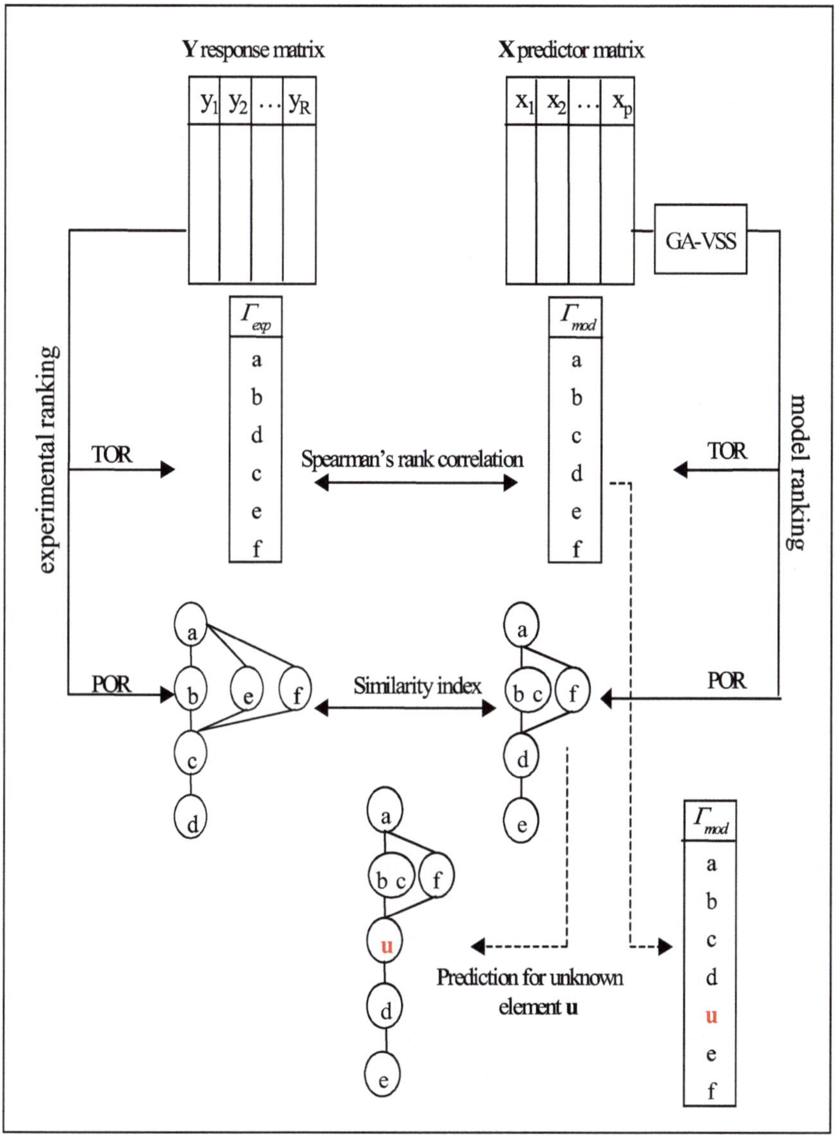

Fig. 2. Scheme of the procedure used for to compare the experimental and model ranking

In the next step, the total order ranking method is applied to the model attributes $x_1, \ldots, x_p$, defining a model ranking, Γ_{mod} and according to that, a model rank is associated to each *i-th* element:

$$\Gamma_i^{mod} \equiv f(x_{i1}, x_{i2}, \ldots, x_{ip}) \rightarrow rank_i^{mod} \tag{11}$$

The correlation between the two rankings (Γ_{exp}, Γ_{mod}) can then be evaluated by Spearman's rank correlation coefficient r_S, according to the following expression:

$$r_s = \frac{6 \cdot \sum_{i=1}^{n} r_i^2}{n^3 - n} \qquad -1 \le r_s \le +1 \tag{12}$$

where r_i is the rank difference for the element i in the two rankings and n is the total number of elements. Fig. 2 shows the procedure used to compare the total experimental ranking and the total model ranking.

Ranking predictions

Once the "goodness of fit" of the model ranking has been verified, predictions can be performed for new elements. The experimental ranking of new compounds that have not yet been investigated experimentally can be estimated by the ranking model; from the set of model attributes $\{x_{u1},\ldots, x_{up}\}$ describing any unknown element u, prediction of the experimental ranking of element u can be performed on the basis of the training set (TS) elements:

$$f\{x_{u1}, x_{u2}, \ldots, x_{up}\} \xrightarrow{\text{training set}} rank_u \tag{13}$$

Thus, in a first step the ranking of the unknown element u is predicted with respect to the training set elements and, in the second step, the experimental responses are predicted. In this case, despite the regression models, the ranking model provides not a single response value but an interval.

To explain ranking predictions, a directed connectivity operator C is introduced. Being s and t two diverse elements in a ranking, and N the set of integer numbers, then the connectivity operator $C(s,t)$ is defined as follows:

$$if \quad s,t \in TS \;\; and \;\; s \neq t \rightarrow C(s,t) \in N \tag{14}$$

$$\begin{aligned} &C(s,t) = 0 \quad iff \quad s \text{ is incomparable with } t \; (s \| t) \\ &C(s,t) \in \mathrm{N}^{+} \quad iff \quad s \text{ is above } t \\ &C(s,t) \in \mathrm{N}^{-} \quad iff \quad s \text{ is below } t \end{aligned} \tag{15}$$

The term "*iff*" means: if and only if and is a typical mathematical idiom. The operator *C(s,t)* has the following properties:

$$\begin{aligned}
&C(s,t) = k \quad \text{iff} \quad 0 \le |k| \le L-1 \\
&C(s,t) = -C(t,s) \quad \rightarrow \quad \text{antisymmetry} \\
&C(s,t) = p \text{ and } C(t,z) = q \Rightarrow C(s,z) = p+q \quad \text{if } p,q > 0 \rightarrow \text{transitivity}
\end{aligned} \tag{16}$$

where the k absolute value is the topological distance between the two elements s and t in the Hasse diagram, i.e. the shortest path length in the diagram, and L the number of levels in the ranking, i.e. the maximum distance in the diagram. According to the first property, the operator is an integer number, taking a value equal to the path length between s to t. If s is above t, and is located in the level immediately above t then $C(s,t)$ takes a value equal to 1. The maximum length of a ranking, which in the Hasse diagram is the maximum number of lines in the longest chain, is equal to L-1, L being the number of ranking levels. If no path exists between s and t, meaning that s and t are incomparable ($s \| t$), then $C(s,t)$ equals 0. Reflecting the ranking order relation properties, the connectivity operator has antisymmetry and transitivity properties. Thus, through the connectivity operator, predictions of the experimental ranking of any unknown element u can be performed looking for the two elements s and t which satisfy the following conditions:

$$min_s C(s,u) > 0 \quad \text{and} \quad min_t C(u,t) > 0 \quad \text{and} \quad min\ [y_s - y_t] > 0 \tag{17}$$

where s and t are the two elements connected (comparable) to u, i.e. $C(s,u) > 0$ (with s above u) and $C(u,t) > 0$ (with u above t), located on the shortest path, and whose experimental difference value constitutes the smallest positive interval. Moreover, $C(s,u)$ represents the u-above rank radius and $C(u,t)$ the u-below rank radius, whereas $C(s,t)$ is the u rank diameter.

A numerical example for partial ranking model is here provided to better explain the prediction calculation. For the sake of simplicity, let us consider an experimental ranking developed on two experimental attributes y_1 and y_2; Table 1 shows their numerical values. Fig. 3 shows the resulted experimental Hasse diagram together with the ranking model developed on the training set composed by 9 elements {*a*, *b*, *c*, *d*, *e*, *f*, *g*, *h*, *i*}, described by an arbitrary set of independent attributes.

Table 1. Numerical values of the experimental attributes

Element	y_1	y_2
a	180	400
b	150	420
c	130	240
d	140	270
e	90	190
f	100	230
g	120	200
h	90	235
i	82	88

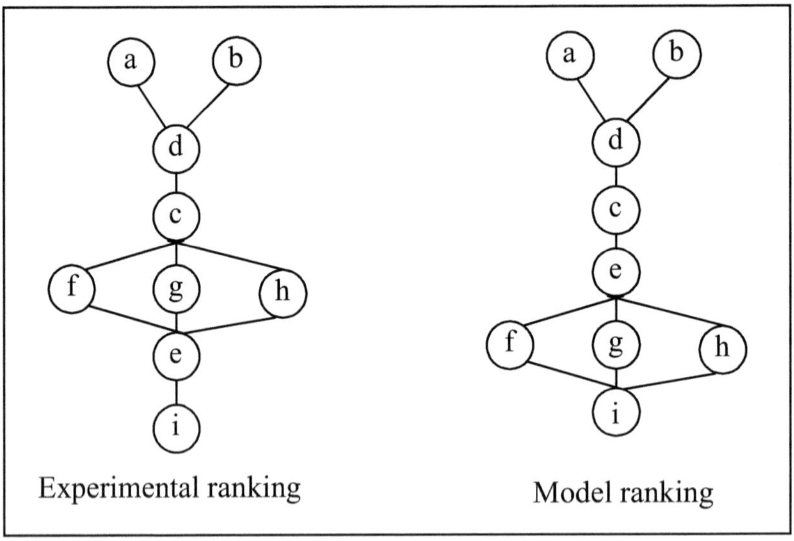

Fig. 3. Experimental and model Hasse diagram

The model's agreement with the experimental ranking is S(**E**,**M**) = 0.92; T(0,0) = T(0,1) = T(1,1) = 0.91.

Fig. 4 shows the model ranking projection of the new unknown element *u* in the model ranking diagram.

To predict the experimental response intervals of the unknown element *u*, a search is made for the element pair located on the shortest path from *u* and with an experimental value difference that constitutes the smallest positive interval.

Firstly, an examination is made of the elements comparable to *u* and located on a path length equal to 1. The experimental values y_1 and y_2 of elements *e*, *f*, *g* and *h* are taken into account and the differences between *e* (located above *u*) and *f*, *g* and *h* (located below *u*) are investigated.

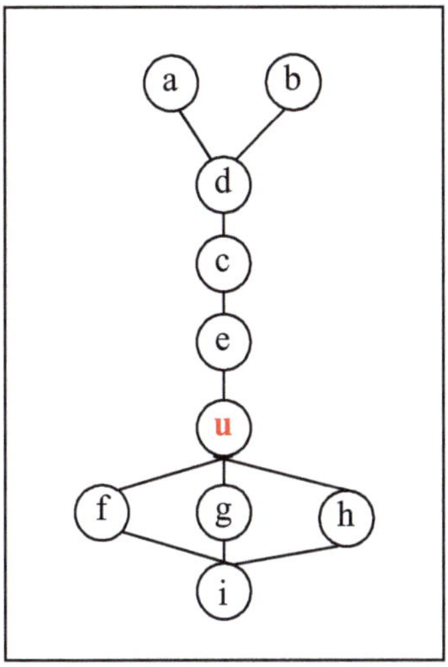

Fig. 4. Projection of the unknown element u in the model ranking diagram

As far as concerns the experimental attribute y_I, on the basis of u location in the model the following intervals can be evaluated (Note that in chapter by Carlsen, p. 163 the intervals are only implicitly used):

$$\begin{aligned} y_{eI} - y_{fI} &= 90 - 100 = -10 \\ y_{eI} - y_{gI} &= 90 - 120 = -30 \\ y_{eI} - y_{hI} &= 90 - 90 = 0 \end{aligned} \tag{18}$$

All three intervals are rejected, as they are not positive interval. Thus the elements located on a path length (relatively to element u) equal to 2 are examined. For example, the elements c and i are considered and the following intervals examined:

$$\begin{aligned} y_{cI} - y_{fI} &= 130 - 100 = 30 \\ y_{cI} - y_{gI} &= 130 - 120 = 10 \\ y_{cI} - y_{hI} &= 130 - 90 = 0 \\ y_{eI} - y_{iI} &= 90 - 82 = 8 \\ y_{cI} - y_{iI} &= 130 - 82 = 48 \end{aligned} \tag{19}$$

The smallest positive interval for y_I is the one provided by elements e and i, thus the experimental value y_I of the unknown element u is predicted as:

$$y_{i1} \leq y_{u1} \leq y_{e1} \quad \Rightarrow \quad 82 \leq y_{u1} \leq 90 \tag{20}$$

It can be observed that element *e* and *i* satisfy all the conditions required to perform a ranking prediction, i.e.:

$$C(e,u) = 1\,(>0) \quad \text{and} \quad C(u,i) = 2\,(>0) \quad \text{and} \quad min[y_e - y_i] = 8\,(>0) \tag{21}$$

In the same way the following intervals are provided for y_2:

$$\begin{aligned} y_{e2} - y_{f2} &= 190 - 230 = -40 \\ y_{e2} - y_{g2} &= 190 - 200 = -10 \\ y_{e2} - y_{h2} &= 190 - 135 = -45 \end{aligned} \tag{22}$$

All the three intervals are rejected, and again the intervals provided by elements *c* and *i* located at a length path equal to 2 are examined:

$$\begin{aligned} y_{c2} - y_{f2} &= 240 - 130 = 10 \\ y_{c2} - y_{g2} &= 240 - 130 = 40 \\ y_{c2} - y_{h2} &= 240 - 235 = 5 \\ y_{e2} - y_{i2} &= 190 - 88 = 102 \\ y_{c2} - y_{2i} &= 240 - 88 = 152 \end{aligned} \tag{23}$$

The smallest positive interval for y_2 is the one provided by elements *c* and *h*, thus the experimental value y_2 of the unknown element *u* is predicted as:

$$y_{h2} \leq y_{u2} \leq y_{c2} \quad \Rightarrow \quad 235 \leq y_{u2} \leq 240 \tag{24}$$

Elements *c* and *h* satisfy all the conditions required to perform a ranking prediction, i.e.:

$$C(c,u) = 2\,(>0) \quad \text{and} \quad C(u,h) = 1\,(>0) \quad \text{and} \quad min[y_c - y_h] = 5\,(>0) \tag{25}$$

According to the position of the unknown element *u* in the model ranking, four different cases can be identified, each characterized by specific prediction:

- u is located in a chain $\rightarrow y_{tr} \leq y_{ur} \leq y_{sr}$
- u is a minimal $\rightarrow y_{ur} \leq y_{sr}$
- u is a maximal $\rightarrow y_{tr} \leq y_{ur}$
- u is isolated $\rightarrow y_{ur} = ?$

s and t being two elements in the Hasse diagram respectively located above and below u. In particular, for case 2, being u a minimal, its rank is predicted to be smaller than the lowest value of the comparable elements ranked above; thus, the rule is the following:

$$C(s,u) = 1 \quad \text{and} \quad min_s(y_s) \tag{26}$$

which means that the estimated interval of u is open on the left and only the first shell of neighbourhoods above is taken into account.

Moreover, for case 3 where u is a maximal, there is no comparable element above, and its rank is predicted to be larger than the highest value of the comparable elements ranked below; thus, the rule is:

$$C(u,t) = 1 \quad \text{and} \quad max_t(y_t) \tag{27}$$

which means that the estimated interval of u is open on the right and only the first shell of neighbourhoods below is taken into account.

In the last case u is an isolated element, i.e. it is not comparable with any of the elements of the training set, thus its rank cannot be predicted by the model ranking developed.

Prediction uncertainty

According to the proposed prediction calculation procedure, it is clear that the actual distance between the two elements s and t, which satisfies the prediction conditions for any unknown element u, is crucial, and the larger the distance the larger the potential uncertainty in the prediction. Thus, a first topological measure of the prediction precision is provided by the connectivity operator $C(s,t)$ previously defined: the precision decreases for increased $C(s,t)$.

$$1 \leq C(s,u) \leq L-1 \quad \text{and} \quad 1 \leq C(u,t) \leq L-1 \tag{28}$$

Moreover a normalised distance measure for each prediction from the upper and lower limits of the interval can be evaluated according to the expression:

$$D_u^{sup} = \frac{C(s,u) - 1}{L - 2} \quad \text{and} \quad 0 \le D_u^{sup} \le 1 \tag{29}$$

$$D_u^{inf} = \frac{C(u,t) - 1}{L - 2} \quad \text{and} \quad 0 \le D_u^{inf} \le 1 \tag{30}$$

s and t being the two elements which, satisfying the prediction conditions are selected to predict the experimental interval of the unknown element u. D_u^{sup} and D_u^{inf} give a measure of the normalised rank uncertainty, above and below respectively. Note that if u is a priority element (maximal) C(s,u) is not defined, as no element exists above u, thus D_u^{sup} is not defined and only D_u^{inf} can be evaluated. Analogously, if u is a minimal element C(u,t) is not defined, as no element exists below u, thus D_u^{inf} is not defined and only D_u^{sup} can be evaluated.

Another way to measure prediction uncertainty is to evaluate the experimental interval width of the prediction on the r-th experimental attribute:

$$Ry_{ur} = \frac{y_{sr} - y_{tr}}{max_{y_r} - min_{y_r}} \qquad 0 \le Ry_{ur} \le 1 \tag{31}$$

where y_{sr} and y_{tr} are the experimental values of *s* and *t* for the *r-th* attribute respectively, and max_{y_r} and min_{y_r} the maximum and minimum experimental values of the *r-th* attribute. The greater the width, the greater the uncertainty. For maximal and minimal elements Ry_{ur} is not defined, as their estimated interval is an open interval. Therefore, D_u^{sup} and D_u^{inf} measure the normalised rank uncertainty of the estimated interval, above and below respectively, whereas Ry_{ur} measures the experimental uncertainty.

Model validation

Further verification of model ranking applicability can be obtained by applying the described ranking prediction procedure to the training set elements initially used to develop the model. This results in the creation of a number of modified data sets from which the elements will be deleted from the data one by one. For each element of the training set the experimentally derived intervals are calculated from the experimental ranking; the other training set elements are then used to calculate the experimental intervals of that element from the experimental ranking. In the same way, the model calculated intervals are obtained by using the other training set elements to calculate the model intervals of the each element from the model ranking. Thus, it is similar to a leave – one – out cross validation procedure (LOO technique), where each element is taken away, one at a time and the response for the deleted element is calculated from the model. Thus, given n objects, n reduced models have to be calculated. This technique is particularly important as this deletion scheme is unique and the predictive ability of the different models can be compared accurately.

Once having obtained the experimentally derived intervals and the calculated intervals, they are compared to establish the model ranking quality.

On comparing two intervals, six different cases, illustrated in Figure 5, can be identified.

As A and B are respectively the lower and upper values of the experimental interval, and C and D those of the model interval, Cases 1 and 2 represent disjoint intervals; Cases 5 and 6 intervals contained one in the other, and Cases 3 and 4 partially overlapped intervals.

Analysing one experimental attribute at a time, for each *i-th* element the disagreement δ_{ir} between its experimentally derived interval (A-B) and its model calculated interval (C-D) on the *r-th* attribute is calculated, assuming the worst case, according to the following expressions:

- Case 1: $\delta_{ir} = |\mathrm{D} - \mathrm{A}|$
- Case 2: $\delta_{ir} = |\mathrm{B} - \mathrm{C}|$
- Case 3, 4, 5, 6: $\delta_{ir} = |\mathrm{C} - \mathrm{A}| + |\mathrm{D} - \mathrm{B}|$

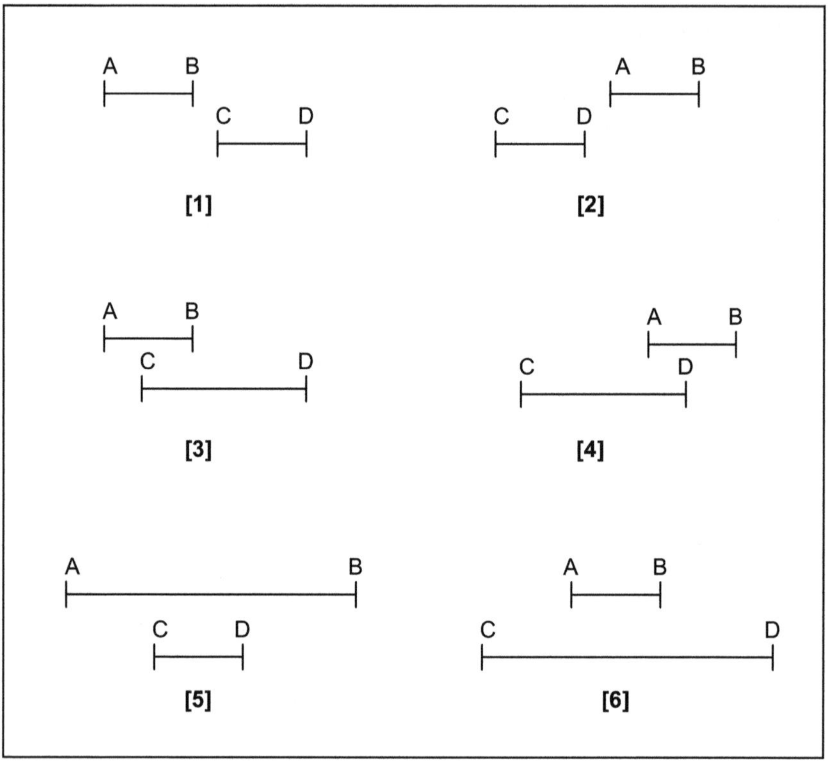

Fig. 5. Interval comparison

A standardised interval disagreement for the *i-th* element on the *r-th* attribute is then derived as:

$$\delta_{ir}^{*} = \frac{\delta_{ir}}{max_{y_r} - min_{y_r}} \tag{32}$$

max_{y_r} and min_{y_r} being the maximum and minimum values of the *r-th* attribute respectively.

The average disagreement between the experimental and the model calculated intervals is then calculated:

$$\bar{\delta}_r = \frac{\sum_{i=1}^{N} \delta_{ir}^{*}}{N} \tag{33}$$

and a measure of the ranking model quality, as far as concerns the *r-th* attribute is calculated as:

$$Q_r = 1 - \bar{\delta}_r \tag{34}$$

The overall ranking model quality, i.e. taking into account all the R responses, can be evaluated by the following expressions:

$$Q_T = \frac{\sum_{r=1}^{R} Q_r}{R} \qquad Q_G = \sqrt[R]{Q_1 \cdot \ldots \cdot Q_R} \qquad Q_M = min\{Q_r\} \tag{35}$$

Q_T being the arithmetic mean of all the R attributes of the ranking model represents the least demanding parameter for evaluating overall model ranking quality. Instead the geometric mean Q_G is a more severe parameter, able to display models not able to reproduce a correct experimental ranking for only a few attributes. The most demanding evaluation parameter of model quality is Q_M, which assumes minimum quality among the R, calculated as the representing overall model quality. This procedure for evaluating model ranking quality is based on ranking interval comparison. Moreover, as the metric scale is usually seen as a “stronger” property than the ordinal scale, it is of interest to measure the loss of information due the replacement of the original “quantitative” information with rank orders. Thus, being the quantitative experimental values intervals with equal lower and upper values, they are compared with the experimentally derived intervals (A-B), and for each *r-th* attribute the standardised interval disagreement ${}^{0}\delta^{*}_{ir}$ is calculated the same way, as described above. The arithmetic mean of the average disagreement between the quantitative experimental values and their derived intervals on the *r-th* attribute provides a measure of the uncertainty increase due to the replacement of a metric scale with an ordinal scale and is calculated as:

$$\tilde{\delta}_r = \frac{\sum_{i=1}^{N} \delta^{*}_{ir}}{N} \tag{36}$$

This quantity is as greater as more loss of information.

Partial order ranking QSAR model for similarly acting phenylurea herbicides

Natural environments and ecosystems are not exposed to individual chemicals but to complex multi-component mixtures of chemicals of various origins. Nevertheless, most ecotoxicological research and chemical regulation focus on hazard and exposure assessment of individual chemicals only and the chemical mixtures in the environment are ignored to a large extent. Therefore, there is the need for developing risk assessment procedures no longer restricted to single toxicants and instead considering combined effects resulting from multiple chemical exposures. The predictive mixture toxicities approaches imply that the chemical composition of the mixture of interest in known. Two different concepts, termed Concentration Addition and Independent Action, are thought of being more generally applicable and allow calculating expected mixture toxicity on the basis of known toxicities of the single components of a mixture. As these two concepts are based on opposite assumption with respect to the similarity of the mechanism of action of the individual components, the first step in mixture risk assessment procedure is to evaluate the similarity of the mechanism of action of the individual components (Faust et al. 2001; Faust et al. 2003).

The similarity in mode of action of different toxicants is frequently described by analysis of their Effect-Concentration curves, where the concentrations of the toxicants that are estimated to cause a predefined effect are plotted. The Hasse diagram technique has been proposed as an useful tool to compare and rank chemical toxicities, not limited to one single level of biological response (as usually by EC_{50}), but taking into account more than one single response at the same time. The Hasse diagram can give information regarding the similarity of action comparable to the different EC profiles. Then, since it is not practically possible experimentally to generate all the necessary input information for the risk assessment of the chemicals, the second step in mixture risk assessment procedure is to obtain part of the information concerning the chemicals fate and effect in the environment by models. The development of efficient and inexpensive technologies for effective risk assessment and to predict physical, chemical and biological properties of new compounds is now driven by the requirements of Commission Directive 93/67/EEC on Risk Assessment for New Notified Substances and Commission Regulation (European Commission (EC) No. 1488/94 on Risk Assessment for Existing Substances (EEC 1993; EEC 1994). In the EU, the proposed system for registration, evaluation and authorization of chemicals, REACH system is likely to have important applications for the development and application of QSARs for predicting chemical toxicity (EC 2002).

Quantitative Structure - Activity Relationships (QSARs) are estimation methods developed and used to predict certain effects or properties of chemical substances, which are primarily based on the structure of the chemicals. The development of QSARs often relies on the application of statistical methods such as multiple linear regression (MLR) or partial least squares regression (PLS). However, since toxicity data often include uncertainties and measurements errors, when the aim is to point out the more toxic and thus hazardous chemicals and to set priorities, order models can be used as alternative to statistical methods such as multiple linear regression.

Toxicity experimental data

The analysed data set consists of 12 congeneric phenylureas previously selected and studied in a EU project on mixture toxicity (Prediction and assessment of the aquatic toxicity of mixtures of chemicals, PREDICT project) (Gramatica et al. 2001; Backaus et al. 2004). These chemicals, frequently found in surface waters where aquatic organisms are exposed to mixture of them, share a common specific mechanism of action (inhibitors of the photosynthetic electron transport). They were tested for toxicity on freshwater algae *Scenedesmus vacuolatus* by the research group of Bremen University, coordinator of the EU project: Bridging Effect Assessment of Mixtures to Ecosystem Situations and Regulation (BEAM) (Backhaus et al. 2003). The dependent variables selected for describing their toxicity were the reproduction inhibition responses with 4 concentrations (μmol/L) provoking 1% (EC_{01}), 10% (EC_{10}) 50% (EC_{50}), 90% (EC_{90}) effect, respectively. Table 2 shows the EC toxicity values of the 12 phenylureas (Scholze et al. 2001). The Fig. 6 reports the effect-concentration curves of the 12 phenylureas, numbered as in Table 2.

Table 2. Toxicity data of 12 phenylureas

			Log(1/EC)			
ID	Substance	CAS	EC01	EC10	EC50	EC90
1	Buturon	3766-60-7	1.877	0.897	0.111	-0.390
2	Chlorbromuron	13360-45-7	3.244	2.059	1.222	0.824
3	Chlortoluron	15545-48-9	2.523	1.576	0.815	0.332
4	Diuron	330-54-1	3.071	2.223	1.538	1.109
5	Fenuron	101-42-8	0.927	0.137	-0.633	-1.214
6	Fluometuron	2164-17-2	2.030	0.772	-0.173	-0.849
7	Isoproturon	34123-59-6	2.226	1.363	0.642	0.166
8	Linuron	330-55-2	3.155	1.990	1.056	0.463
9	Metobromuron	3060-89-7	1.430	0.630	-0.019	-0.490
10	Metoxuron	19937-59-8	2.320	1.209	0.319	-0.249
11	Monolinuron	1746-81-2	2.058	0.920	0.007	-0.575
12	Monuron	150-68-5	2.569	1.367	0.402	-0.212

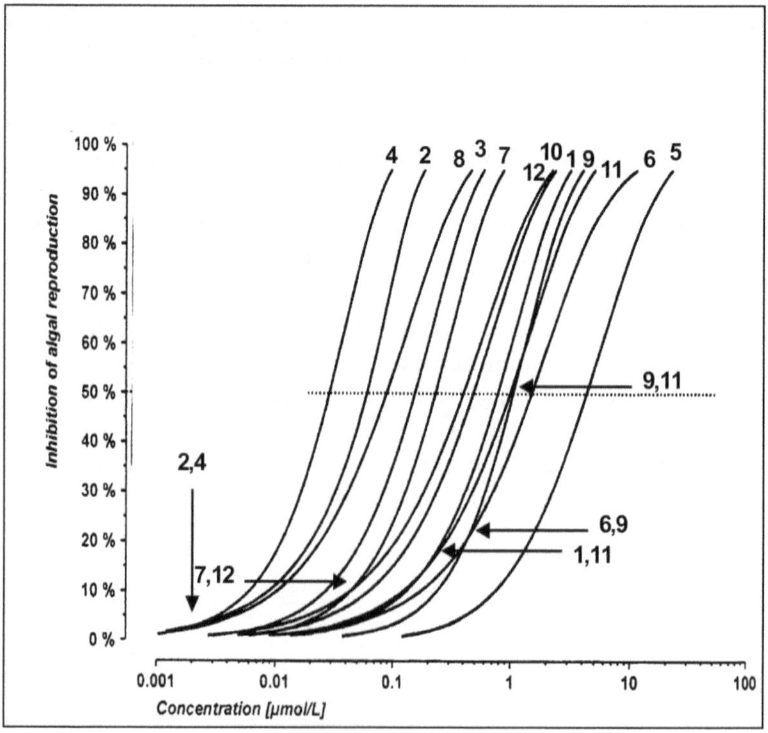

Fig. 6. Effect-Concentration curves of 12 phenylurea herbicides. Crossing of lines are indicated

Molecular descriptors

The chemical structures of the phenylureas have been described with more than 1500 molecular descriptors, in order to catch all the structural information.

The molecular descriptors have been calculated by the Dragon program (Todeschini et al. 2004) on the basis of the minimum energy molecular geometries optimized by HyperChem package (HYPERCHEM 1995) (PM3 semiempirical method). In this study the following sets of molecular descriptors have been calculated: constitutional descriptors, topological descriptors (Bonchev 1983; Devillers et al. 2000), walk and path counts, connectivity indices (Kier et al. 1986), information indices, Moreau-Broto 2D-autocorrelations (Moreau et al. 1980a; Moreau et al. 1980b; Broto 1984), edge adjacency indices (Estrada 1995), BCUT descriptors (Pearlman et al. 1998; Pearlman 1999), topological charge indices (Gálvez et al. 1994; Gálvez et al. 1995), eigenvalue based indices (Balaban et al. 1991), Randić molecular profiles (Randić 1995; Randić 1996), geometrical descriptors, radial distribution function descriptors (Hemmer et al. 1999), 3D-MoRSE descriptors (Schuur et al. 1996; Schuur et al. 1997), WHIM descriptors (Todeschini et al. 1994; Todeschini et al. 1997), GETAWAY descriptors (Consonni et al. 2002), functional group counts and atom centred fragments. Definitions and further information regarding all these molecular descriptors can be found in the 'Handbook of Molecular Descriptors' (Todeschini et al. 2000).

Experimental ranking

The Hasse Diagram Technique applied on the four toxicity responses of algae reproduction inhibition with the concentrations provoking different levels of biological response (EC_{01}, EC_{10}, EC_{50} and EC_{90}) is shown in Fig. 7.

It is a quite simple diagram arranged in seven levels. It identifies two maximals: chlorbromuron (2) and diuron (4), highlighted as the most toxic. They are incomparable since some contradictions exist among their EC values: diuron is more toxic than chlorbromuron on EC_{10}, EC_{50} and EC_{90} levels, but it is less toxic on EC_{01} Nevertheless, taking into account the minor reliability of the EC_{01} value, diuron is certainly the most toxic of this data set: actually, it is included in priority lists of organic pollutants for aquatic systems. Fenuron (5) is the less toxic substance in this set of herbicides, as it is characterised by the lowest effect of all the concentration values.

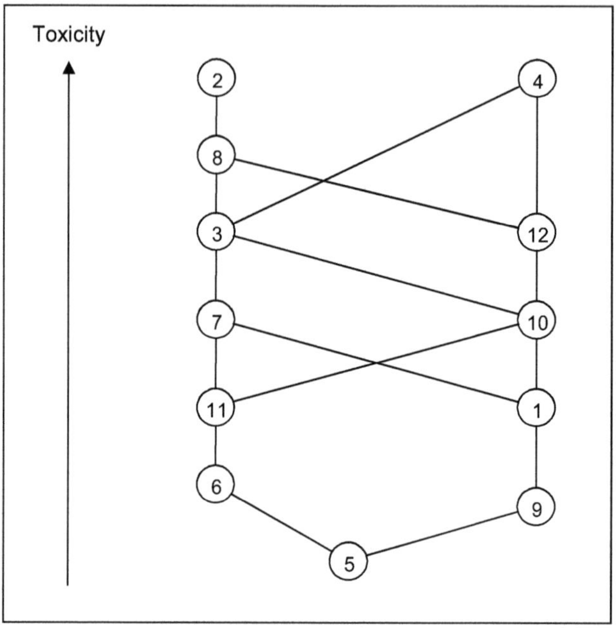

Fig. 7. Experimental Hasse diagram of 12 phenylureas

The Hasse diagram is characterised by a few number of incomparabilities: 18 over 132 comparisons, highlighting that not too many contradictions in the EC values exist when chemicals have a common mechanism of action. Actually, the phenylureas are all inhibitors of photosynthesis. It is interesting to compare the Hasse diagram with the EC curves of Fig. 6: the shapes and slopes of the EC curves appear quite similar, but they are not strictly parallel (as usually happen with similarly acting chemicals); the majority of the intersections between the curves (highlighted in Figure 6) correspond to the Hasse diagram incomparabilities. These differences in EC curve shape and slope and, similarly, these incomparabilities in Hasse diagram may be interpreted in terms of different toxicokinetic properties and/or indicate different binding behaviour (Backaus et al. 2004).

The analysis performed by the Hasse diagram not only allows to rank the phenylureas according to their overall toxicity values and to provide a priority list, but also highlights in a clear and simple way the EC curves crossing.

Model ranking

The correlations between the overall toxicity of the considered chemicals and the molecular descriptors have been estimated by the partial ranking Hasse Diagram Technique (HDT). However as an exhaustive search for the best ranking models within a wide set of descriptors requires extensive computational resources and is time consuming, given the extremely high number of possible descriptor combinations, the Genetic Algorithm (GA-VSS) approach has been used as the variable selection method. Starting from a population of 100 random models with a number of variables equal to or less 3, the algorithm has explored new combinations of variables, selecting them by a mechanism of reproduction/mutation similar to that of biological population evolution. The models based on the selected subsets of variables have been tested and evaluated by maximising the similarity index S(**E**,**M**). All of the calculations have been performed by the in-house software *RANA* for variable selection for WINDOWS/PC (Todeschini et al. 2003).

The best model obtained is a very simple model, based on two variables: a 3D-MoRSE descriptor of signal 9 weighted by atomic mass (Mor09m) and a functional group count accounting for the number of substituted aromatic carbons (nCaR). The maximal elements of the experimental Hasse diagram are the more toxic element (priority elements), whereas the minimal elements are the less toxic. According to the model Hasse diagram, the more toxic elements are those with a greater number of substituted aromatic carbons and with a greater value of Mor09m. The model Hasse diagram is shown in Fig. 8: it is arranged on ten levels and characterized by 61 comparable pairs of elements and 10 contradictions. The two model descriptor values are illustrated in Table 3.

The diagram points out chlorbromuron (2) as maximal element, being characterized by the highest nCaR value (nCaR = 3) and the highest Mor09m value (Mor09m = -0.242).

Diuron (4), experimentally the most toxic chemical has the same value of nCaR and very close value of Mor09m (Mor09m = -0.335). Fenuron (5) and metobromuron (9) are identified as minimal elements, the former is characterised by the lowest nCaR value (nCaR = 1), the latter by the lowest Mor09m value (Mor09m = -1.332).

The agreement degree between experimental and model diagrams is quite satisfactory (S(E,M) = 89.4). The Tanimoto indices have been calculated (Sørensen et al. 2003):

$$T(0,0) = 98.1 \qquad T(0,1) = 85.2 \qquad T(1,1) = 80.0 \qquad (37)$$

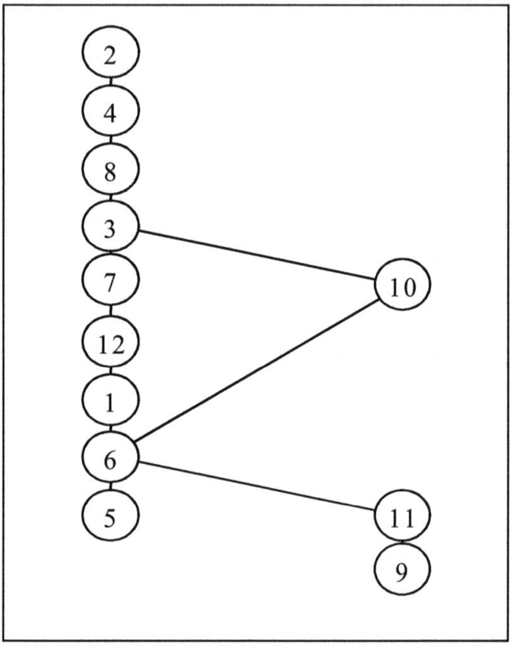

Fig. 8. Model Hasse diagram developed with Mor09m and nCaR descriptors

The "goodness of fit" of the partial ranking model calculated by the similarity index is lower than that calculated by T(0,0) but higher than the one by T(0,1) and T(1,1), confirming that the similarity index S(**E**,**M**) is a reasonable compromise between the over optimistic and the over pessimistic evaluation provided by T(0,0), and T(0,1), T(1,1), respectively.

Interval estimation

The experimental ranking of each chemical has been estimated according to the procedure described above. The calculated intervals have been compared to the corresponding experimentally derived intervals, obtained by deleting each chemical from the experimental ranking diagram; and using the remaining training set elements to calculate the experimental intervals of the deleted element from the experimental ranking diagram.

Analysing one experimental response at a time, for each chemical the standardised disagreement δ_{ir} between its experimentally derived interval and model-calculated interval has been calculated. The experimentally derived intervals and the calculated intervals for Log(1/EC_{01}), Log(1/EC_{10}),

Log(1/EC_{50}), Log(1/EC_{90}), together with the corresponding standardised disagreements are illustrated in Table 4, 5, 6 and 7, respectively.

Table 3. Model descriptors value for 12 phenylureas

ID	Substance	Mor09m	nCaR
1	Buturon	-0.973	2
2	Chlorbromuron	-0.242	3
3	Chlortoluron	-0.661	3
4	Diuron	-0.335	3
5	Fenuron	-1.102	1
6	Fluometuron	-1.077	2
7	Isoproturon	-0.803	2
8	Linuron	-0.517	3
9	Metobromuron	-1.332	2
10	Metoxuron	-0.990	3
11	Monolinuron	-1.136	2
12	Monuron	-0.935	2

Table 4. Experimental Log(1/EC_{01}) interval estimation (bold fonts indicate disjoint intervals)

Response: Log(1/EC_{01})		Experimental		Calculated		
ID	Substance	Min	Max	Min	Max	δ^*_{EC01}
1	Buturon	1.430	2.226	2.030	2.569	0.407
2	Chlorbromuron	> 3.155	-	> 3.071	-	0.036
3	Chlortoluron	2.320	3.071	2.320	3.155	0.036
4	Diuron	> 2.569	-	3.155	3.244	0.253
5	Fenuron	-	< 1.430	-	< 2.030	0.259
6	Fluometuron	0.927	2.058	0.927	1.877	0.078
7	Isoproturon	2.058	2.523	**2.569**	**3.155**	0.493
8	Linuron	2.569	3.244	2.523	3.071	0.095
9	Metobromuron	0.927	1.877	-	< 2.058	0.078
10	Metoxuron	2.058	2.523	2.030	2.523	0.012
11	Monolinuron	2.030	2.226	**1.430**	**2.030**	0.344
12	Monuron	2.320	3.071	**1.877**	**2.226**	0.556

Table 5. Experimental Log(1/EC_{10}) interval estimation (bold fonts indicate disjoint intervals)

Response: Log(1/EC_{10})		Experimental		Calculated		
ID	Substance	Min	Max	Min	Max	δ^*_{EC10}
1	Buturon	1.209	0.630	0.772	1.367	0.563
2	Chlorbromuron	> 1.990	-	> 2.223	-	0.112
3	Chlortoluron	1.363	1.990	1.363	1.990	0.000
4	Diuron	> 1.576	-	1.990	2.059	0.198
5	Fenuron	-	<0.630	-	< 0.772	0.068
6	Fluometuron	0.137	0.920	0.137	0.897	0.011
7	Isoproturon	0.920	1.576	1.367	1.576	0.214
8	Linuron	1.576	2.059	1.576	2.223	0.079
9	Metobromuron	0.137	0.897	-	< 0.920	0.011
10	Metoxuron	0.920	1.363	0.772	1.576	0.173
11	Monolinuron	0.772	1.209	**0.630**	**0.772**	0.278
12	Monuron	1.209	1.990	0.897	1.363	0.450

Table 6. Experimental Log(1/EC_{50}) interval estimation (bold fonts indicate disjoint intervals)

Response: Log(1/EC_{50})		Experimental		Calculated		
ID	Substance	Min	Max	Min	Max	δ^*_{EC50}
1	Buturon	-0.019	0.319	-0.173	0.402	0.109
2	Chlorbromuron	> 1.056	-	> 1.538	-	0.222
3	Chlortoluron	0.642	1.056	0.642	1.056	0.000
4	Diuron	> 0.815	-	1.056	1.222	0.111
5	Fenuron	-	< -0.173	-	< -0.173	0.000
6	Fluometuron	-0.633	0.007	-0.633	0.111	0.048
7	Isoproturon	0.111	0.815	0.402	0.815	0.134
8	Linuron	0.815	1.222	0.815	1.538	0.146
9	Metobromuron	-0.633	0.111	-	< 0.007	0.048
10	Metoxuron	0.111	0.402	-0.173	0.815	0.321
11	Monolinuron	-0.173	0.319	**-0.019**	**-0.173**	0.298
12	Monuron	0.319	1.056	0.111	0.642	0.287

Table 7. Experimental Log(1/EC_{90}) interval estimation (bold fonts indicate disjoint intervals)

Response: Log(1/EC_{90})		Experimental		Calculated		
ID	Substance	Min	Max	Min	Max	δ^*_{EC90}
1	Buturon	-0.490	-0.249	-0.849	-0.212	0.170
2	Chlorbromuron	> 0.463	-	> 1.109	-	0.278
3	Chlortoluron	0.166	0.463	0.166	0.463	0.000
4	Diuron	> 0.332	-	0.463	0.824	0.056
5	Fenuron	-	< -0.849	-	< -0.849	0.000
6	Fluometuron	-1.214	-0.575	**-0.575**	**-0.390**	0.355
7	Isoproturon	-0.390	0.332	-0.212	0.332	0.077
8	Linuron	0.332	0.824	0.332	1.109	0.123
9	Metobromuron	-1.214	-0.390	-	< -0.575	0.080
10	Metoxuron	-0.390	-0.212	-0.849	0.332	0.432
11	Monolinuron	-0.849	-0.249	-0.490	-0.390	0.215
12	Monuron	-0.249	0.463	-0.390	0.166	0.189

As example the interval disagreement for the Buturon between its experimentally derived interval and its model calculated interval on the Log(1/EC_{90}) is calculated as follows:

$$\delta_{Buturon\ EC90} = \left|-0.849 + 0.490\right| + \left|-0.212 + 0.249\right| = 0.396 \quad (38)$$

The standardised interval disagreement is then derived according to equation 1.32 as:

$$\delta^*_{Buturon\ EC90} = \frac{0.396}{(1.109 + 1.214)} = 0.170 \quad (39)$$

Overall model quality

By comparing the experimentally derived intervals with the calculated ones, an average disagreement has been calculated on each response:

$$\bar{\delta}_{Log(1/EC01)} = 0.221 \ \bar{\delta}_{Log(1/EC10)} = 0.180 \ \bar{\delta}_{Log(1/EC50)} = 0.144 \ \bar{\delta}_{Log(1/EC90)} = 0.164 \quad (40)$$

The average disagreement between the quantitative experimental values and their derived intervals has been calculated:

$$\tilde{\delta}_{\text{Log(1/EC01)}}=0.262 \quad \tilde{\delta}_{\text{Log(1/EC10)}}=0.270 \quad \tilde{\delta}_{\text{Log(1/EC50)}}=0.235 \quad \tilde{\delta}_{\text{Log(1/EC90)}}=0.223 \tag{41}$$

The uncertainty increase due to the replacement of a metric scale with an ordinal scale, calculated as arithmetic mean on all the four experimental attributes, is equal to 0.248.

For each response, the model quality has been evaluated by complement of the average disagreement between experimental and calculated intervals (Q_r):

$$Q_{\text{Log(1/EC01)}}=0.779 \qquad Q_{\text{Log(1/EC10)}}=0.820 \tag{42}$$
$$Q_{\text{Log(1/EC50)}}=0.856 \qquad Q_{\text{Log(1/EC90)}}=0.836$$

The overall ranking model quality, i.e. taking into account all the four responses, has been evaluated from the above parameters by arithmetic mean (Q_T), geometric mean (Q_G) and by the minimum value obtained on the four responses (Q_M):

$$Q_T=0.823 \qquad Q_G=0.822 \qquad Q_M=0.779 \tag{43}$$

The present case study reveals that partial order ranking and its modelling by structural molecular descriptors provides an attractive alternative to conventional QSAR modelling tools. The method appears, from a mathematical point of view, robust and transparent. It is thus possible using partial ranking techniques to develop ranking models and it is suggested that ranking models have a general potential in the area of risk assessment of environmentally hazardous chemicals. However, further analyses of the proposed method appear appropriate to investigate validation techniques suitable for ranking models and to evaluate the potential of ranking models for QSAR modelling.

Conclusions

Being based on elementary methods of Discrete Mathematics, ranking methods are a very useful and simple tool of QSAR modelling for exposure analyses and risk assessment, not to substitute conventional statistics but to supplement them. A complete procedure to perform a ranking model has been here proposed, based on the following main steps: experimental and model ranking development, comparison of the experimental and model rankings to evaluate model reliability, and finally interval estimations to provide experimental ranking from the ranking model obtained. In order to allow processing of data described by a wide set of variables the Genetic Algorithm (GA-VSS) approach has been proposed as the variable selection method. It is worthwhile to highlight that the procedure proposed can be located between fitting and predictive approaches, since the interval estimation and the model validation appear combined in one step. In fact, the model calculated intervals are obtained by deleting one element at a time from the model ranking, and using the remaining training set elements to calculate the model intervals of the deleted element from the model ranking. Thus, it is similar to a leave – one – out cross validation procedure (LOO technique), where each element is taken away, one at a time and the response for the deleted element is calculated from the model. In ranking model searching, the validation is not performed during the evolutionary optimisation procedure, but the model predictive ability is simulated once the model has been defined. The approach proposed seems, from a mathematical point of view well grounded. However, further analyses of the interval estimation procedure as well as of the uncertainty evaluation are required. Moreover, one of the main theoretical aspect not yet fully investigated concerns the search for validation techniques suitable for ranking models.

Acknowledgement: The present work was sponsored by the Commission of the European Communities (R&D project BEAM EVK1-1999-00012)

References

Backhaus T, Altenburger R, Arrhenius A, Blanck H, Faust M, Finizio A, Gramatica P, Grote M, Junghans M, Meyer W, Pavan M, Porsbring T, Scholze M, Todeschini R, Vighi M, Walter H, Grimme LH (2003) The BEAM-project: prediction and assessment of mixture toxicities in the aquatic environment. Continental Shelf Research 23:1757-1769

Backaus T, Faust M, Scholze M, Gramatica P, Vighi M, Grimme LH (2004) Joint Algal Toxicity of Phenylurea Herbicides is Equally Predictable by Concentration Addition and Independent Action. Environmental Toxicology and Chemistry 23:258-264

Balaban AT, Ciubotariu D, Medeleanu M (1991) Topological Indices and Real Vertex Invariants Based on Graph Eigenvalues or Eigenvectors. J Chem Inf Comput Sci 31:517-523

Bath PA, Morris CA, Willet P (1993) Effects of Standardization on Fragment-Based Measures of Structural Similarity. J Chemom 7:543-550

Bonchev D (1983) Information Theoretic Indices for Characterization of Chemical Structures. Research Studies Press: Chichester, UK

Broto P, Moreau G, Vandycke C (1984) Molecular Structures: Perception, Autocorrelation Descriptor and SAR Studies. Autocorrelation Descriptor, Eur J Med Chem 19:66-70

Brüggemann R, Zelles L, Bai QY, Hartmann A (1995) Use of Hasse Diagram Technique for Evaluation of Phospholipid Fatty Acids Distribution as Biomarkers in Selected Soils. Chemosphere 30:1209-1228

Brüggemann R, Bartel HG (1999) A Theoretical Concept to Rank Environmentally Significant Chemicals. J Chem Inf Comput Sci 39:211-217

Brüggemann R, Pudenz S, Carlsen L, Sørensen PB, Thomsen M, Mishra RK (2001a) The use of Hasse Diagrams as a Potential Approach for Inverse QSAR. SAR and QSAR in Environmental Research 11:473-487

Brüggemann R, Halfon E, Welz G, Voigt K, Steinberg CEW (2001b) Applying the Concept of Partially Ordered Sets on the Ranking of Near-Shore Sediments by a Battery of Tests. J Chem Inf Comput Sci 41:918-925

Carlsen L, Sørensen PB, Thomsen M (2001) Partial Order Ranking-based QSAR's: estimation of solubilities and octanol-water partitioning. Chemosphere 43:295-302

Carlsen L, Sørensen PB, Thomsen M, Brüggemann R (2002a) QSAR's Based on Partial Order Ranking, SAR and QSAR in Environmental Research 13:153-165

Carlsen L, Lerche DB, Sørensen PB (2002b) Improving the Predicting Power of Partial Order Based QSARs through Linear Extensions. J Chem Inf Comput Sci 42:806-811

Commission Regulation (EC) No 1488/94 laying down the principles for assessment of risks to man and the environment of existing substances in accordance with Council Regulation (EEC) No 793/93

Consonni V, Todeschini R, Pavan M (2002) Structure / Response Correlation and Similarity / Diversity Analysis by GETAWAY Descriptors. Part 1. Theory of the Novel 3D Molecular Descriptors. J Chem Comput Sci 42:693-705

Devillers J, Balaban AT (2000) Topological Indices and Related Descriptors in QSAR and QSPR. Gordon & Breach: Amsterdam, The Netherlands

EC 2002. White Paper on the Strategy for a new Future Chemicals Policy. Brussels: European Commission. Available: http://europa.eu.int/comm/environment/chemicals/whitepaper.htm.

Estrada E (1995) Edge Adjacency Relationships and a Novel Topological Index Related to Molecular Volume. J Chem Inf Comput Sci 35:31-33

European Commission Directive 96/67EEC of 20 July 1993. Laying down the principles for assessment of risks to man and the environment of substances notified in accordance with Council Directive 67/458/EEC

Faust M, Altenburger R, Backaus T, Boedeker W, Gramatica P, Hamer V, Scholze M, Vighi M, Grimme LH (2001) Predicting the Joint Algal Toxicity of Multi-component s-Triazine Mixtures at Low-effect Concentrations of Individual Toxicants. Aquatic Toxicology 56:13-32

Faust M, Altenburger R, Backaus T, Blanck H, Boedeker W, Gramatica P, Hamer V, Scholze M, Vighi M, Grimme LH (2003) Joint Algal Toxicity of 16 Dis-

similarly Acting Chemicals is Predictable by the Concept of Independent Action. Aquatic Toxicology 63:43-63
Gálvez J, Garcìa R, Salabert MT, Soler R (1994) Charge Indexes. New Topological Descriptors. J Chem Inf Comput Sci 34:520-525
Gálvez J, Garcìa-Domenech R, De Julián-Ortiz V, Soler R (1995) Topological Approach to Drug Design. J Chem Inf Comput Sci 35:272-284
Gramatica P, Vighi M, Consolaro F, Todeschini R, Finizio A, Faust M (2001) QSAR Approach for the Selection of Congeneric Compounds with a Similar Toxicological Mode of Action. Chemosphere 42:873-883
Goldberg DE (1989) Genetic Algorithms in Search, Optimization and Machine Learning. Addison-Wesley, Massachusetts, MA
Halfon E, Reggiani MG (1986) On Ranking Chemicals for Environmental Hazard. Environ Sci Technol 20:1173-1179
Halfon E (1989) Comparison of an Index Function and a Vectorial Approach Method for Ranking of Waste Disposal Sites. Environ Sci Technol 23:600-609
Halfon E, Brüggemann R (1998) On Ranking Chemicals for Environmental Hazard. Comparison of methodologies. Proceedings of the Workshop on Order Theoretical Tools in Environmental Sciences 11-48
Harrington EC (1965) The Desirability Function, Industrial Quality Control. 21:494-498
Hemmer MC, Steinhauer V, Gasteiger J (1999) Deriving the 3D Structure of Organic Molecules from Their Infrared Spectra. Vibrat Spect 19:151-164
Hendriks MMWB, Boer JH, Smilde AK, Doorbos DA (1992) Multicriteria Decision Making. Chemom Intell Lab Syst 16:175-191
Hocking RR (1976) The Analysis and Selection of Variables in Linear Regression. Biometrics 32:1-49
HYPERCHEM Rel 4 for Windows (1995) Autodesk Inc Sausalito CA. USA
Keller RH, Massart DL (1991) Multicriteria decision making: a case study, Chemom Intell Lab Syst 175-189
Kendall MG (1948) Rank Correlation Methods Charles Griffin & Co, London. 195 pp 202-204
Kier LB, Hall LH (1986) Molecular Connectivity in Structure-Activity Analysis. Research Studies Press - Wiley, Chichester (UK) pp 262
Leardi R, Boggia R, Terrile M (1992) Genetic Algorithms as a Strategy for Feature Selection. Journal of Chemometrics 6:267-281
Leardi R (1994) Application of Genetic Algorithms to Feature Selection Under Full Validation Conditions and to Outlier Detection. J Chemom 8:65-79
Leardi R (1996) Genetic Algorithms in Feature Selection. In Genetic Algorithms in Molecular Modeling. Principles of QSAR and Drug Design. Vol 1 (Devillers, J., ed). Academic Press, London (UK) pp 67-86
Lewi PJ, Van Hoof J, Boey P (1992) Multicriteria Decision Making Using Pareto Optimality and PROMETHEE Prefernce Ranking. Chemom Intell Lab Syst 16:139-144

Luke BT (1994) Evolutionary Programming Applied to the Development of Quantitative Structure-Activity Relationships and Quantitative Structure-Property Relationships. J Chem Inf Comput Sci 34:1279-1287

Massart DL, Vandeginste BGM, Buydens LMC, De Jong S, Lewi PJ, Smeyers-Verbeke J (1997) Handbook of Chemometrics and Qualimetrics: Part A, Amsterdam, Chapter 26 pp 783-803

Miller AJ (1990) Subset Selection in Regression. Chapman & Hall, London (UK) pp 230

Moock TE, Grier DL, Hounshell WD, Grethe G, Cronin K, Nourse JG, Theodosious J (1998) Similarity Searching in the Organic Reaction Domain. Tetrahedron Computer Methodology 1:117-128

Moreau G, Broto P (1980a) The Autocorrelation of a Topological Structure: A New Molecular Descriptor. Nouv J Chim 4:359-360

Moreau G, Broto P (1980b) Autocorrelation of Molecular Structures, Application to SAR Studies. Nouv J Chim 4:757-764

Pearlman RS, Smith KM (1998) Novel Software Tools for Chemical Diversity. In 3D QSAR in Drug Design - Vol. 2; Kubinyi H, Folkers G, Martin YC, Eds.; Kluwer/ESCOM: Dordrecht, The Netherlands pp 339-353

Pearlman RS (1999) Novel Software Tools for Addressing Chemical Diversity. Internet Communication http://www.netsci.org/Science/Combichem/feature08.html

Randić M (1995) Molecular Shape Profiles. J Chem Inf Comput Sci 35:373-382

Randić M (1996) Quantitative Structure-Property Relationship - Boiling Points of Planar Benzenoids. New J Chem 20:1001-1009

Rogers DJ, Tanimoto TT (1960) A computer Program for Classifying Plants. Science 132:1115-1118

Scholze M, Boedeker W, Faust M, Backaus T, Altenburger R, Grimme LH (2001) A general best-fit method for concentration response curves and the estimation of low-effect concentration. Environ Toxicol Chem 20:448-457

Schuur J, Gasteiger J (1996) 3D-MoRSE Code - A New Method for Coding the 3D Structure of Molecules. In: Software Development in Chemistry - Vol 10 (J. Gasteiger Ed). Fachgruppe Chemie-Information-Computer (CIC) Frankfurt am Main Germany

Schuur J, Gasteiger J (1997) Infrared Spectra Simulation of Substituted Benzene Derivatives on the Basis of a 3D Structure Representation. Anal Chem 69:2398-2405

Sørensen PB, Brüggemann R, Carlsen L, Mogensen BB, Kreuger J, Pudenz S (2003) Analysis of Monitoring Data of Pesticide Residues in Surface Waters Using Partial Order Ranking Theory. Environmental Toxicology and Chemistry 22:661-670

Todeschini R, Lasagni M, Marengo E (1994) New Molecular Descriptors for 2D- and 3D-Structures, Theory. J Chemom 8:263-273

Todeschini R, Gramatica P (1997) 3D-Modelling and Prediction by WHIM Descriptors. Part 5. Theory Development and Chemical Meaning of WHIM Descriptors. Quant Struct-Act Relat 16:113-119

Todeschini R, Consonni V, Mauri A, Pavan M (2004) MobyDigs: software for regression and classification models by genetic algorithms in Nature-inspired methods in chemometrics: genetic algorithms and artificial neural networks (R. Leardi Ed.), Chapter 5, Elsevier pp 141-167

Todeschini R, Consonni V, Mauri A, Pavan M (2004) DRAGON, rel. 5 for Windows; Talete srl: Milano Italy

Todeschini R, Consonni V (2000) Handbook of Molecular Descriptors, Wiley-VCH, Weinheim Germany p 667

Todeschini R, Consonni V, Mauri A, Pavan M (2003) RANA for Windows; Talete srl: Milano Italy

Wehrens R, Buydens LMC (1998) Evolutionary optimization: a tutorial, TrAC, Trends in Analytical Chemistry. 17(4):193-203

4 Decision support

The very idea behind partial order is to compare objects and by doing this to learn from them. Hence, it is obvious that partial order may also be useful in decision support. Indeed partial orders play implicitly an important role in many established multi-criteria decision aid models. Being aware that there is important literature about this topic, the motto of the chapters presented here is to let speak the partial order alone - as far as possible.

In the chapter of Simon et al., the water management strategies of the urban area Berlin-Potsdam are analyzed. Strategies and their expected success are evaluated on the basis of hydrological and hydrochemical indicators. Hence, a multi-attribute problem evolves and it is analyzed by means of the Hasse diagram technique (or more recently used: "partial order ranking"). Instead of indicators the different strategies ("scenarios") may qualitatively be evaluated by experts who had designed the strategies. Beside the background information about complex decision problems the reader will learn something about similarity of Hasse diagrams and about the concept of antagonistic indicators.

In the second chapter of Brüggemann et al. the approach of utility functions and of PROMETHEE is compared with that of HDT. As example chemicals are serving. The renewed interest for evaluation of chemicals, initiated by the European Commission's proposal for a new system for assessing and regulating chemicals, REACH, may support our decision. The main methodological tool is the use of Monte Carlo simulations and the comparison of the Monte Carlo simulations with the rank probability of chemicals. The comparison, albeit restricted to only two competing methods, suggests the advantageous partial order in decision problems.

Aspects of Decision Support in Water Management: Data based evaluation compared with expectations

Ute Simon[1*], Rainer Brüggemann[1], Stefan Pudenz[2], Horst Behrendt[1]

[1] Leibniz-Institute of Freshwater Ecology and Inland Fisheries
Müggelseedamm 310, D-12587 Berlin, Germany

[2] Criterion Evaluation and Information Management,
Mariannenstrasse 33, D-10999 Berlin, Germany

[*]e-mail: simon@igb-berlin.de

Abstract

In the cities of Berlin and Potsdam nine water management strategies (scenarios) were evaluated with respect to their ecological effects to the system of surface water. Scenarios were generated by combining different water management measures such as wastewater and storm water treatment. Indicators were qualitatively modelled as well as quantitatively evaluated by experts' knowledge. For decision support Hasse Diagram Technique (HDT) was used. The scenario modular structure increases the transparency of the evaluation process and brought up the question whether time and work consuming calculation of data by mathematical models is needed or experts' knowledge is sufficient for evaluation. To clarify this question, the results of two evaluation examples were compared: (a) data based and (b) experts expectations. Beyond the concept of antagonistic indicators the similarity-profile is introduces as a new tool to compare HDT evaluation results. Our study revealed that in the present investigation evaluation by expert knowledge is not satisfactory. The shift in the type of indicators from state to pressure and the effect of up scaling from local to regional may be the reason.

Introduction

In a research project about sustainable water management in the cities of Berlin and Potsdam (Germany), an interdisciplinary working group, including ecologists, landscape architects and civil engineers developed a framework to evaluate water management strategies (Steinberg et al. 2002, Weigert & Steinberg 2002). As an evaluation tool the Hasse Diagram Technique (Halfon & Reggiani 1986, Brüggemann et al. 2001 and 2003) was applied. Altogether nine water management strategies (scenarios) were evaluated with respect to their ecological effects to the system of surface water. The scenarios are considered of being composed of different modules describing measures for (A) hydrological boundary conditions, (B) waste water treatment, and (C) management of storm water. The modular structuring of scenarios follows the idea of Saaty (1994) to handle complex problems by dividing them into smaller, manageable compartments. While progressing the evaluation process, however, in our research project an unexpected side effect occurs. Members of the working group start arguing about being able to predict the evaluation result, even without using any modelled data. One reason was the modular structure of the scenarios, by which the transparency of the evaluation problem is increased and by which the impression might be given to know already which scenario will be the best.

To clarify the question whether indicator values based on calculated data by mathematical models are needed or solely knowledge of experts is sufficient to evaluate our water management strategies, we analyzed the results of both approaches, the data based evaluation and the evaluation by expert knowledge. The question about the need of modelled data is closely related to two topics, which are of general importance in every decision process: the efficient use of project resources - data modelling is time and work consuming - and the acceptance of the evaluation result. Stakeholders will hardly approve results, which distinctly disagree with their expectations (Lahdelma et al. 2000). The comparison of the evaluation result based on data calculated by the model MONERIS (Behrendt et al. 1999 and 2002) with the evaluation result based on data representing the experts' expectations was carried out by the HDT originated tools of antagonistic indicators and by a similarity-profile. The similarity-profile we introduce as a new approach to compare the evaluation results, namely the structures of Hasse Diagrams (HD) in a detailed and objective way.

Methods

Research Area

Object of research is the complex surface water system of the cities of Berlin and Potsdam (Fig. 1). To evaluate the ecological effects of the water management strategies on the surface waters, not only the evaluation of each indicator representing a certain scenario characteristic is of interest, but also where these patterns appear. Thus, to detect local effects of the scenarios, the water system has been split into 14 sections, each of which contributes its own characteristics to the decision procedure.

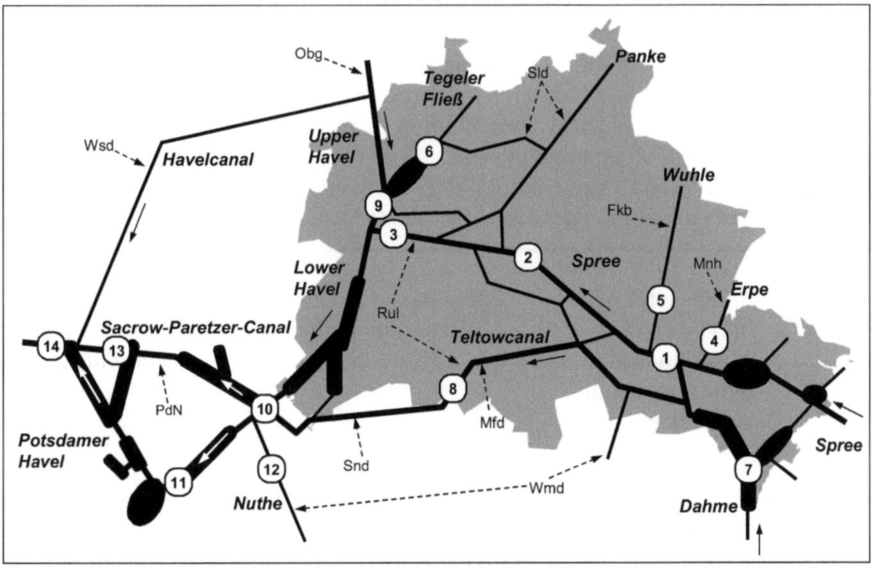

Fig. 1. Schematic diagram of the surface water system of Berlin and Potsdam. River sections: (1) Spree Köpenick (including Dahme), (2) Spree Mühlendamm, (3) Spree Sophienwerder, (4) Erpe (Neuenhagener Mühlenfließ), (5) Wuhle, (6) Inflow to lake Tegeler See, (7) Dahme Schmöckwitz, (8) Teltowkanal, (9) Upper Havel, (10) Lower Havel, (11) Havel Caputh, (12) Nuthe Babelsberg, (13) Sacrow-Paretzer-Kanal, (14) Havel Ketzin. Waste water treatment plants: Obg=Oranienburg, Sld=Schönerlinde, Fkb=Falkenberg, Mnh=Münchehofe, Rul=Ruhleben, Snd=Stahnsdorf, Mfd=Marienfelde, Wmd=Waßmannsdorf, Wsd=Wansdorf, PdN=Potsdam Nord. Dashed lines show wastewater pipe lines. Shaded area = administrative border of the city of Berlin

Water Management Strategies

Altogether, nine water management strategies, 1a, 1, 2, 3, 4, 5, 6i, 6ii and 6iii, in the following called scenarios, were evaluated. Each scenario consists of three modules comprising measures for: (A) hydrological boundary conditions, in particular the amount of water flowing into the area and its nutrient concentrations; (B) wastewater treatment, including the technical equipment of the wastewater treatment plants (wwtp), as well as the spatial and quantitative distribution of purified waste water; and (C) quality and quantity of storm water discharge into river section. The current state represented by scenario 1a is the reference for all other scenarios. The measures, belonging to each scenario are summarised in Table 1. A more detailed description can be found in Simon et al. (2004a, 2004b).

Table 1. Water management strategies. Abbreviations of wastewater treatment plant names are given in Fig. 1. Example how to read Table 1: Scenario 2 includes the following measures: Module (A) reduced amount of water flowing towards Berlin, carrying same nutrient concentration as in the current state. Module (B) technical upgrade of all operating wwtps. Three Wwtps, namely Falkenberg (Fkb), Marienfelde (Mfd) and Oranienburg (Obg) are assumed to be shut down. Module (C) current state of storm water discharge into the surface waters

Abbr. of scenarios	Measures of module (A): hydrological boundary conditions	Measures of module (B): wastewater treatment		Measures of module (C): entry of storm water
		Purification technique	shut down of wwtps	
1a	current state (average of the years 1993-1997)			
1	reduced amount of water			
2		technical upgrade	Fkb, Mfd, Obg	
3	reduced amount of water and lower nutrient concentrations	advanced waste water treatment (micro-filtration)		emission 50% reduced
4			Mfd, Odg	
5			Fkb, Mfd, Obg	
6i		alternative sanitary technique		
6ii			Mfd, Obg, Mnh, Snd	
6iii			Mfd, Obg, Mnh, Sld	

Indicators of data based evaluation

In the first example, the data based evaluation, the nine scenarios (Table 1) are characterised by a set of four indicators. For better recognition the „dat" subscript is added to the indicator abbreviations:

Q_{dat}: Reduction of the discharge in river sections
P_{dat}: Difference of phosphorus concentration from target concentration
N_{dat}: Concentration of total nitrogen
S_{dat}: Short-term pollution of surface waters by storm water

Each of these indicators gets numerical values separately for the 14 river sections. The Q_{dat}, P_{dat} and N_{dat} indicators have been calculated with the model MONERIS, which is described in (Behrendt et al. 1999 and 2002). These indicators are metric quantities. Although the quantitative calculation of the S_{dat} indicator values is included in the MONERIS modelling, the S_{dat} indicator is evaluated qualitatively. The reason is that quantitative effects turned out to be not significant within the uncertainty of the model. The S_{dat} indicator is evaluated best, if there is no direct influence of the river sections by storm waters at all. The reduction of emissions by storm water events of about 50% (SenSUR 1999) is evaluated middle and the present state is evaluated worse. Thus the S_{dat} indicator is considered as an ordinal quantity. The simultaneous consideration of quantities of different scaling levels (metric together with ordinal ones) is one of the core advantages of HDT. Note, that for consistent orientation of indicators, here high values always represent a bad evaluation. Consequently, a high value in one of the measures implies automatically a rather high rank (bad evaluation). As each of the 14 river sections is evaluated separately, a large matrix of 9 scenarios multiplied by 4 indicators multiplied by 14 river sections equals 504 entries is obtained, which we would like to introduce as the data based evaluation matrix.

Indicators of evaluation by experts' knowledge

The modular structure of the scenarios as described in section Water Management Strategies facilitates to predict the ecological effects of the measures within each module, at least as an ordinal quantity. Therefore, an evaluation solely based on the knowledge of experts, here by members of the project group, becomes possible. To transform the experts' expectations into a data matrix, indicators for qualitative evaluation of the measures within each module are defined (see below). For consistent comparison of evaluation results, high indicator values again represent a bad evaluation. Note, that interactions among measures cannot be considered. Indicators representing the experts' expectations are labelled with the „exp" index. The indicator values defined by members of our project group are shown in Table 2.

Indicators to evaluate measures of module (A): Hydrological boundary conditions:

Ha_{exp}: Amount of water entering the research area. The indicator is evaluated in two classes. A good evaluation of the present state (indicator value 0) and a worse evaluation in case of scenarios 1 to 6iii, comprising a reduced amount of water entering Berlin (indicator value 1).

Hq_{exp}: Quality of water entering the research area. Scenarios 1a, 1 and 2 represent the present state and got a bad evaluation (indicator value 1), whereas the scenarios 3 to 6iii are evaluated better (indicator value 0) because of lower nutrient concentration due to an improved technical standard of wwtps in the catchments area upstream of Berlin.

Table 2. Evaluation matrix based on expectations. High values are representing a bad evaluation

Indicator/ Scenario	P_{exp}	N_{exp}	S_{exp}	Ha_{exp}	Hq_{exp}
1a	2	2	1	0	1
1	2	2	1	1	1
2	1	2	1	1	1
3	0	1	0	1	0
4	0	1	0	1	0
5	0	1	0	1	0
6i	0	0	0	1	0
6ii	0	0	0	1	0
6iii	0	0	0	1	0

Indicators to evaluate measures of module (B): waste water treatment:

P_{exp}: phosphorus emission of the wwtps. Scenarios 1a, and 1 get the highest indicator value (2), representing the worse evaluation. Scenario 2 comprises technically upgraded wwtps with a reduction of phosphorus emissions. Consequently it is given the indicator value 1. Scenarios 3 to 6iii are evaluated best (indicator value 0). Advanced wastewater treatment and alternative sanitary technique will reduce phosphorus emissions of the plants significantly.

N_{exp}: nitrogen emission of the wwtps. Scenarios 1a, 1 and 2 are evaluated equivalently worse (indicator value 2). Scenarios 3 to 5 assuming technical upgrade of all wwtps is evaluated middle (indicator value 1). Scenarios suggesting alternative sanitation technique (6i to 6iii) are evaluated best (indicator value 0). Due to separation of urine and faeces, the discharge of nitrogen into the surface water will be drastically reduced.

Indicators to evaluate measures of module (C): short term pollution of surface waters by storm water

S_{exp}: A bad evaluation (indicator value 1) is given to scenarios 1a, 1 and 2, representing the present state. Scenarios 3 to 6iii are evaluated better (indicator value 0), as storm water events are reduced for about 50%, according to the Sewage Disposal Plan (SenSUR 1999).

Hasse Diagram Technique and the concept of antagonistic indicators

The Hasse Diagram Technique (HDT) is a method to sort options (here scenarios) evaluative with respect to all indicator values simultaneously, however without aggregation of indicators. The HDT evaluation is based on a simple ≤ comparison of the options indicator values within every single indicator. For consistent evaluation, all indicator values have to be oriented uniformly: for instance, high values always have to represent a bad evaluation. More technical details can be found for example in Brüggemann and Carlsen, p. 61 and in references Halfon & Reggiani (1986), Brüggemann et al. (2001, 2003) and Brüggemann and Drescher-Kaden (2003). The evaluation result is depicted in a so-called Hasse Diagram (HD). Connective vertical lines show that the indicator values of the options will simultaneously increase (upwards) or decrease (downwards). Note that the evaluation of options is only deduced following exactly one vertical direction. Options not being connected with a sequence of vertical lines are not comparable with each other because of antagonistic indicators. For explanation let us consider two incomparable objects: There is at least one pair of indicators in which one indicator is better evaluated with respect to one option and worse in the other. The other indicator is evaluated in the reverse sense. Thus the incomparability among objects indicates differences in their profile of characteristic properties and can be analysed by the HDT-originated tool of antagonistic indicators, which formalizes the set of advantages and disadvantages with respect to each indicator. Note, that more than two indicators can be necessary to explain the complete separation of any of two objects or group of objects. The reason is possible overlapping of the antagonistic indicator intervals. Overlapping indicator intervals can explain the incomparability among objects only to a certain percentage. Consequently, more than two antagonistic indicators are needed to explain total separation of objects (Simon 2003).

By automated identification of antagonistic indicators with the WHASSE software, immanent conflicts in the evaluation matrix can be discovered in a convenient way, and thus the advantages and disadvantages of each option under discussion can be named. The precise knowledge about antagonisms supports the stakeholders' decision process as further discussions can focus on these immanent evaluation conflicts. The methodologically strategy how to solve these conflicts is one of the most crucial steps of the evaluation process (Strassert 1995).

Similarity of Hasse Diagrams

Similarity indices are well known in statistical literature and also discussed in this book by Pavan et al., p. 181, especially their S(E,M)-index. Mostly similarity indices, however only provide highly aggregated information, and for that reason they imply a lost of information. For detailed comparison of HDT results, visualized by Hasse Diagrams (HD), we introduce a new tool, the similarity-profile. By the similarity-profile the structural accordance and discordance between any two HDs can be described in detail. As explained in the chapter by El-Basil, p. 3 and in chapter Brüggemann and Carlsen, p. 61, Hasse Diagrams are graph theoretical structures. Therefore the comparison of evaluation results is not only to relate one object to other ones, but also to investigate the graph as a whole. In that sense we are speaking of a structure of an evaluation result.

Our similarity-profile is adapted from an approach proposed by Sørensen et al. (2004), see also the more general discussion about correlation in chapter by Sørensen et al., p. 259. The relation of each option to another one is written down in a matrix, separately for both diagrams. Altogether four possible relations can occur:

- $>$ scenario x is evaluated better than scenario y.
- $<$ scenario x is evaluated worse than scenario y.
- $\sim$ scenario x is equivalent to scenario y and
- $||$ scenario x is incomparable to scenario y.

Consequently, maximal 16 combinations of relations can be found if two diagrams are compared with each other. The similarity-profile however describes four different kind of relations:

(1) Parallel relations of options in both diagrams, such as < <, > > and ~ ~. E.g., the parallel relation < < means that scenario x < scenario y in HD_1, and scenario x < scenario y in HD_2. Parallel order relations indicate a similar ranking of options in the two compared Hasse Diagrams and thus there is no evaluation conflict.
(2) Indifferent relations of options in both diagrams, such as > ~, < ~, and ~ >, ~ <. While in one Hasse Diagram the options are evaluated equivalent, in the other Hasse Diagram the options are ranked. Thus „indifferent" shows a difference in the evaluation, but not a conflict as strong as „anti-parallel".
(3) Anti-parallel relations of options in both diagrams, such as < > and < >. They show contrary rankings of options in two Hasse Diagrams, and thus discover a strong evaluation conflict.
(4) Uncertain relations of options in both diagrams arise, when an option is incomparable to others in at least one of the diagrams. Uncertainty includes: || <, || >, || ~, || || and < ||, > ||, ~ ||. To generate an incomparability, there must be at least one pair of antagonistic indicators. For that reason „uncertainty" expresses also a strong conflict of the compared evaluation results.

The similarity-profile can be generated by counting all relations of each of the four groups and can be visualized by a bar plot.

Evaluation Results

The Hasse Diagrams, visualising the results of the two evaluation examples are shown in Fig. 2. By the data based evaluation (Fig. 2, left diagram) the three scenarios 4, 6i and 6ii are identified as favourable, whereas by expectations of experts (Fig. 2, right diagram) four scenarios results as best possible solutions. These are 1a, 6i, 6ii and 6iii. However the scenarios 6i, 6ii and 6iii are equivalent, i.e. they have got an identical evaluation in all indicators. The incomparability between the winner scenarios within each evaluation example can be explained by analysing the antagonistic indicators, revealing the advantages and disadvantages of each scenario (Fig. 3, left hand side). In case of the data based evaluation two reasons were identified to cause the incomparability: (1) thematic antagonisms occur because different indicators such as phosphorus concentration (P_{dat}) and discharge reduction (Q_{dat}) are involved. And (2) spatial antagonisms, as different river sections such as the tributaries Erpe (section no. 4) and Wuhle (section no. 5) are affected. In contrast, in the evaluation based on experts expectations incomparability is only caused by thematic antagonisms: The higher amount of water entering the research area (Ha_{exp}) is identified of being the only advantage of scenario 1a, whereas the four indicators phosphorus (P_{exp}), nitrogen (N_{exp}), short term pollution (S_{exp}) and the quality of water entering the research area (Hq_{exp}) are evaluated better in the scenarios 6i, 6ii and 6iii (Fig. 3, right hand side).

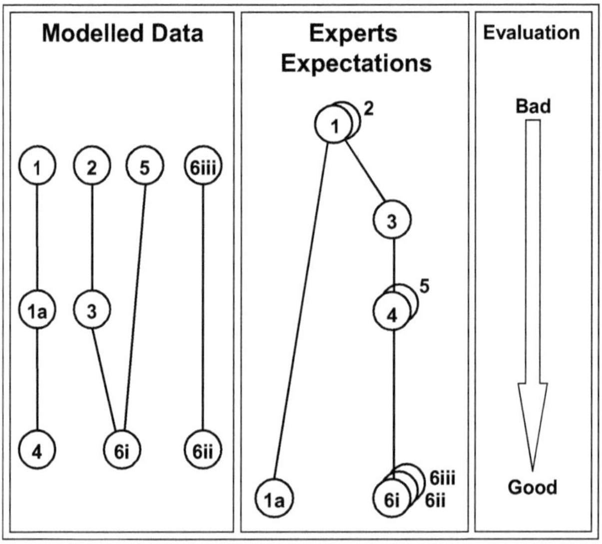

Fig. 2. Results of the two evaluation examples. Circles symbolize scenarios. For description of scenario abbreviations see Table 1. Segments of circles symbolize equivalent evaluation of scenarios. Left diagram: result based on modelled data. Right diagram: result based on experts knowledge

Beyond differences between both evaluation examples concerning the scenarios, which are evaluated, the best, optical inspection of both Hasse Diagrams reveals further obvious dissimilarities. These structural differences can be described in more detail and objectivity by the similarity-profile (Fig. 4). When the structure of both HD's are compared, only few parallel relations (about 17%) can be found, indicating total agreement in the evaluation of scenarios in both evaluation examples. There are also only few indifferent relations (about 2.7%), showing differences in the evaluation of scenarios: while in one HD the scenarios are ranked, they are evaluated equivalent in the other HD. Indifferent relations can be addressed to the evaluation result based on experts knowledge (Fig. 2, left diagram). There are no anti-parallel relations, which would indicate severe evaluation conflicts because of converse ranking of options in both HD's. However, there is a clear dominance of uncertainties (about 80%), discovering severe disagreements in how the scenarios are ranked in both evaluation examples. This high amount of uncertainties is caused by incomparability among scenarios and can be addressed to the existence of three hierarchies in the databased Hasse Diagram. These differences, however, need to be traced back in more detail. They discover conflicts between modelled data and experts expectation and therefore they will reduce the acceptance of the evaluation result.

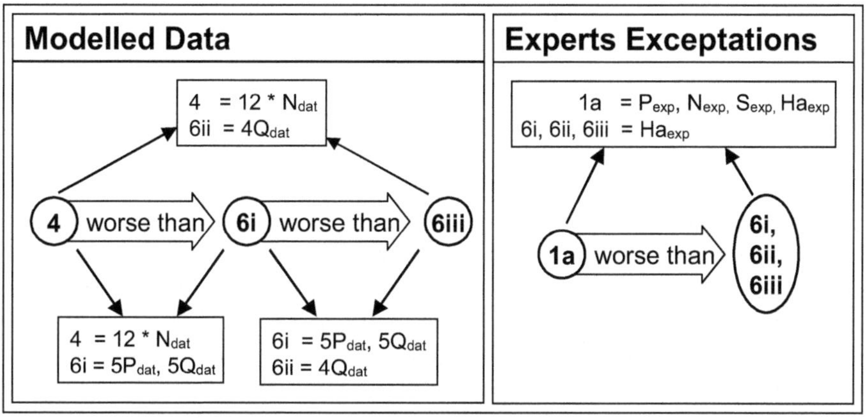

Fig. 3. Antagonistic indicators of favourable scenarios. Note that here only the better-evaluated indicators of the antagonisms are given. They represent the advantage of one of the two incomparable options, which implies to be the disadvantage of the other one. An example how to read the graphic: In the first evaluation example scenario 6ii is incomparable to 6i (||), because in 6ii the indicators phosphorus load and discharge reduction, both concerning the river section Wuhle ($5P_{loc}$, $5Q_{loc}$) are evaluated better than in 6i. In contrast, in 6i the indicator discharge reduction concerning the river section Erpe ($4Q_{loc}$) is evaluated better than in 6ii

Discussion

In our study, the comparison of the evaluation results based on modelled data and based on experts expectations, prove the need to calculate data by a mathematical model to obtain sufficiently detailed and precise results. The insufficiency of evaluation based on expectations can be mainly addressed to two topics: (1) A shift in the type of indicators from pressure to state and (2) a shift in the geographical scale from local to regional.

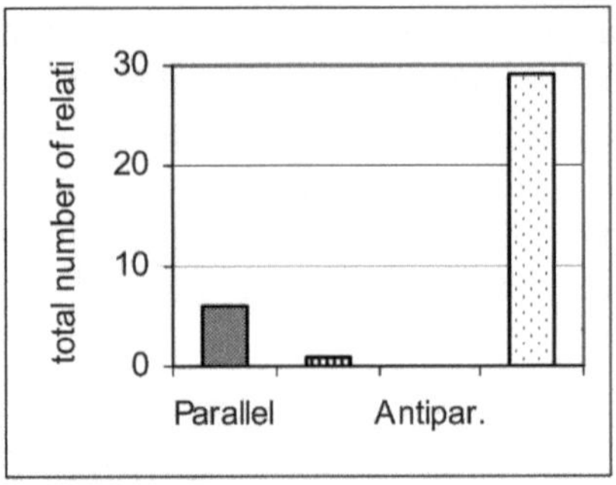

Fig. 4. Similarity-profile of the evaluation examples

According to the P-S-R-approach of the OECD, indicators can be classified into three basic groups (OECD 1994): Pressure indicators (P) are describing the causing factors such as emissions of technical assets. State indicators (S) represent the present state, for example the trophic state of an aquatic ecosystems. Response indicators (R) are mapping reactions of the society to a certain problem. In our case study, all the data input by experts' knowledge can be characterized as pressure indicators. For example: indicators characterizing the impact of waste water treatment plants on the surface waters, only provide information about emission of phosphorus and nitrogen and the amount of waste water which is discharged into river sections. In contrast, by using the model MONERIS, the final nutrient concentrations and discharges of the river sections are calculated. Thus the original input data representing pressure indicators are transformed into state indicators describing resulting effects. In addition, by the modelling of data the interactions between the original pressure indicators can be implemented. The final nutrient concentration for example, results from multiple sources such as initial level of water pollution and emissions of several different pathways such as wastewater treatment and storm water.

When the evaluation is based on expert knowledge, only pressure indicators can be relevant. Whereas precise information about emissions into the surface waters can be available, the prediction or estimation of the resulting concentrations of substances as well as the final discharge is almost impossible. However, under certain conditions it might be manageable to predict state indicators sufficiently. For example if there is only one source of emission and only one river section. In our case study this for instance was true for the river section Wuhle. Experts with precise local knowledge were able to predict the effects of the shut down of the wwtp Falkenberg precisely, including the resulting degradation of the water quality. If advanced wastewater treatment is assumed, the discharge of sewage from the wwtp into the river Wuhle will actually cause a dilution of the nutrient concentration. However, having more than one source of emission and a complex system of surface water including tributaries, a precise prediction of effects (state indicators) is impossible to derive from pressure indicators alone.

Beyond the type of indicators (pressure-state-respond) also different geographical scales are addressed when data based evaluation is compared with that of experts' knowledge. Data based evaluation can be referenced to a local scale. Thus spatial effects can be detected in more detail, even though the definition of the river sections is determined on the one side by the official monitoring program, surveying the quality and quantity of the flowing waters, and on the other side by the need to observe the influence of tributaries. In contrast, data based on information about emissions (pressure indicators) can be only addressed to the directly affected river sections. Consequently the evaluation will be incomplete with respect to the entire system of surface waters. Alternatively, emissions can be summarily evaluated, which rather equals a regional scale. Thus up scaling is taking place, because as stated before it is not possible to detect spatially differentiated effects by expectations if a large system of surface water is investigated.

The topics discussed above showed that the decision whether effort of time and manpower to model data is legitimated, or experts' expectations are sufficient for evaluation, largely depends on the complexity of the problem. Complexity might be related to the geographical extend of the study, e.g., a complex system of surface water, or might be caused by the variety of influencing variables, such as social or political interests to be represented by indicators. In our study the complexity of a local referenced evaluation required data calculated by a mathematical model. Therefore the effort to model data can be legitimated by the advantage of precise information. Experts' expectations would provide insufficient information to evaluate the effects of water management strategies with respect to the surface water system of the cities of Berlin and Potsdam. The corresponding loss of information is expressed in the similarity-profile high number of uncertainties. In some cases there can be a better efficiency in using project resources such as time, manpower and knowledge, if the evaluation is solely based on experts knowledge. The evaluation of single river sections such as the river Wuhle, which is briefly described above, is an example. As expectations provide sufficient precise results, modelling of data cannot be legitimated by an increase of information. However, the efficiency in using project resources will be not detectable in the similarity index, as both evaluation results, data based and expectations, should be in good agreement.

Between the unambiguous extremes of a total preference of modelled data and expectations respectively, there is a wide range where both approaches increasingly conform with each other with respect to the agreement of their results and efficiency of project resources respectively. However, this range might include that the choice of the method is not adequate to the problem. For example: evaluation on a regional scale is inappropriate to decide about local referenced water management strategies, independently from the methodological question whether modelled data or experts expectations should be the data base. Again, such a spatial inadequacy is not detectable by a tool such as the similarity profile, as data based evaluation as well as expectations both might provide sub-optimal results. The specific problem of finding the adequate scale and method including the generation of a complete set of indicators and a broad variety of options cannot be supported by methodological tools but has to be solved discursive by the stake holders and public respectively. The latter is concerned to the topic of participation (Lahdelma et al. 2000, Munda 2004, De Marchi & Ravetz 2001). Methodological tools such as the similarity-profile, however can help to analyze and to explain discrepancies between experts expectations and modelled data. Mediation between both evaluation results can be important to increase both, transparency and acceptance of decisions.

Conclusions and Prospect

The comparative study of evaluation results based on modelled data and obtained by experts expectations respectively, revealed that in the present investigation the evaluation by expert knowledge is not satisfactory. Even though there are agreements in both results, such as scenarios which are identified as potential winners in both approaches, the dominance of differences (disagreements) prove the need of modelled data to obtain sufficiently detailed and precise results. The insufficiency of evaluation based on expectations can be mainly addressed to two topics: A shift in the type of indicators from pressure to state and a shift in the geographical scale from local to regional.

The modular structure of the water management strategies (scenarios) facilitates to solve the complex problem by split it into manageable parts (Saaty 1994). Consequently transparency is increased as demanded by the "Lokale Agenda". The model MONERIS and the Hasse Diagram Technique respectively support this strategy. MONERIS, which is of modular structure as well, facilitates to adapt the input data to changing conditions such as adding or modification of scenarios. Furthermore in practical application the model is holding excellent balance between generalization and detailed information. HDT is providing convenient tools for data analysis.

It is somehow paradox, that the good transparency supported by modular scenarios gives the impression that expectations will offer sufficient evaluation results, superfluous the need to model data. Anyway, discrepancies between expected and modelled evaluation results need to be removed, as stakeholders will hardly accept an evaluation result, which extensively disagrees with their expectations. HDT-originated analysis tools such as the antagonistic indicators and the similarity profile proved to be helpful in such conflicts. Thus beyond the application of HDT in the field of multicriteria decision aid (MCDA) the approach might be a helpful tool to mediate the whole decision process.

Acknowledgements

We thank the Deutsche Bundesstiftung Umwelt (DBU, AZ 12953) for financial support, and Mr. Behrendt and Mr. Opitz for modelling the data. We like to thank Ms. Gertraud Fendt for improving the language.

References

Behrendt H, Eckert B, Opitz D (1999) The Havel river a source of pollution for the Elbe river - the retention funktion of regulated river stretches. In: Senate Department of Urban Development (Senatsverwaltung für Stadtentwicklung, Umweltschutz und Reaktorsicherheit), Ed. Zukunft Wasser, Dokumentation zum 2. Berliner Symposium Aktionsprogramm Spree/Havel 2000 der Senatsverwaltung für Stadtentwicklung, Umweltschutz und Reaktorsicherheit, Berlin: 33-39

Behrendt H, Huber P, Kornmilch M, Opitz D, Schmoll O, Scholz G, Uebe R (2002) Estimation of the Nutrient Inputs into River Basins - Experience from German Rivers. Regional Environmental Changes 3:107-117

Brüggemann R, Welzl G, Voigt K (2003) Order Theoretical Tools for the Evaluation of Complex Regional Pollution Patterns. J Chem Inf Comp Sc 43:1771-1779

Brüggemann R, Drescher-Kaden U (2003) Introduction into model supported evaluation of environmental chemicals (Einführung in die modellgestützte Bewertung von Umweltchemikalien). Berlin, Springer-Verlag: 1-513

Brüggemann R, Halfon E, Welzl G, Voigt K, Steinberg C (2001) Applying the Concept of Partially Ordered Sets on the Ranking of Near-Shore Sediments by a Battery of Tests. J Chem Inf Comp Sc 41(4):918-925

Brüggemann R, Bücherl C, Pudenz S, Steinberg C (1999) Application of the concept of Partial Order on Comparative evaluation of environmental chemicals. Acta hydrochim Hydrobiol 27:170-178
De Marchi B, Ravetz JR (2001) Participatory approaches to environmental policy. Concerted Action EVE, Policy Research Brief 10:1-25
Halfon E, Reggiani MG (1986) On Ranking Chemicals for Environmental Hazard. Environ Sci & Technol 20:1173-1179
Lahdelma R, Salminen P and Hokkanen J (2000) Using Multicriteria Methods in Environmental Planning and Management. Environmental Management 26:595-605
Munda G (2004) Social multi-criteria evaluation: methodological foundations and operational consequences. European Journal of Operation Research 158:662-677
OECD (Ed.) Environmental indicators - OECD Core Set, Paris, 1994
Saaty TL (1994) How to Make a Decision: The Analytic Hierarchy Process. Interfaces 24:19-43
SenSUR (Senatsverwaltung für Stadtentwicklung, Umweltschutz und Reaktorsicherheit) (Ed.) (1999) Abwasserbeseitigungsplan. Berlin: 1-108
Simon U, Brüggemann R, Pudenz S (2004) Aspects of Decision Support in Water Management – Example Berlin and Potsdam (Germany) I: Spatially Differentiated Evaluation. Water Research 38:1809-1816
Simon U (2003) Multikriterielle Bewertung wasserwirtschaftlicher Maßnahmen aus gewässerökologischer Sicht - Beispiel Berlin. Dissertation. Tenea. Berlin: 1-130
Simon U, Brüggemann R, Pudenz S (2004) Aspects of Decision Support in Water Management - Berlin and Potsdam (Germany) II: Improvement of management strategies. Water Research 38:4085-4092
Sørensen PB, Gyldenkærne S, Lerche DB, Brüggemann R, Thomsen M, Fauser P and Mogensen BB (2004) Probability approach applied for prioritisation using multiple criteria. Cases: Pesticides and GIS. In: Sørensen PB, Carlsen L, Mogensen BB, Brüggemann R, Luther B, Pudenz S, Simon U, Halfon E, Bittner T, Voigt K, Welzl G, Rediske F (Ed.): Order Theoretical Tools in Environmental Sciences - Proceedings of the 5. Workshop, National Environmental Research Institute, NERI Technical Report No. 479, Denmark: 121-137
Steinberg C, Weigert B, Möller K, and Jekel M (Ed.) (2002) Nachhaltige Wasserwirtschaft - Entwicklung eines Bewertungs- und Prüfsystems. Berlin, Erich Schmidt: 1-311
Strassert G (1995) Das Abwägungsproblem bei multikriteriellen Entscheidungen - Grundlagen und Lösungsansatz unter besonderer Berücksichtigung der Regionalplanung. Frankfurt am Main: Peter Lang, Europäischer Verlag der Wissenschaften: 1-111
Weigert B, Steinberg C (2002) Sustainable development - assessment of water resource management. Wat Sci Techn 46:6-7

A Comparison of Partial Order Technique with Three Methods of Multi-Criteria Analysis for Ranking of Chemical Substance

Rainer Brüggemann*[1], Lars Carlsen[2], Dorte B. Lerche[3] and Peter B. Sørensen[3]

[1] Leibniz-Institute of Freshwater Ecology and Inland Fisheries
Müggelseedamm 310, D-12587 Berlin-Friedrichshagen, Germany

[2] Awareness Center
Hyldeholm 4, Veddelev, DK-4000 Roskilde, Denmark

[3] The National Environmental Research Institute
Department of Policy Analysis, Frederiksborgvej 399, DK-4000 Roskilde, Denmark

*Corresponding author: Phone: +49 30 64181666, Fax. +49 30 64181663 and e-mail: brg@igb-berlin.de

Abstract

An alternative to time-consuming risk assessments of chemical substances could be more reliable and advanced priority setting methods. Hasse Diagram Technique (HDT) and/or Multi-Criteria Analysis (MCA) provide an elaboration of the simple scoring methods. The present chapter evaluates HDT relative to two MCA techniques. The main methodological step in the comparison is the use of probability concepts based on mathematical tools such as linear extensions of partially ordered sets and Monte Carlo simulations. A data set consisting of 12 High Production Volume Chemicals (HPVCs) is used for illustration.

It is a paradigm in this investigation to claim that the need of external input (often subjective weightings of criteria) should be minimized and that the transparency should be maximized in any multicriteria prioritisation. This study illustrates that the Hasse Diagram Technique (HDT) needs least external input, is most transparent and is therefore the least subjective of the techniques studied. However, HDT has some weaknesses if there are criteria, which exclude each other. In such cases weighting is needed. Multi-Criteria Analysis (i.e. Utility function approach and PROMETHEE as examples) can deal with such mutual exclusions because their formalisms to quantify preferences allow participation e.g. weighting of criteria. Consequently MCA include more subjectivity and loose transparency. The recommendation, which arises from this study, is that a first step in decision-making is to run HDT and as a second step possibly to run one of the MCA algorithms.

Introduction

Many chemical substances have adverse effects on human health and the environment. Thus, within the European Union the risk from chemicals on the European market are to be assessed and the use successively regulated if necessary (EEC, 1993). The suggested introduction of the REACH scheme reflects the increased interest in ranking of chemicals. (see EEC, 2001 and also chapter by Carlsen, p. 163). At present, due to the lack of data, only a rather limited number of risk assessments have been adopted. Considering, the 100106 existing chemical substances registered in Europe (EEC, 1996), an efficient alternative to the present risk assessment procedure is needed. With the OMNIITOX programme, (see Larsen et al. 2004), a broad study is ongoing, where some simple sorting schemes are compared and where the Hasse Diagram Technique (HDT) is compared to other very simple ranking schemes, like EDIP, EURAM, Priofactor method etc. (see Larsen 2004, for further references) The present chapter does the other way round: It compares Hasse Diagram Technique with methods which usually are considered as well equipped and more sophisticated decision analysis tools. As examples we select MAUT (Schneeweiss 1991) (in its simplest version) and PROMETHEE (Brans, Vincke 1985, Brans et al. 1986).

One of the MCA methodologies, the Utility Function (MAUT), covers the general principle of scoring methods. The conclusion from the comparison of HDT with the Utility Function can therefore to some degree be transferred to the scoring methods mentioned above. The Utility Function is sometimes also referred to as the Index Function or the Quality Function.

When priority setting methods are compared or evaluated the following two criteria should be kept in mind (see for a very good compilation of relevant criteria this book):

- Subjectivity/ Transparency. Political or subjective considerations should only be involved when strictly necessary. It appears that the transparency of a methodology is often inversely related to the degree of subjective consideration involved.
- Precision. The result needs to be precise enough to allow useful decisions.

The main methodological step in the comparison is to use probability concepts based on mathematical tools like linear extensions of partial ordered sets and Monte Carlo simulations. Finally, a first step in the identification of a general mathematical scheme to evaluate the ranking methods is considered.

A data set consisting of 12 High Production Volume Chemicals (HPVC) is used for illustration (EEC, 2000).

Survey of the Ranking Methods

Hasse Diagram Technique

Hasse Diagram Technique (HDT) is based on partial order theory (see this book, chapter by El-Basil, p. 3 and Brüggemann et al. 2001). The HDT appears as a simple method, which a priori includes "$\leq$" as only mathematical relation. As the HDT is explained in chapter by Brüggemann and Carlsen, p. 61, we establish here only the main notation:

X: a set of chemicals. Here this set contains 12 chemicals.

IB: is a set of characterizing descriptors, i.e. the information basis of the evaluation.

$q_j(x_i)$: the value of the jth descriptor $\in$ IB of the ith chemical $\in$ X

If all descriptors for a substance x_1 are equal to the corresponding descriptors for the substance x_2, i.e. $q_j(x_2) = q_j(x_1)$, the two substances will have identical rank and will be considered as equivalent, $x_1 \sim x_2$. (For more details concerning equivalence, see Brüggemann, Bartel 1999).

The concept of Linear Extensions is of specific importance in this chapter (for details see the Chapter by Brüggemann and Carlsen, p. 61).

To illustrate the principle a simple Hasse diagram is constructed for five substances $x_1, x_2, .., x_5$ using two descriptors in Fig. 1.

Five substances and two descriptors

Substance	q_1	q_2
x_1	5	5
x_2	5	0
x_3	3	2
x_4	1	4
x_5	1	1

Hasse Diagram

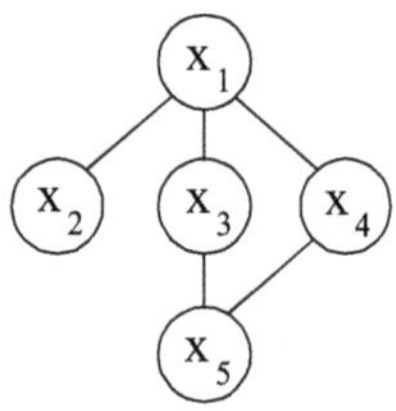

Total set of linear extensions written as linear sequences (LE- no are arbitrarily given)

LE_1: $x_5<x_3<x_4<x_2<x_1$
LE_2: $x_5<x_3<x_2<x_4<x_1$
LE_3: $x_5<x_2<x_3<x_4<x_1$
LE_4: $x_2<x_5<x_3<x_4<x_1$
LE_5: $x_5<x_4<x_3<x_2<x_1$
LE_6: $x_5<x_4<x_2<x_3<x_1$
LE_7: $x_5<x_2<x_4<x_3<x_1$
LE_8: $x_2<x_5<x_4<x_3<x_1$

Fig. 1a. HDT exemplified by five objects x_1, x_2,…,x_5

Ranking frequency from the total set of linear extensions

Rank	Substances				
	x_1	x_2	x_3	x_4	x_5
5	8	0	0	0	0
4	0	2	3	3	0
3	0	2	3	3	0
2	0	2	2	2	2
1	0	2	0	0	6

Ordinal ranking probabilities and the average ranks

Rank	Substances				
	x_1	x_2	x_3	x_4	x_5
5	1	0	0	0	0
4	0	0.25	0.375	0.375	0
3	0	0.25	0.375	0.375	0
2	0	0.25	0.25	0.25	0.25
1	0	0.25	0	0	0.75
Average	5	2.5	3.125	3.125	1.25

Fig. 1b. Calculation of the Ranking frequencies (left table) and ordinal ranking probabilities (right table) and of averaged ranks

In Fig. 1 it is illustrated that the substance x_1 is related to the highest rank since, x_1, obviously is placed above all other substances in all the linear extensions. On the other hand, the substance x_3 is equally related to three ranking levels, as it may be located

- below x_5
- between x_5 and x_3 and
- between x_3 and x_1

without hurting the already given order relations.

For the substance x_3 the ordinal ranking probability for rank no. 5 is equal to 0 and for rank no. 4 it is 0.375. Note that one can apply symmetry considerations (i.e. in mathematical terms, analyze the automorphism group of a Hasse diagram (see Schröder 2003)): For example x_3 "sees" the same order theoretical environment as x_4. Therefore the ordinal ranking probabilities of these two objects (see chapter by Brüggemann and Carlsen, p. 61) are the same and consequently all derived quantities.

When all possible linear extensions are found it is also possible to calculate the averaged ranks of the substances in a partially ordered set (Winkler, 1982 and 1983; Lerche, Sørensen, 2003, Lerche et al. 2003). The averaged rank is simply the average of the ranks in all the linear extensions. Using Fig. 1 as an example, the averaged rank of the substance

$$\mathrm{Rkav}(x_2) = 4 \cdot 0.25 + 3 \cdot 0.25 + 2 \cdot 0.25 + 1 \cdot 0.25 = 2.5 \qquad (1)$$

For the substance x_3 the averaged rank analogously is found to be 3.125. Therefore in the linear rank, x_3 will be given a higher rank than x_2. In the example given in Fig. 1 the most probable sequence of ranks is:
$x_1 > x_3 = x_4 > x_2 > x_5$.

The HDT does not need any additional knowledge which goes beyond the data matrix, given by set X and set IB. Therefore the extent of external input, which we would like to call EXT, is 0. Formally we write $EXT_{HDT} = 0$.

Utility Function Approach

In contrast to HDT, when using the Utility Function, each (normalized to the scale [0,1] and then called instead of q_i now p_i) descriptor is given a weight indicating the relative importance of that particular descriptor. More complex assumption, how a descriptor can be transformed into an "individual" preference function can be found by Schneeweiss (1991), where also the axiomatic foundation of this method is discussed. Instead of the simultaneous consideration of the descriptors the function

$$\Gamma(x_i) = \sum_i g_i \cdot p_i(x_i) \qquad (2)$$

whereby $p_i(x_i)$ are the normalized descriptors for element x_i and g_i are the individuals' weights reflecting the agreed mutual importance of the single descriptors. i is ranging from 1 to m. Thus, the utility function induces the ranking of chemicals according to their $\Gamma(x_i)$ - values, i.e. substances are totally ordered according to the Γ-function.

Considering the descriptors used in this paper, it is thus necessary to decide if, for example, the bioaccumulation expressed by log K_{ow} is more or less important than acute toxicity for fish expressed by LC_{50}. Weighting of the descriptors is a way to take into account additional external judgements. It can be seen as formalism to introduce the participation principle in the ranking procedure.

Thus, beyond the "internal" information given by the data values of the p_i's, at least m additional information in form of weights is needed. This additional information, which is classically the result of political, ethical, sociological preferences we call "external" knowledge. Thus $EXT_{Utility}$ = m. This is important information since the participation principle seams to play a more and more important role in modern environmental policy-making.

PROMETHEE

The Preference Ranking Organisation METHod for Enrichment Evaluations (PROMETHEE) was originally developed as a tool in operational research, up to now it has only proved limited applications in environmental sciences (Le Teno 1999, Poschmann et al. 1998, Drechsler 2004).

In contrast to the utility function approach the various descriptors of each substances are not aggregated. However a preference function pr_i will be constructed based on the mutual comparison with respect to one descriptor q_i (note that we return to the original attributes q_i) of any two substances. The preference function needs -as in the utility function approach- weights and information about the significance of numerical differences between the descriptor values of two substances, here called Δq^0. Any difference $q_j(x_i)$ - $q_i(x_k)$ will be assigned to the preference function pr_i. i and k are ranging from 1 to m. In the simplest case the preference function may be formulated as follows:

$$pr_j(x_i,x_k)=\begin{cases} 1 \ \text{ if } (q_j(x_i)-q_j(x_k))\geq \Delta q_j^0 \\ (1/\Delta_j^0)\cdot(q_j(x_i)-q_j(x_k)) \ \text{ if } 0<(q_j(x_i)-q_j(x_k))<\Delta q_j^0 \\ 0 \ \text{ if } (q_j(x_i)-q_j(x_k)) \quad \leq 0 \end{cases} \qquad (3)$$

By equation 1 the ith, k^{th} entry of a matrix is calculated. Thus a number of m × m matrices, corresponding to the number of descriptors, are formed, as any chemical is to be compared with any other in dependence of q_i. Therefore m additional parameters, describing the shape of the preference function $pr_j(x_i,x_k)$ are needed.

From the local matrices pr the "global" matrix PR can be written as follows:

$$PR = (1/m) \cdot \Sigma\, g^{PROMETHEE}{}_i \cdot pr_i \tag{4}$$

Therefore, once again m weights are needed (see below and for more details Lerche et al. 2002). The outcome of PROMETHEE is usually divided into two parts: By PROMETHEE I a dominance, $d\pi$, and a subdominance indicator, $sd\pi$, is found. The dominance indicator describes how much a substance is preferred over all other. If "preference" means: necessity to analyse this chemical more carefully as this chemical may be a hazardous one, then $d\pi_i$ is the sum over the entries of the ith row of PR, whereas the subdominance indicator $d\pi_j$ (sum over the jth column of PR) describes how much all other substances are preferred over some specific substance. In PROMETHEE II just the difference $\Gamma(i) = d\pi_i - sd\pi_i$ is formed. By the quantity $\Gamma(i)$ the substance i can be ranked, i.e. a total order is formed.

This means: In total (in the simple form, given by equation 1) 2·m external parameters are needed.

$$EXT_{PROMETHEE} = 2 \cdot m \tag{5}$$

Data

Data for 12 HPVC are found in the IUCLID database (EEC, 2000). Four characteristic properties (descriptors of molecular/environmental properties) are used to describe the environmental impact of the pesticides, see Table 1.

The pesticides, their CAS[1] no., the abbreviations used in figures and tables and the descriptors are given in Table 2. The data in Table 2 thus form the decision matrix, which is the basis for both HDT and the MCAs.

[1] CAS: Chemical Abstracts Service

Table 1. Descriptors (syn. attributes or indicators used in this study)

Descriptor	abbreviation	name	orientation	remark
q1	PV	production volume (Exposure)	1	as the higher PV the higher the risk
q2	log K_{ow}	n-Octanol-Water partition coefficient (accumulation)	1	as the higher the log K_{ow} the higher the bio-accumulation
q3	LC_{50}	Toxicity	-1	as the higher LC_{50} the less the hazard
q4	BD	Biodegradation	-1	as the higher BD the less the persistance

The Hasse Diagram

Fig. 2 shows the Hasse diagram for the 12 substances included in this investigation. It is convenient to arrange the substances in levels and to avoid crossings of the lines (see also chapter by Brüggemann and Carlsen, p. 61). The Hasse diagram in Fig. 2 is arranged into four levels. In the highest level (corresponding to the highest environmental impact) there are four substances (malathion (MAL), linuron (LIN), 1-chloro-4-nitrobenzene (CNB) and thiram (THI)) and in the lowest there is only one substance (chlormequat chlorid (CHL)). In the present study four descriptors for twelve substances are used.

The Hasse diagram (Fig. 2) reveals two groups of substances and an isolated substance. These groups are also called hierarchies and motivate for the analysis of antagonisms (see chapter by Simon et al., p. 221). The largest group includes eight and the small nontrivial group includes three substances. The production volume and log K_{ow} are denoted "antagonistic" indicators because they are capable of separating the Hasse diagram in two groups. Malathion (MAL) is an isolated substance because it has the lowest LC_{50} and at the same time the fastest biodegradation.

Using all linear extensions the probability distribution of ranks for each substance can be found. Table 3 gives the ordinal probability for the individual substances to occupy a certain rank. The ranking probabilities are spread out in the interval of possible ranking positions. Clearly the probabilities of isolated objects, like malathion (MAL) (compare Fig. 2) will be spread equally out over the whole interval providing an averaged rank of 6.5.

Table 2. Data on production volume (PV), acute toxicity for fish (LC_{50}), the n-octanol-water partitioning coefficient (log K_{ow}) and bio-degradation (BD) for 12 pesticides

CAS No.	Abbr.	Name	PV*	LC_{50} (mg/l)	log K_{ow}	BD (%/day)
100-00-5	CNB	1-chloro-4-nitrobenzene	4	1.5	2.6	0.2
100-01-6	4NA	4-nitroaniline	2	35	1.4	0
100-02-7	4NP	4-nitrophenol	1	7	1.9	0.1
1912-24-9	ATR	atrazin	2	4.3	2.5	0.5
999-81-5	CHL	chlormequat chlorid	2	80	-2.2	1
333-41-5	DIA	diazinon	1	2.6	3.3	0
60-51-5	DIM	dimethoate	2	7.5	0.7	0
26761-40-0	LIN	ethofumesate	1	11	2.7	0.4
1071-83-6	GLY	glyphosate	2	52	0.002	0.3
34123-59-6	ISO	isoproturon	2	3	2.5	30
121-75-5	MAL	malathion	3	0.04	2.7	100
137-26-8	THI	thiram	2	0.3	1.7	0

*(1 = 5.000 – 10.000 tons/year, 2 = 10.000 – 50.000 tons/year, 3 = 50.000 – 100.000 tons/year and 4 = 100.000 – 500.000 tons/year)

The highest probability for a position is only 0.553 for chlormequat chlorid (CHL) on rank no. 1. The averaged rank is presented in the HDT-column in Table 4.

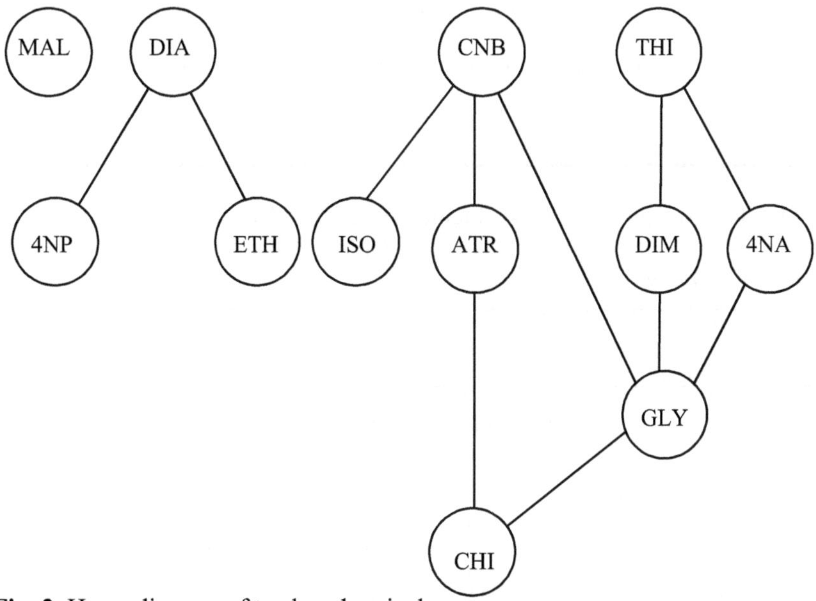

Fig. 2. Hasse diagram of twelve chemicals

Table 3. The probability distribution of ranks when using all linear extensions (HDT)

Sub-stance/ Rank	12	11	10	9	8	7	6	5	4	3	2	1
CNB	0,285	0,228	0,177	0,132	0,092	0,055	0,025	0,007	0	0	0	0
4NA	0	0,059	0,107	0,141	0,161	0,163	0,147	0,116	0,074	0,031	0	0
4NP	0	0,023	0,043	0,061	0,077	0,091	0,102	0,111	0,118	0,123	0,125	0,125
ATR	0	0,031	0,059	0,084	0,106	0,123	0,136	0,141	0,135	0,114	0,072	0
CHL	0	0	0	0	0	0	0,002	0,011	0,042	0,118	0,273	0,553
DIA	0,250	0,205	0,164	0,127	0,095	0,068	0,045	0,027	0,014	0,005	0	0
DIM	0	0,059	0,107	0,141	0,161	0,163	0,147	0,116	0,074	0,031	0	0
ETH	0	0,023	0,043	0,061	0,077	0,091	0,102	0,111	0,118	0,123	0,125	0,125
GLY	0	0	0	0	0,010	0,041	0,094	0,162	0,228	0,259	0,207	0
ISO	0	0,026	0,049	0,068	0,085	0,098	0,107	0,112	0,114	0,114	0,114	0,114
MAL	0,083	0,083	0,083	0,083	0,083	0,083	0,083	0,083	0,083	0,083	0,083	0,083
THI	0,382	0,263	0,169	0,100	0,053	0,024	0,008	0,002	0	0	0	0

Table 4. The most probable rank obtained by using all linear extensions (HDT) and the rankings obtained when using the Utility Function and PROMETHEE with equal weightings ($g_j = 0.25$)

Rank	HDT All linear extensions	HDT average rank	The Utility Function	Utility Γ	PROMETHEE	PRO.[a)] Γ
12	CNB	10.71	CNB	0.963	CNB	29.40
11	THI	10.71	ATR	0.782	ATR	10.95
10	DIA	9.75	THI	0.760	THI	7.91
9	4NA	7.28	DIA	0.742	DIA	7.02
8	DIM	7.28	ISO	0.713	ISO	2.73
7	MAL	6.50	DIM	0.692	DIM	-0.01
6	ATR	5.93	ETH	0.687	ETH	-0.10
5	ISO	5.09	4NP	0.664	MAL	-0.63
4	4NP	4.88	MAL	0.639	4NP	-3.32
3	ETH	4.88	4NA	0.638	4NA	-4.95
2	GLY	3.84	GLY	0.520	GLY	-16.24
1	CHL	1.69	CHL	0.331	CHL	-32.75

a): PROMETHEE will be abbreviated by PRO

Results of Multi-Criteria Analysis

The Utility Function

To illustrate the methodological differences between HDT and the two MCA, the rank using equal weights ($g_j = 0.25$) is calculated (Table 4, column "The Utility Function"). Compared with the rank obtained when using the HDT average rank, the most significant differences are atrazin (ATR) that changes from position no. 6 to no. 11 and 4-nitroaniline (4NA) that changes from position no. 9 to 3. The change in position of 4-nitroaniline (4NA) can probably be ascribed to the rather low LC_{50}. When using a metric method, such as the Utility Function, the actual numbers obviously influences the result more than when using a non-metric method, such as HDT. Thus, if a substance, like 4-nitroaniline (4NA) has a descriptor with a very low value, the ranking result may vary significantly. Note that there are other examples of inversion like 4-nitrophenol (4NP) and malathion (MAL): They switch their positions, when the ranking by HDT is compared with that of the Utility function.

If a set of weights is chosen it will result in one definite rank. If another set of weights is selected another rank will occur. To illustrate the range of possible ranks the influence of the weights is estimated by a Monte Carlo Simulation. In the Monte Carlo Simulation random weights (from 0 to 1) are repetitively chosen and the utility, Γ, for each substance and the linear rank is calculated (number of runs = 10000) (Table 5a). The average utility, $\Gamma_{average}$, is calculated and based on the linear ranks; the ranking probabilities are calculated and given in Table 5.

Comparing the ranking probabilities for the Utility Function with the ordinal ranking probabilities found by HDT it appears that the probabilities for the Utility Function are less spread out and clearly higher. As an illustrative example the probability for 1-chloro-4-nitrobenzene (CNB) on rank no. 12 is as high as 0.944. Fig. 3 exemplifies how the ordinal HDT ranking probabilities are more spread out than those for the Utility Function. It can thus be seen that the non-metric approach (HDT) generally supplies more ranking possibilities than the variation of weights. Note that not all linear extensions can necessarily be considered as result of a Utility function approach. Because of the metric nature of the Utility function many different sets of weights may mapped onto the same linear extension, which explains the different distributions found by HDT and Utility function approach respectively.

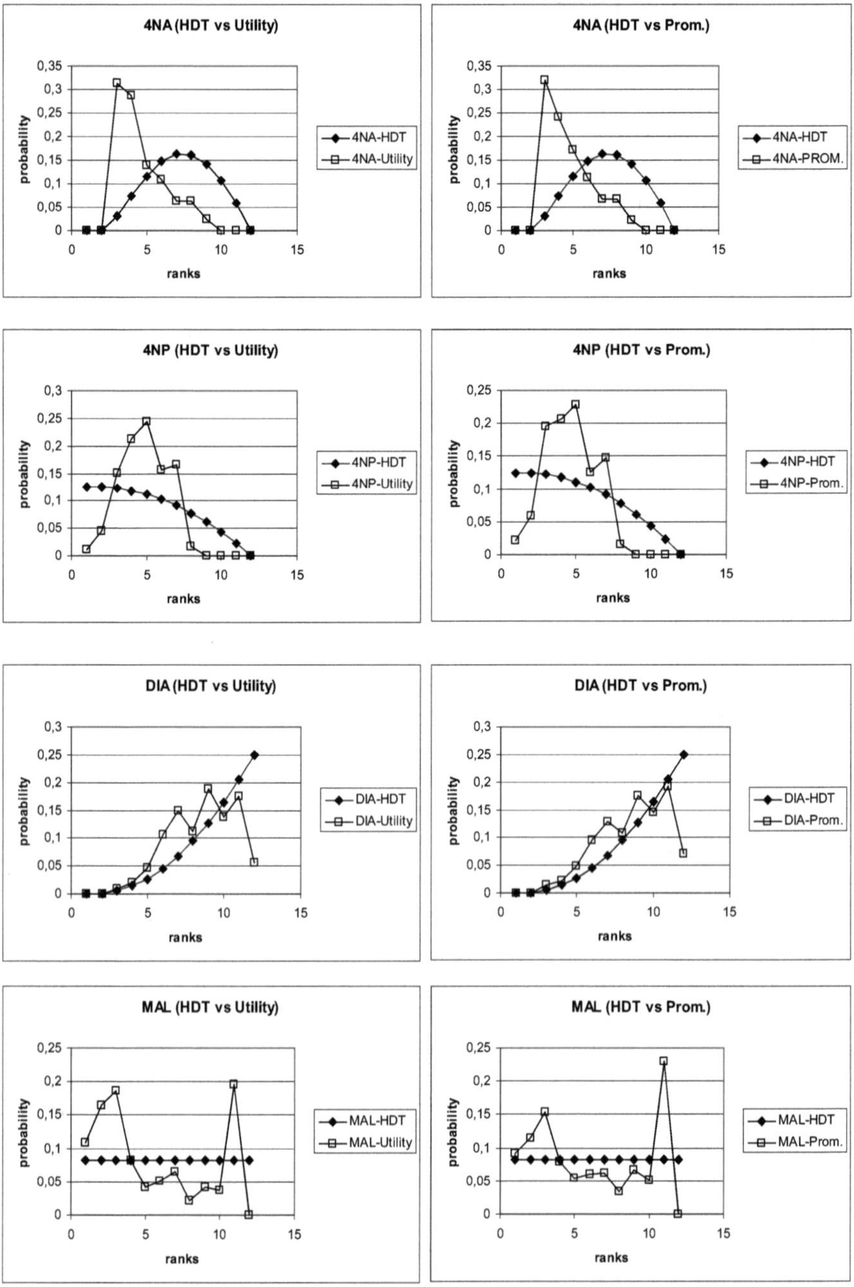

Fig. 3. Probability distributions of 4 of the twelve chemicals to demonstrate the differences in the three approaches. For a full description, see (Lerche et al.)

PROMETHEE

The Γ-values derived from PROMETHEE and the ranks deduced from them are shown in Table 4. The calculation was performed using equal weights ($g_j = 0.25$), like for the Utility function and $\Delta q^0{}_j = 0.5 \cdot \text{std}(q_j)$ ($\text{std}(q_j)$: standard deviation of q_j) taken from the data matrix (Table 2). Compared with the ranking based on the Utility Function no significant changes in positions are observed, only malathion (MAL) and 4-nitrophenol (4NP) changes position. As for the Utility Function, the weights of the descriptors play a crucial role for the ranking of the substances in PROMETHEE. The possible rankings, which can be obtained due to the variation of the weights, are again illustrated by a Monte Carlo Simulation using random weights. The simulation result is shown in Table 5b (number of runs = 10000). Comparing the ranking probabilities for PROMETHEE with the ordinal probabilities found by HDT then the ranking probabilities are less spread out and clearly higher. As an example the probability for 1-chloro-4-nitrobenzene (CNB) on rank no. 12 is as high as 0.929. Compared with the Utility Function the pattern seems similar.

Table 5. The distributions of ranking probabilities when using a Monte Carlo simulation on the weights for a: the Utility Function and b: PROMETHEE

Table 5a. Utility Function

Substance \ Rank	12	11	10	9	8	7	6	5	4	3	2	1
CNB	0.944	0.055	0.001	0	0	0	0	0	0	0	0	0
4NA	0	0	0	0.023	0.064	0.064	0.109	0.139	0.288	0.312	0	0
4NP	0	0	0	0	0.017	0.165	0.156	0.244	0.212	0.151	0.044	0.012
ATR	0	0.424	0.477	0.086	0.012	0.001	0	0	0	0	0	0
CHL	0	0	0	0	0	0	0.008	0.003	0.011	0.010	0.093	0.876
DIA	0.055	0.176	0.137	0.188	0.112	0.149	0.106	0.046	0.021	0.009	0	0
DIM	0	0	0	0.164	0.118	0.150	0.187	0.200	0.152	0.030	0	0
ETH	0	0	0.007	0.091	0.150	0.123	0.233	0.179	0.140	0.054	0.018	0.004
GLY	0	0	0	0	0	0.023	0.013	0.030	0.038	0.225	0.671	0
ISO	0	0	0.015	0.145	0.341	0.179	0.118	0.115	0.057	0.022	0.009	0
MAL	0	0.196	0.038	0.042	0.022	0.065	0.051	0.042	0.082	0.187	0.165	0.109
THI	0.001	0.148	0.325	0.261	0.163	0.081	0.021	0.001	0	0	0	0

Table 5b. PROMETHEE

Substance \ Rank	12	11	10	9	8	7	6	5	4	3	2	1
CNB	0.929	0.070	0	0	0	0	0	0	0	0	0	0
4NA	0	0	0	0.021	0.067	0.068	0.113	0.171	0.242	0.319	0	0
4NP	0	0	0	0	0.015	0.148	0.126	0.229	0.206	0.195	0.059	0.021
ATR	0	0.368	0.510	0.107	0.014	0.001	0	0	0	0	0	0
CHL	0	0	0	0	0	0	0.010	0.005	0.014	0.016	0.074	0.880
DIA	0.070	0.191	0.146	0.176	0.109	0.128	0.095	0.049	0.022	0.015	0	0
DIM	0	0	0	0.136	0.127	0.127	0.197	0.190	0.181	0.040	0	0
ETH	0	0	0.008	0.104	0.144	0.103	0.231	0.164	0.140	0.070	0.029	0.007
GLY	0	0	0	0	0	0.028	0.014	0.034	0.053	0.164	0.707	0
ISO	0	0	0.017	0.109	0.320	0.231	0.117	0.102	0.063	0.026	0.015	0
MAL	0	0.230	0.052	0.067	0.034	0.062	0.060	0.054	0.079	0.154	0.115	0.092
THI	0.001	0.141	0.266	0.280	0.169	0.105	0.036	0.002	0	0	0	0

Discussion

The Generality of and additional information obtained from HDT

The range of possible ranks for the Monte Carlo version of the MCA methods are all within the range of the minimum and maximum limits as defined by the Hasse diagram (cf. Fig. 3 and Tables 5a and 5b). In general the distribution of the ranking probabilities is narrower and the probabilities are higher for the two Multicriteria Decision Tools, examined here. HDT can thus be characterised as being the most conservative and due to the neglecting of weights the least subjective method. Evaluating the findings one should consider three cases.

- Case C: There is coincidence of HDT with all other two methods, i.e. the ranking is approximately the same when applying the different methods
- Case W: There are different results, between HDT and the other two methods, and the other two methods coincide rather well, i.e. HDT delivers a rank for a substance, which differs from that of the other methods. Utility function approach, and PROMETHEE deliver approximately the same ranks.

- Case S: There is no coincidence between HDT and the other two methods. However among the Utility function - and PROMETHEE - approach there are also conflicting results, i.e. each of the three methods deliver another rank for a substance.

In case C, the general coincidence, there is no further discussion needed. In principle the result of ranking does not depend on weights, and other subjective information.

In case S, the result depends on the algorithm used, on weights, on the use of preference functions; discrepancies to HDT are due to the specific choices within the MCAs. Therefore in the case of uncertainties in the selection of weights, preference functions, it seems that HDT is transparent displaying the ranking interval. (HDT may be on the safe side in priority setting exercises.)

Case W: This is the worst case to be interpreted. In general terms it can be stated that this situation arises when a metric algorithm (Utility function approach, PROMETHEE) is to be compared with a non-metric (HDT) one. It can be expected that isolated substances or substances only loosely connected with the others in the Hasse diagram will be candidates for the case W, because then the ranking depends severely on the kind how participation is included (for example by weightings). If the number of comparable compounds is counted (for example ATR has 2, DIM 3, MAL 0 THI 4 comparable compounds, etc) then the case W is found for such substances, which have a low number of connections to others. The cases C and S are associated with substances having a higher number of connections. Note that a loose connection of a substance does not necessarily lead to a broad distribution over the diverse ranks if metric algorithms are applied. It may be possible

- that for example by a weighted sum some single linear extensions can not be realized (because of the restricted functional form)
- that in the case of a weighted sum, the dominant part of the space of weights leads nevertheless to a small variation of the ranks (see below and the chapter by Voigt and Brüggemann, p.327) and
- the conclusions based on the linear extensions of HDT can be found for many different datasets, if they lead to the same Hasse diagram.

Table 6 discloses the single compounds belonging to the three cases C, S and W, respectively.

If the three cases are identified, how do we come to decisions?

Case C: This does not lead to a problem at all, because the results are coincident. Then HDT is most convenient, because it does not need to find weights, reference substances etc.

Case S: HDT will display an extra ranking. So what to do? Once again, HDT shows that obviously the ranking result depends on the kind how preferences are formulated. Therefore -if a decision is needed- HDT warns that the actual decision is severely depending on the algorithmic realization due to the two methods (Utility function approach, PROMETHEE).

Case W: In these cases case the numerical values of the descriptors allow a rather unique decision by MCA. The discrepancy with HDT arises from a high amount of incomparabilities. Therefore the selection of weights can, but must not strongly influence the results.

Summarizing:

In sum we conclude that in case C a limited uncertainty is observed for both the HDT and the MCA, whereas in case S the ranking uncertainty is large for the MCA. Case W is related to substances where the uncertainty of the rankings done by HDT are large and where those of the MCA are small.

The classification of the 12 substances is given in Table 6 with respect to the cases C (Coincidence among all four methods), S (the rank of a substance depends on the method and therefore on the formalisms to include participation. HDT, as a method based purely on the data matrix seems to be on the safe side) and case W ("Worst" case, discrepancies arise because of the different approaches (non-metric (HDT) vs. metric (the other three methods)).

Table 6. Summary of the cases C, S, W

Cases	Substances[a]	Remarks
C	CNB, CHL, (DIA), GLY, (DIM)	More or less the same ranking results
S	4NA, (DIA), (DIM), ETH, THI	Different rankings at all
W	ISO, 4NP, ATR, MAL	HDT differs from the results of the others

[a] Substance identifiers in parentheses mean that the classification may not be convincing; such substances may also be listed in different classes

For the MCA methods the uncertainty associated with the input data has a significant influence on the results, whereas it has been demonstrated that HDT is relatively robust to uncertainties on the input data (Sørensen et al. 2000). This is due to the fact that HDT is not concerned with the metric value, but simply the relative difference between the descriptors. Thus, HDT demonstrates to be more reliable, even though, or actually because it is a less specific method. The transparency of HDT is regarded to be higher than that of the MCAs, since only a minimum of, if any external input is needed.

The issue of subjectivity in ranking methods

When comparing priority setting methodologies it is important to identify additional external information. For all the methods considered a choice is made by choosing the descriptors. The choice of descriptors is however often based on key parameters from risk assessment schemes or environmental fate models, which makes it less subjective. In the present chapter exposure (given by production volume), aquatic toxicity, bioaccumulation and persistence were chosen as descriptive parameters. For HDT the selection of the descriptors is the main contribution of subjectivity. Additionally, some indirect weighting can be added if the data are separated into classes.

For the MCAs another level of subjectivity is added when the descriptors are weighted or weights are introduced elsewhere in the algorithm. Then the more important descriptors have to be identified. This often raises more debate than the actual choice of the descriptors. It is easy to imagine discussions concerning the relative importance for the ranking of, for example, persistence and bioaccumulation or even between exposure and effect. The choice of the weights might thus be considered more subjective than the choice of descriptors. In HDT the descriptors are not weighted. A priori we argue that this makes the method more scientifically based than the MCAs. Note, however that HDT can be extended by a stepwise aggregation of descriptors. The principle of stepwise inclusion of weights is explained in the chapter "Information System and Databases" (Voigt and Brüggemann; this book) and is called METEOR (Method of Evaluation by Order theory). As the number of possible Hasse diagrams is finite if weights (varying continuously between 0 and 1) are introduced, the infinite number of points in a space of weights is mapped onto finite subspaces characterized by one and only one Hasse diagram. A first attempt to describe METEOR from a systematic point of view was published by Simon et al., 2005.

Compared with the Utility Function approach, PROMETHEE contain an additional level of subjectivity: On top of the choice of the weights for the descriptors PROMETHEE needs information on delta zero value.

Even though not analysed in the present paper, MCA methods, like the Analytical Hierarchy Process (AHP) (Saaty 1994), are available, which rely solely on participation. If an estimation of EXT is done for AHP, then approximately it is found: $EXT_{AHP} = n^2 \cdot m + m^2$. The number n counts the objects and as before m is the number of descriptors, i.e. criteria as the concept "descriptor" does not match the algorithm of AHP.

The various ranking methods are built on sets of assumptions. In the Utility Function it is assumed that there is a numerical (often linear) relation among the individual descriptors. When ranking is made based on PROMETHEE it is assumed that all preferences are linearly comparable. The HDT do not assume anything about linearity among the descriptors. It only assumes one can find a common orientation of the descriptors.

Conclusion

In addition to providing the most general rank, HDT further provides additional information on levels, groups and the importance of the descriptors. It is rather insensitive to data uncertainty and seems methodologically transparent. This makes HDT suitable as an individual ranking tool, but also as an additional tool for MCAs. Nevertheless, if scientifically well-founded information or agreement can be obtained on weights and on the delta zero (Δq_j^o) and the assumptions on the relation among the descriptors are fulfilled, the MCAs do give a more specific rank than the HDT.

The difference between the HDT and the MCA also illustrates the concept of scientifically based ranking versus public participation. The study reveals that if a ranking of high scientific degree is desired, HDT is to be preferred, whereas if some room for public participation or negotiation among parties is desired one of the MCA methods might be a better choice.

Based on the conclusion the following tired approach for obtaining the most scientific rank is suggested. A priori a HDT analysis should be performed. The result can be thus considered as the scientific base result. On top of this analysis, if needed, MCAs can be undertaken allowing the use of additional external information on the ranking that are not sufficiently clarified by HDT.

References

Brans JP, Vincke PH (1985) A Preference Ranking Organisation Method (The PROMETHEE Method for Multiple Criteria Decision - Making). Management Science 31:647-656

Brans JP, Vincke PH, Mareschal B (1986) How to select and how to rank projects: The PROMETHEE method. European Journ Oper Research 24:228-238

Brüggemann R, Bartel HG (1999) A Theoretical Concept to Rank Environmentally Significant Chemicals. J Chem Inf Comp Sc 39:211-217

Brüggemann R, Halfon E, Welzl G, Voigt K, Steinberg CEW (2001) Applying the Concept of Partially Ordered Sets on the Ranking of Near-Shore Sediments by a Battery of Tests. J Chem Inf Comp Sci 41:918-925

Drechsler M (2004) Model - based conservation decision aiding in the presence of goal conflicts and uncertainty. Biodiversity and Conservation 13:141 - 164

European Commission. Council Regulation (EEC) No 793/93. Official Journal No. L84 05/04/1993

European Commission (1996) Technical Guidance Documents in support of Directive 93/67/EEC on risk assessment of new notified substances and Regulation (EC) No. 1488/94 on risk assessment of existing substances (Part I, II, III and IV). The European Commission: Luxembourg

European Commission. The white paper on Chemical Strategy, COM (2001) 88. European Commission: Luxembourg

European Communities. IUCLID CD-ROM Year (2000) Edition, Public data on high volume chemicals, EUR 19559EN. European Communities: Luxembourg

Larsen HF, Birkved M, Hauschild M, Pennington MW and Guinée JB (2004) Evaluation of Selection Methods for Toxicological Impacts in LCA - Recommendations for OMNIITOX. Int J LCA 9:307-319

Le Teno, JF (1999) Visual Data Analysis and Decision Support Methods for Non-Deterministic LCA. Int J LCA 4:41-47

Lerche D, Brüggemann R, Sørensen PB, Carlsen L, Nielsen OJ (2002) A Comparison of Partial Order Technique with three Methods of Multicriteria Analysis for Ranking of Chemical Substances. J Chem Inf Comp Sc 42:1086-1098

Lerche D, Sørensen PB (2003) Evaluation of the ranking probabilities for partial orders based on random linear extensions. Chemosphere 53:981-992

Lerche D, Sørensen PB, Brüggemann R (2003) Improved Estimation of the Ranking Probabilities in Partial Orders Using Random Linear Extensions by Approximation of the Mutual Probability. J Chem Inf Comp Sc 53:1471-1480

Poschmann C, Riebenstahl C, Schmidt-Kallert E (1998): Umweltplanung und -bewertung. Klett-Perthes, Gotha

Saaty TL (1994) How to Make a Decision: The Analytical Hierarchy Process. Interfaces 24:19-43

Schneeweiss C (1991) Planung 1 – Systemanalytische und entscheidungstheoretische Grundlagen. Springer-Verlag Berlin

Schröder B (2003) Ordered sets - an introduction. Birkhäuser, Boston

Simon U, Brüggemann R, Mey S, Pudenz S (2005) METEOR - application of a decision support tool based on discrete mathematics. MATCH Commun. Math. Comput. Hem. 54:623-642

Sørensen PB, Mogensen BB, Carlsen L, Thomsen M (2001) The influence on partial order ranking from input parameter uncertainty, definition of a robustness parameter. Chemosphere 41:595-601

Winkler PM (1982) Average height in a partially ordered set. Discrete Mathematics 39:337-341

Winkler PM (1983) Correlation among partial orders (1983) Siam J Alg Disc Meth 4:1-7

5 Field, Monitoring and Information

This section focuses at similar examples as described in the chapters in section 2, i.e., sampling sites, river sections, landscape regions, etc. Some methodological steps are presented and the reader will see that the problem of chemical pollution has another important aspect, namely that of information.

The chapter by Sørensen et al., concerned with monitoring results of chemicals in Danish small streams is mainly a methodological paper. Thus, partial order induced by simultaneous inclusion of more than one attribute can be seen as an expression of rank correlation. A slim Hasse diagram indicates that the attributes correlate; a broad Hasse diagram indicates an anti-correlation. Hence, similarity among Hasse diagrams indicates some degree of correlation between groups of attributes. The reader may learn how to make use of different attribute groups by constructing "agreement diagrams" and "conflict diagrams" and he will see that by Monte Carlo type simulations a confidence analysis can be obtained.

The application of partial order in an empirical context is hampered by the problem of possibly insignificant numerical differences. In the chapter by Helm this problem is carefully analyzed and a rounding algorithm is proposed. The German Environmental Specimen Bank serves as an illustrative example, including samples representing terrestrial, limnic and marine environments. Beside others the samples are characterized by their content of contaminants. Here the consideration is concerned with samples from German rivers. Cluster analysis, Principal component analysis together with an analysis based on partial orders allows a comprehensive view about the pollution pattern. The concept of averaged ranks is applied to finally obtain a well readable Hasse diagram, indicating that the river Elbe is stronger contaminated than the river Rhine.

In the chapter of Myers and Patil a further aspect is considered. Here the potential habitat suitability is assessed for species groupings of the vertebrate fauna. This chapter introduces once again many important methodological ideas: How can we take use of the charm of Hasse diagrams if the number of objects is very high? Here the authors present ideas to overcome this difficulty. The concept of Rank range runs is represented for 184 watersheds according to species richness. The final result will select some of the habitats having a superior status. Hence, this chapter is not only informative for biologists and ecologists, but also for decision makers.

The last chapter comes back to chemicals, i.e. it considers 12 high production volume chemicals and how information can be obtained from the EU database IUCLID. A brief description of the Hasse diagram technique is given. It may be worth to mention the "four-point-program", which can be considered as some kind of protocol applying partial orders in empirical context. The reader will see that - as in other chapter - the prioritization is not uniquely found. Hence, an extension of partial order theory is shown, which is called METEOR (Method of Evaluation by Order Theory). The central point is that within METEOR a stepwise aggregation of indicators is performed which finally leads to a linear rank. The advantage above other Multi-criteria support systems is that the introduction of additional knowledge can be better controlled, hence the transparency of the decision making process is enhanced.

Developing decision support based on field data and partial order theory

Peter B. Sørensen*, Dorte B. Lerche and Marianne Thomsen

Department of Policy Analysis,
The National Environmental Research Institute (NERI),
Frederiksborgvej 399, Postbox 358, DK-4000 Roskilde, Denmark

*Corresponding author: e-mail: pbs@dmu.dk

Abstract

The corner stone in the development of decision support systems is to secure that the partial ordering of descriptors does reflect reality. The similarity between descriptor ranking and field scale data ranking is thus highly critical and this chapter shows how to establish this linkage. The partial order technique is used as a robust and non-parametric similarity quantification method and illustrated using monitoring data of pesticide findings in streams of Denmark. The approach has a general appeal where the consequence of false positives (accidentally identification of a similarity) is critical and/or only rough knowledge exist about relations between the data sets that are going to be analysed for similarity. A simple and transparent mapping of a correlation profile is possible and the software named Po Correlation supports the principle described in this chapter. The principle is an extension of the conventional *Kendalls Tau* that is modified to include ordering using more than two data sets simultaneously and thus being a kind of a multi-variate rank correlation analysis. The multi-variate nature opens up for several measures of discordance that shows different aspects of discrepancy between the data set. A graphical display using Hasse diagrams of respectively concordant and discordant rankings shows how individual objects are respectively correlated and anti-correlated with regard to all the other objects. A testing algorithm using randomized data sets are included in order to test for statistically significance of both similarity and discrepancy.

Introduction

The analysis of similarity is often the corner stone in scientific work or related areas and this chapter will show how the partial order technique can contribute to the toolbox of similarity quantification methods. Ranking will be the methodological basis, so, the concept will be non-parametric, without being based on specific functional relationships including parameterisation of functional parameters. In general a non-parametric procedure on one side is relatively robust as data interpretation, but on the other hand it is a rough approach that may over look important information. The classic discussion of parametric vs. non-parametric data interpretation is relevant here, where the usage of partial orders for data analysis can be characterised as a robust but also rough methodological approach. This problem of method selection is governed by the dilemma between confidence and completeness: Confidence has to be maximised in order to increase the statistically strength of the conclusion. Completeness has to be maximised in order to secure that no important similarity remains undiscovered. The confidence and the completeness form a complementary pair, however. A highly confident analysis can only focus on the most obviously relations missing finer similarity structures and thus violates the completeness. On the other side, if finer structures of the similarity are going to be identified in a more complete analysis then there will be a danger for modelling of noise and thus for violating the confidence. So, there is no easy answer to the problem of selecting the method, but the following good praxis is relevant: (1) If both detailed and valid knowledge about functional relations exist between the data sets then a parametric method will benefit from being most complete and still rather confident due to the relatively well known conditions. (2) If the consequence of false positives (accidentally identification of a similarity) is critical then a non-parametric method will most safe. (3) If only rough knowledge exist about relations between the two data sets then non-parametric methods become attractive. A more detailed description of non-parametric methods can be seen in Conover (1999) and Gibbons (1993).

Partial ordering is useful as technique for rank correlation analysis, where a simple and transparent mapping of a correlation profile is possible. The principle described in this chapter is supported by the software named "Po Correlation" presented by Sørensen et al. (2005) and the content of this chapter is based on that paper. Non-commercial use of the software for research and education is free if reference is given to Sørensen et al., (2005) and can be made available by contacting the first author of this chapter. Two other software products exist for application of Partial

Order Theory in decision support: (1) the software WHASSE (Brüggemann et al. 1999); (2) The software ProRank© (Ver. 1.0) (Pudenz 2004). The software Po Correlation differs from the other applications by focusing on the correlation analysis between two partially ordered sets.

The method is an extended and improved version of the methodology for assessing ranking similarity as first presented by Sørensen et al. (2003). This paper will not made a comprehensive review of ranking correlation methods, however, and for a more general discussion of ranking correlation see e.g. Brüggemann et al. (2001), Conover (1999), Gibbons (1993), Pavan (2003) and Pudenz (1998).

Data background for illustration

The increasing accessibility to environmental data makes it more and more attractive to investigate the similarity between full-scale environmental conditions on one side and either laboratory conditions or human activity on the other side. Such a similarity analysis will typically face the challenge of having high complexity and thus ill-defined knowledge regarding specific relations between the variables in the data sets. The rank correlation is attractive for this type of data as illustrated in the following.

The data set, selected for illustration, is taken from the Danish Monitoring Program (NOVA 2003) and includes pesticides finding during year 2000 in small streams, see Table 1. The data set is made based on 23 sampling stations, each covering a separated catchments area. At each station, 6 water samples were analysed for a series of pesticide active ingredients in the following denoted pesticides. The detection frequency (*DetFreq*) is defined as the frequency for a pesticide to be detected above detection limit in the joint set of measurement from the 23 sample stations. If the set of stations is assumed representative for Danish conditions then the *DetFreq* is a measure for the propagation of a given pesticide in the stream water environment in Denmark. For each station and for each pesticide, the maximum measured concentration level among the 6 single samples is identified. This yields 23 maximum concentration values and the median (*MedMax*) is subsequently calculated characterising the level of contamination. So, the two numbers *DetFreq* and *MedMax* together form an ecotoxicological meaningful way of characterising occurrence by taken into account both propagation (*DetFreq*) and level (*MedMax*) as discussed by Sørensen et al. (2003).

The ranking of pesticides using data of *DetFreq* and *MedMax* together will be compared with two variables for the usage due to the human activity of agriculture in form of recommended dosage level (*Dose* in g/ha) and sprayed area at country scale level (*SpArea* in 1000 ha). The use data is taken from the Danish sales statistics and the reported recommended dosage. The data set is shown in Table 1. Some of the pesticides in the monitoring program has been banned since year 1995 and are thus not used in year 2000. They are identified in Table 1 as: *Dose*=0 and *SpArea*=0.

The primary topic for the correlation analysis is to investigate the coincidence between the ranking of pesticides based on the measured variable set *DetFreq* and *MedMax* on one side *(Set 1)* and the usage variable set *Dose* and *SpArea* on the other side *(Set 2)*. The variables *DetFreq* and *MedMax* are thus denoted the predicted variables while *Dose* and *SpArea* are denoted the predicting variables.

Correlation analysis

A simple fundamental rank correlation measure is *Kendalls Tau* (Kendall, 1938). The principle in *Kendalls Tau* is closely linked to a partial order as explained in the following. For a set of two variables as *e.g.* the variables *DetFreq* and *MedMax* in Table 1, the ranking of two objects (pesticides in Table 1) can be done using either the first or the second variable. If the ranking using the first variable is equivalent with the ranking using the second variable then the ranking is claimed to be concordant. A pair of objects is discarded if at least one of the variables is equal or equivalent. In Table 1, a concordant ranking is seen for the ranking of Id. 13 above Id. 8, where Id. 13 > Id. 8 for both the variables *DetFreq* (76>2 in Table 1) and *MedMax* (220>200 in Table 1). A discordant order (also later simply called ' discordant ranking') appears when there is discordance between the rankings induced by the single variables. The variable pair formed by the Ids. 6 and 13 is an example of a discordant ranking, where *DetFreq* (1<73 in Table 1) and the *MedMax* (380>220 in Table 1) yielding a different ranking of the two objects. The number of concordant rankings is denoted *C* and the number of discordant rankings is denoted *D*. A modified *Kendalls Tau* was suggested by Goodman and Kruskal (1963) as

$$\tau = \frac{C-D}{C+D} \tag{1}$$

Table 1. The data set of pesticide finding in streams and related usage in agriculture used for methodological illustration

		Predicted variables Set 1		Predicting variables Set 2	
Id	Substances	DetFreq (%)	MedMax (ng/l)	Dose (g/ha)	SpArea (1000 ha)
1	2,4_D	2	40	0	0
2	Atrazine	9	30	0	0
3	Bentazone	36	20	523	91
4	Bromoxynil	6	80	383	110
5	Carbofuran	0	0	659	1
6	Chloridazon	1	380	0	0
7	Chlorsulforon	2	30	0	0
8	Cyanazin	2	200	0	0
9	Diclorprop	7	70	847	2
10	Dimethoat	2	40	304	81
11	Ethofumesat	5	90	491	31
12	Fenpropimorph	2	70	477	249
13	Glyphosat	76	220	1172	573
14	Ioxynil	6	30	349	113
15	Isoproturon	40	130	2750	4
16	Maleinhydrazid	1	10	1790	0.3
17	MCPA	20	140	1410	101
18	Mecoprop	17	30	900	13
19	Metamitron	8	90	2098	48
20	Metribuzine	1	50	250	27
21	Metsulfuron methyl	1	10	5	151
22	Pendimethalin	12	40	1368	178
23	Pirimicarb	4	30	135	7
24	Propiconazole	6	20	6837	3
25	Terbuthylazine	33	100	1500	22

The τ value is 1 for complete ranking agreement and –1 to complete disagreement between rankings of the two variables.

The correlation between two variables can also be graphically displayed as a partial order using a Hasse diagram. This is illustrated in the following for the variables *DetFreq* and *MedMax* in Fig. 1. In this diagram two pesticides are ranked if there are no discordance between the ranking of respectively *DetFreq* and *MedMax*. A discordant ranked pair of objects in the Hasse diagram do not have downward connecting lines between the two objects as seen for the object pair formed by the Ids. 6 and 13 in Fig. 1. A more detailed discussion about these relationships can be seen in Brüggemann and Bartel (1999) or else where in this book (see chapter by Brüggemann and Carlsen, p. 61). The value for C and D in case of the ranking correlation between *DetFreq* and *MedMax* is respectively: $C = 169$ and $D = 97$ and $\tau = 0.24$, indicating a positive but not strong correlation.

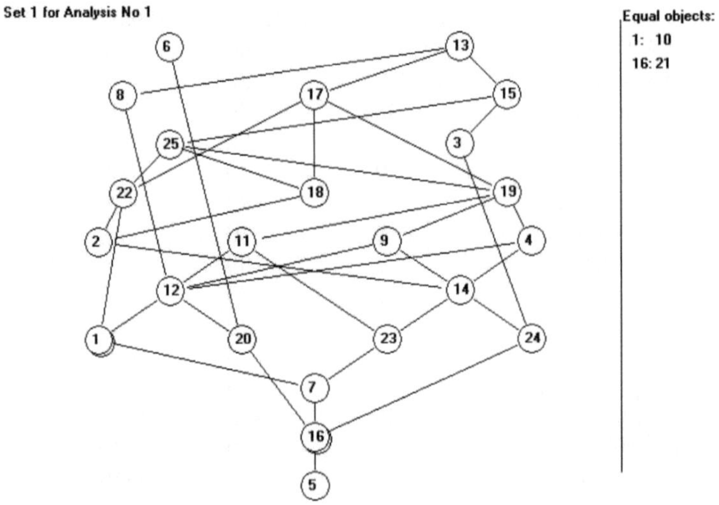

Fig. 1. Hasse diagram using *DetFreq* and *MedMax* as parameters. The numbering is referring to the Id's in Table 1. (Figure drawn using Po correlation)

The value of τ is calculated for all combinations of variable pairs in Po Correlation as shown in Table 2. The correlation in Table 2 shows only the correlation of pairs while more complex correlation formed by combined ranking of several variables will be investigated in the following.

Table 2. τ values for all parameter combinations using the complete data set in Table 1

		Predicted variables		Predicting variables	
		DetFreq	MedMax	Dose	SpArea
Predicted variables	DetFreq	1,00	0,27	0,42	0,24
	MedMax	0.27	1,00	0,07	0,07
Predicting variables	Dose	0,42	0,07	1,00	0,20
	SpArea	0,24	0,07	0,20	1,00

Two partial ordered sets are defined: (1) *Set* 1, composed by *DetFreq* and *MedMax* as shown by the Hasse diagram in Fig. 1; (2) *Set* 2 composed by *Dose* and *SpArea*. Both concordant and discordant rankings are compared between the two sets for all pair of objects. This procedure is illustrated for two pair of objects in Fig. 2.

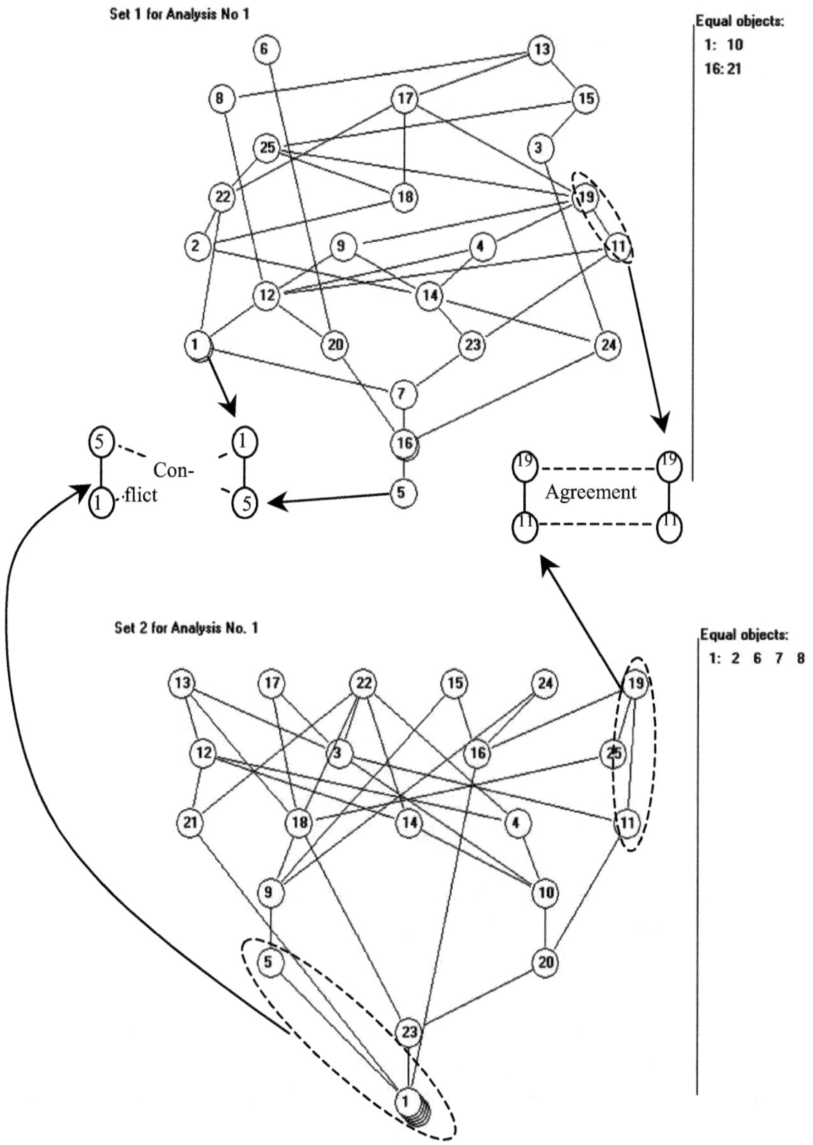

Fig. 2. Example of the comparison between *Set 1* and *Set 2*

In Fig. 2 the *Set* 1 is the partial ordered set for respectively *DetFreq* and *MedMax* and *Set* 2 the partial ordered set for respectively the *Dose* and the *SpArea*. All pesticides in Table 1 are included The similarity of the rankings between the sets is graphically displayed in a agreement diagram in form of a Hasse diagram where all variables in the two sets are applied for ranking in one diagram as defined by Sørensen et al. (2003). This is shown in Fig. 3, where the variables *DetFreq*, *MedMax, Dose* and *SpArea* are included.

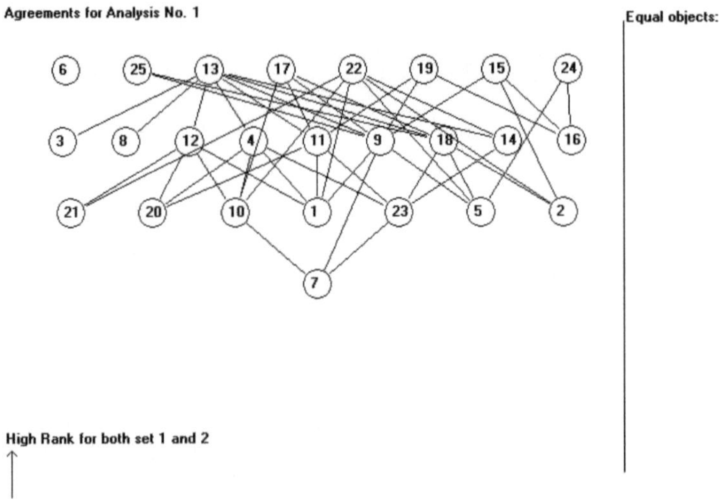

Fig. 3. Graphical displays of the agreements (*agreement diagram*) between *Set* 1 and *Set* 2 (The sets shown in Figure 2). (Figure drawn using *PO correlation*)

The agreement diagram has a complementary diagram denoted the conflict diagram (Sørensen et al. 2003) in where the *Set* 1 parameters are ranked upward while the *Set* 2 parameters are ranked downward (inverse rank). Such a diagram is shown in Fig. 4, where the variables: *DetFreq*, *MedMax,* negative *Dose* and negative *SpArea* are used. No ranking can exist simultaneously in both the agreement diagram and the conflict diagram. These two diagrams show important elements of the correlation profile for each single object in relation to the other objects. The conflict diagram maps the conflicting ranking for each object. In this way Id. 8 is seen to be ranked above several objects in the conflict diagram telling that *Set* 1 tends to rank Id. 8 upward while the *Set* 2 tends to rank Id. 8 downward. The top objects in the conflict diagram having multiple comparisons to other objects are dominated by pesticides banned in 1995 and thus not used in year 2000 (The Ids.: 1, 2, 6, 7, 8). Hence the conflict diagram indicates that the correlation between measured occurrence and usage the year of measurement is damaged by the fact that some of the pesticides have not been used since 1995. It also tells that there are only a few conflicts between the pesticides, which are still in use. This will be analysed later in this chapter.

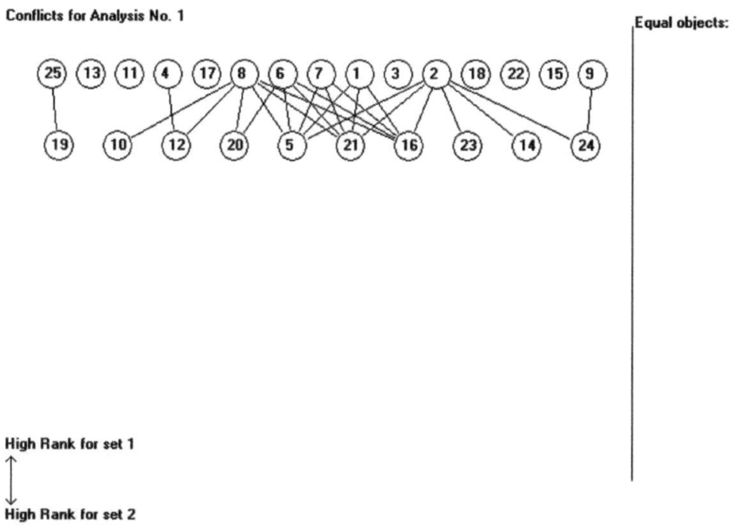

Fig. 4. Graphical displays of the conflicts (conflict diagram) between *Set* 1 and *Set* 2 (The sets shown in Fig. 2). (Figure drawn using Po correlation)

A series of different indexes describes various part of the similarity and a single number is unable to capture all possible information. So, the similarity between partial orders is a multi-dimensional problem and any one-dimensional representation in a single number will discard information. The principle of the modified Tanimoto index as a similarity index, *T(...,...)* and the linkage to other concepts are shown by Sørensen et al. (2003).

The quantification of similarity in Po Correlation is based in the following counting of object pairs covering all basic properties of similarity:

a. Concordant ranked in both sets having the same rank in the two sets (see Figure 2: agreement).
b. Concordant ranked in both sets but having different rank in the two sets (see Figure 2: conflict).

c. Concordant[1] ranked in *Set 1* and discordant[2] ranked in *Set 2*.
d. Concordant ranked in *Set 2* and discordant ranked in *Set 1*.
e. Concordant ranked in *Set 1* and concordant ranked in *Set 2* simultaneously ($e=a+b$).
f. Discordant ranked in both sets simultaneously.
g. Equivalent in *Set 1* and not equivalent in *Set 2*.
h. Equivalent in *Set 2* and not equivalent in *Set 1*.
i. Equivalent in both *Set 2* and *Set 1* simultaneously.

These definitions will be used in the following tables showing the correlation between the two sets. The similarity of *Set* 1 and *Set* 2 is analysed as shown in Table 3.

Table 3. The correlation results between Set 1 and Set 2 as shown in Fig. 2

Comparison n=25	Counting	Probability for larger or equal value
a	82	0.067
b	25	0.993
c	91	0.030
d	72	0.030
e	107	0.980
f	25	0.980
g	2	0.932
h	10	0.932
i	0	1.000
T(0,0) = a/(a+b)	0.77	0.014
T(1,0) = a/(a+b+c)	0.41	0.078
T(0,1) = a/(a+b+d)	0.46	0.078
T(1,1) = a/(a+b+c+d)	0.30	0.124

The number of agreements between *Set* 1 and *Set* 2 is relatively high (a=82) compared with the number of disagreements (b=25). Obviously some positive correlation seems to exist between the predicting and predicted variables as also indicated in Table 2. All the pairs of objects that contribute to the a value are ranked in the agreement diagram, Fig. 3, while all the pairs contributing to b are ranked in the conflict diagram, Fig. 4.

The confidence is estimated as the probability for a randomly formed value to be equal or larger than the actual value. This procedure will be explained below. If the probability estimate is close to zero the actual

[1] concordant ranked objects x, y: $x \perp y$
[2] discordant ranked objects x, y: $x \parallel y$

value is "relatively large" and contrary estimates close to one shows that the actual value is "relatively low". So, the *T* (0,0) value of 0.77 is seen to be relatively large by having a probability of 0.014 for a randomly formed estimate to be at least 0.77. A concordant ranking in one set tends to be discordant in the other set as seen in Table 3, where the *c* and *d* values are significantly high (related probability estimate of 0.030). This is supported by relatively low values for *e* and *f*. This shows some degree of correlation within the two pairs (*DetFreq*, *MedMax*) and (*SpArea*, *MedMax*), which is not reflected in a corresponding correlation between the pairs. *E.g.* there are some positive internal correlation between *DetFreq* and *MedMax* (see Table 2) and the high *c* value indicates that this positive correlation is not reflected completely in the similarity with *Set 2*. This is very important knowledge because it reveals some structure in *Set 1* that is not explained by *Set 2*.

The similarity between *Set 1* and *Set 2* is governed by the following three factors:

1. The value setting of the descriptors, within the sets. This is illustrated in Table 3 where the value setting of both *Dose* and *SpArea* shows several zero values in pairs and thus many equal objects in *Set 2*. This tends to reduce the number of concordant rankings and thus the potential number of rankings, which can be compared with rankings done in *Set 1*.
2. More or less internal correlation within the variable in a set is critical for the number of concordant rankings within each set and thus also for the potential correlation between the two sets.
3. Correlation between the ranking of the predicted descriptors in *Set 1* compared to the ranking of the predicting descriptors in *Set 2*.

Only the third factor is important when the confidence of correlation between *Set 1* and *Set 2* are going to be assessed. So, the challenged is to design a statistical test that can hold the two first factors constant (take them as conditions) and only test the correlation between the two sets of descriptors. Keeping the structure of the Hasse diagram of Set 2 constant as a condition for the test is one way to solve this problem. Such a procedure is shown in the following by using a simple example.

The following simple example will use two, small and arbitrary chosen, partial ordered sets for illustration of the probability estimates, see Fig. 5. Consider two partial ordered sets: *Set 1* and *Set 2*. The ranking of the objects named A, B, C and D is done in *Set 1* using predicted variables (like *e.g., DetFreq* and *MedMax*) and the same objects are ranked in *Set 2* using predicting variables (like *e.g. SpArea, MedMax*). The box in top of Fig. 5

shows the two Hasse diagrams of respectively *Set 1* and *Set 2* and the small table between them shows the actual values for the parameters defined in Table 3.

The concept of testing is based on a fixed Hasse diagram structure for *Set 2*, in where positioning of the objects in this Hasse diagram is varied. This mean that all possible naming of objects are allowed in the Hasse diagram and the procedure is to test all these combinations for similarity to *Set 1*. This yield 24 possible Hasse diagrams in Fig. 5, listed from high correlation towards lower correlation measured positive for a high value for *a* and a low value for *b*. In this way it is seen that *Set 2* belongs to the Hasse diagram, which is among the four best in similarity with *Set 2*. The probability for an *a* value to be equal of larger than the actual *a* value of 3 is seen to be 6/24 and the probability for a *b* value to be equal or larger than the actual value of 0 is 20/24.

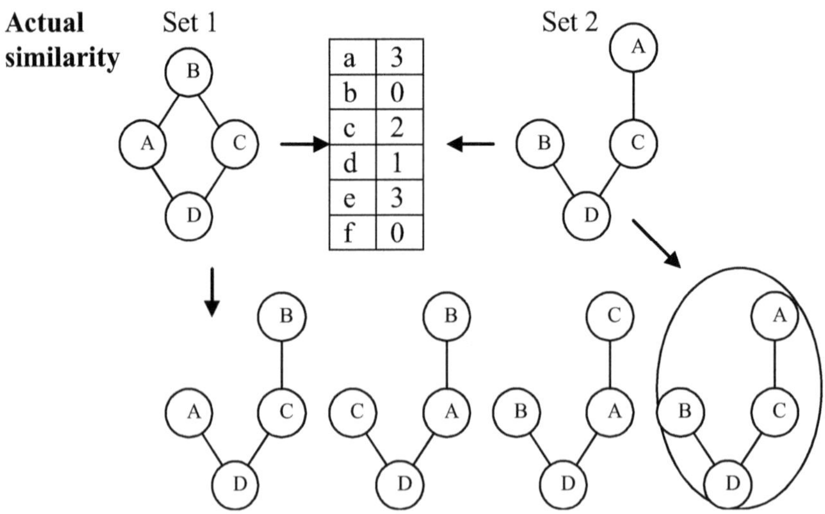

a	4	4	3	3
b	0	0	0	0
c	1	1	1	2
d	0	0	2	1
e	4	4	3	3
f	1	1	0	0

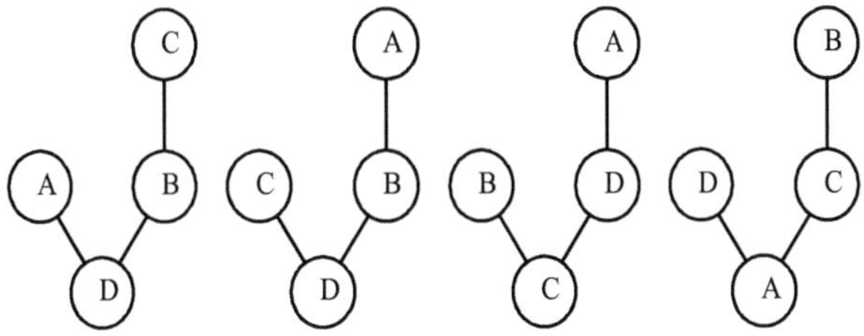

a	3	3	2	2
b	1	1	1	1
c	1	1	2	2
d	0	0	1	1
e	3	4	3	3
f	1	1	0	0

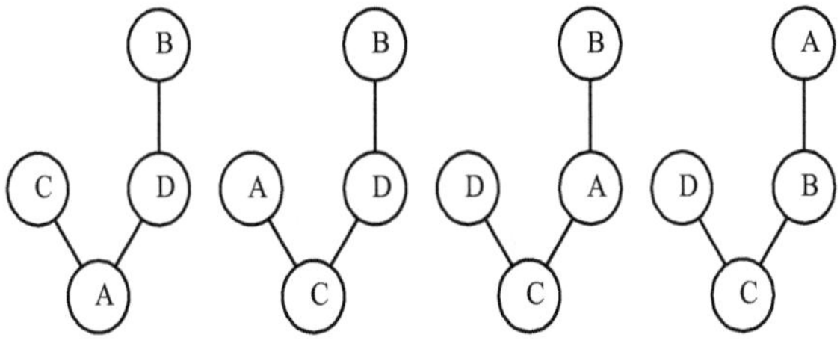

a	2	2	2	1
b	1	1	1	2
c	2	2	2	2
d	1	1	1	1
e	3	3	3	3
f	0	0	0	0

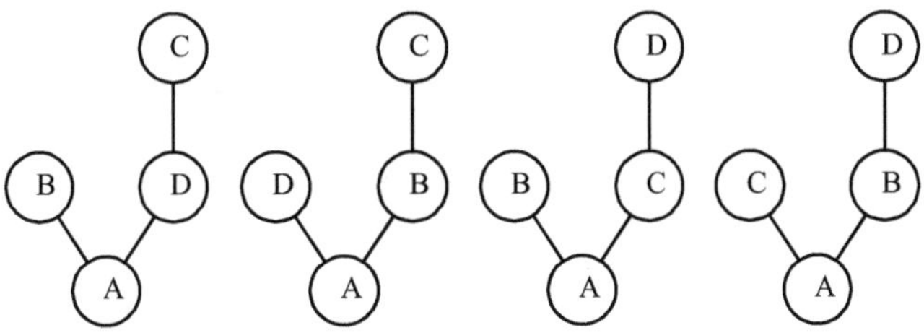

a	1	1	1	1
b	2	2	2	2
c	2	2	2	2
d	1	1	1	1
e	3	3	3	3
f	0	0	0	0

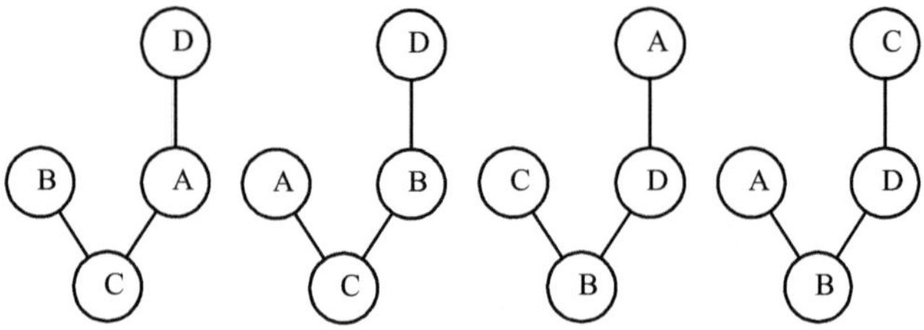

a	1	1	1	1
b	2	2	3	3
c	2	2	1	1
d	1	1	0	0
e	3	3	4	4
f	0	0	1	1

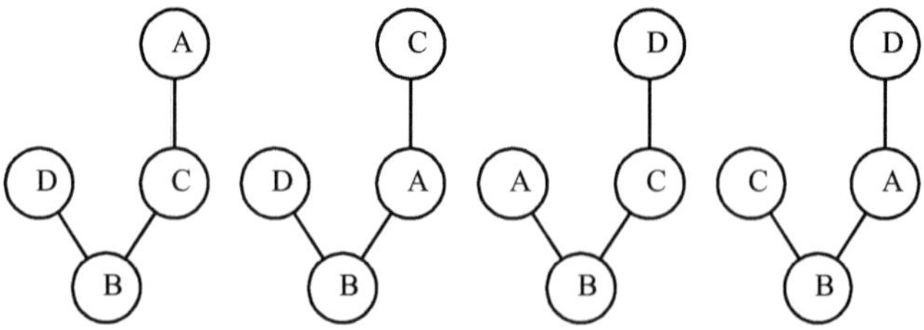

a	0	0	0	0
b	3	3	4	4
c	2	2	1	1
d	1	1	0	0
e	3	3	4	4
f	0	0	1	1

Fig. 5. A simple example of the confidence tests using all possible object combinations in Set 2

This simple example shows that there exist a variety of interrelations between the single similarity parameters. The four objects have six possible ranking relations between each other (generally for n as the number of objects: $n + (n\text{-}1)/ 2$). Every relation can either be a concordant or a discordant ranking (Note: we have neglected equivalence). Thus the vertical sum of the comparison parameters (neglecting e, which just is the sum $a + b$) will always be 6. The total number of discordant rankings are 3 (one in *Set 1* and two in *Set 2*), so the relation $c + d + f = 3$ needs to be valid due to the constancy of the Hasse diagram structure. Similar, the number of concordant rankings in *Set 2* is 4, so $a + b + d = 4$ and for *Set 1* there are 5 concordant rankings yielding $a + b + c = 5$. Another example of interaction is seen for d and e, where $d = 1$ for $e = 4$ and $d = 2$ for $e = 3$. These simple interrelations become more complex when some objects are equivalent in *Set 1* and/or *Set 2*. However, there will still be a close interrelation between the single comparison parameters as seen in Table 3, where the probability estimates for respectively (c, d), (e, f) and (g, h) in pairs are equivalent.

The number of possible Hasse diagram versions for testing of *Set 2* is $n!$ (4!=24 diagrams in Fig. 5). So, it will never be computational realistic to test more than about 10 objects using the outlined method directly and a method of approximation is applied in order to solve this problem. The full number of possible *Set 2* versions is replaced in this procedure by a random sampled subset, see Fig. 6. Every random sample is found by mixing up the object Ids in *Set 2*. In other words, first id. 1 is by random choice interchanged with *e.g.* id. 5. This procedure is repeated for all the other object id.'s 2, 3, ...n ending up with a Hasse diagram, where the object naming is randomly distributed. The correlation between *Set* 1 and the original *Set 2* is first calculated yielding the "actual" correlation result (*AC*). Then subsequently a series of randomly formed Hasse diagrams are generated as explained above and a comparison with *Set 1* generates a correlation estimate (*RC*). A sum (*sum* in Fig. 6) counts the number of times that *RC*>*AC* is true out of totally *I* randomly formed Hasse diagrams. The ratio *I*/*sum* is an estimate for the probability for a randomly formed correlation to exceed the actual correlation.

The value of *I* needs to be high enough to secure a robust probability estimate, however, the analysis has an upper limited for a meaningful increment of the *I* value around the factorial value for the number of objects (n!). For higher values of *I* only limited additional information will be gained by further increasing *I* value. However, in case of 25 objects as in Table 3, the factorial value is $1.6 \cdot 10^{25}$ and thus far above any realistic value for *I*. Different values of *I* are tested for the data set in Table 1 and the results is shown in Fig. 7.

Original data table

Substances	Set 1		Id	Set 2	
	DetFreq (%)	MedMax (ng/l)		Dose (g/ha)	SpArea (1000ha)
2,4_D	2	40	1	0	0
Atrazine	9	30	2	0	0
Bentazone	36	20	3	523	91
-	-	-	-	-	-
-	-	-	-	-	-
Terbuthylazine	33	100	25	1500	22

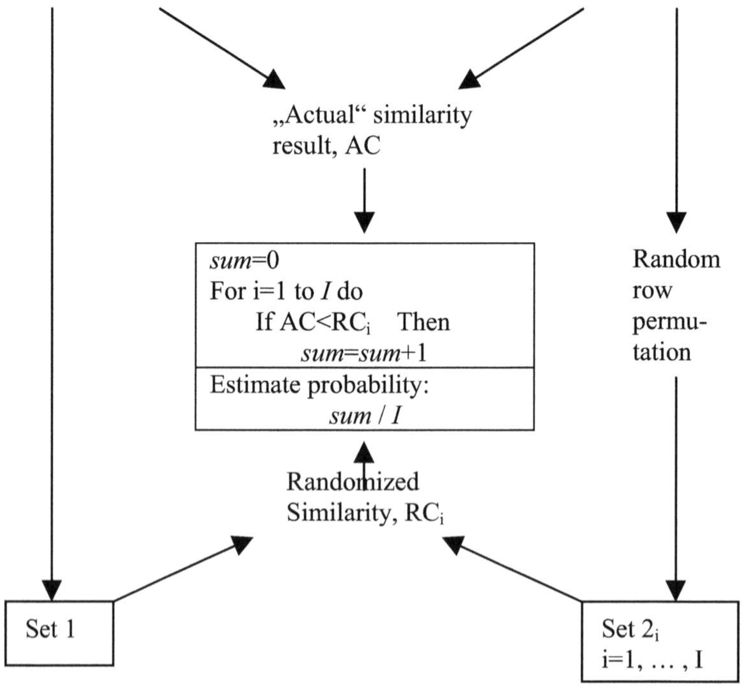

Fig. 6. Significance testing principle. The probability is estimated for a randomly formed similarity value to be of higher or equal the true similarity between Set 1 and Set 2. A randomised version of Set 2 denoted Set 2_i is formed by mixing the Ids (names) for the rows in the data table of Set 2.

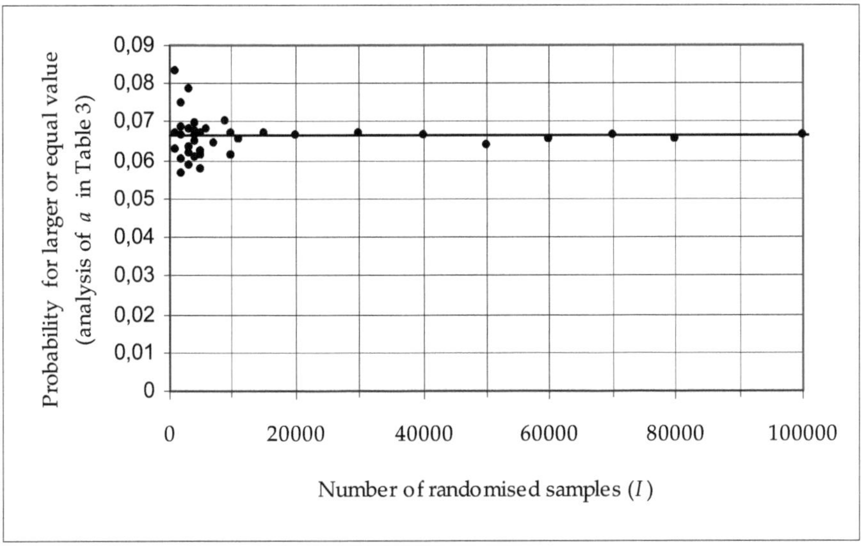

Fig. 7. The principle shown in Fig. 6 applied for different values of I and for testing a=82 from Table 3. The line shows an estimate of the "true" probability using $I=10^6$

The probability estimate is graphically shown in Po Correlation, in form of a significance plot, see Fig. 8. This figure shows two numbers for the correlation quality and every point is formed based on *Set* 2_i for i=1, 2..,I. The quality number on the x-axis is T (0,0) as a measure for the goodness of correlation as discussed by Sørensen et al. (2003). The quality number on the y-axis describes the total completeness of tested correlation. This quality number is calculated as the ratio between the number of rankings, which are included as either an agreement or a conflict ($a + b$ in Table 3), and the total number of possible ranking relations in the data set ($n{\cdot}(n-1))/2$. The highlighted circle is the actual correlation estimate. In Fig. 8 the actual estimate is located in the high end of the point cluster in the direction of T (0,0), which indicates some degree of correlation. On the y-axis the actual estimate is located close to the lower edge of the point cluster showing some misfit between the comparability in *Set* 1 and in *Set* 2. This was also seen in Table 3 as discussed above, where quite high probability is seen (0.980) for randomly formed e values to equal or larger than 107.

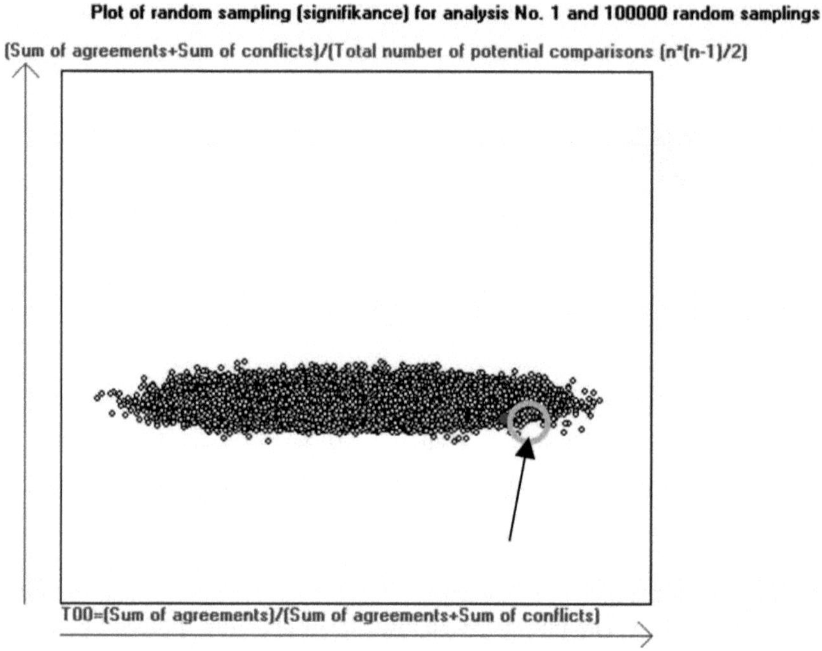

Fig. 8. Significance plot based on *Set 2*$_i$ for i=1,2,..,I (I=100000) using the data set in Table 1. The highlighted circle is equal to the actual value from Table 3. (Figure drawn using Po Correlation)

In the following the banned pesticides, which have not been used in a 4-5 years period before the measurements was sampled (Ids. 1, 2, 6, 7, 8) are excluded from the data set and the correlation analysis is repeated. The results for the τ correlation are shown in Table 4. The efficiency of the product between *Dose* and *SpArea* as correlation variable is tested. This product may be an effective variable, having unit of used amount per year (kg/year). However, it has been shown for the Swedish data that this product is far from being complete (Sørensen et al. 2003). The correlation in Table 4 has changed substantial compared with Table 2. The correlation between the predicted and predictive variables has improved. The product *Dose·SpArea* has the best correlation to both *DetFreq* and *MedMax*.

The correlation between *Set 1* and *Set 2* is recalculated for the data set without the banned pesticide and the results are displayed in Table 5.

The numbers in Table 5 are smaller compared to Table 3 due to a smaller number of pesticides. However, a much more confident positive correlation is seen having only 3 conflicting rankings and thus a T (0,0) value of 0.94.

Table 4. τ values for all parameter combinations using the data set in Table 1, where the banned and thus not used pesticides, Ids. 1, 2, 6, 7 and 8, are excluded

		Predicted variables		Predicting variables		
		DetFreq	MedMax	Dose	SpArea	Dose+SpArea
Predicted variables	DetFreq	1.00	0.49	0.41	0.14	0.46
	MedMax	0.49	1.00	0.22	0.21	0.43
Predicting variables	Dose	0.41	0.22	1.00	-0.22	0.17
	SpArea	0.15	0.22	-0.22	1.00	0.61
	Dose+SpArea	0.46	0.43	0.17	0.61	1.00

Table 5. Correlation results between *Set* 1 and *Set* 2, where the banned pesticides having the Ids. 1, 2, 6, 7 and 8 are withdrawn from the analysis leaving 19 pesticides in the correlation analysis.

Comparison n=19	Value	Probability for larger or equal value
a	51	0.003
b	3	1.000
c	91	0.273
d	20	0.273
e	54	0.815
f	25	0.786
g	1	1.000
h	0	1.000
i	0	1.000
T(0,0) = a/(a+b)	0.94	0.000
T(1,0) = a/(a+b+c)	0.35	0.003
T(0,1) = a/(a+b+d)	0.69	0.003
T(1,1) = a/(a+b+c+d)	0.31	0.006

The significance plot also shows an improved confidence compared to Fig. 8, as graphical displayed in Fig. 9. A more clear separation between the actual correlation (highlighted circle) and the cluster of points is seen. It is also seen in Fig. 9 that the actual correlation is placed in the centre of the cluster in the direction on the y-axis. This indicates absence of information within the sets, which is not reflected in the correlation between the two sets. The latter is also seen in Table 4, where the probability level for a higher *e* value has moved away from unity (0.980 in Table 3 down to 0.815 in Table 4).

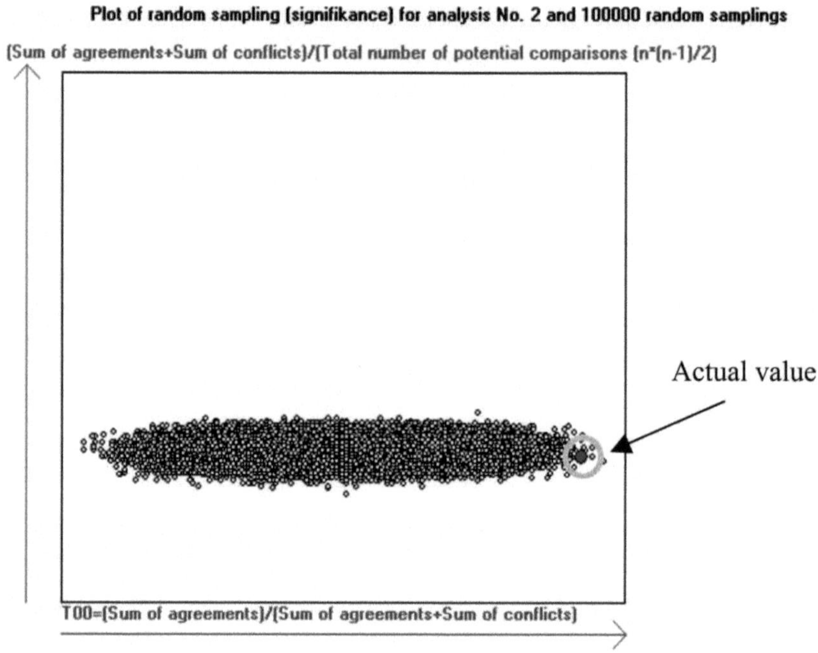

Fig. 9. Significance plot formed by random testing for the data set (*I*=100000), where the banned pesticides are neglected (Ids. 1, 2, 6, 7, 8). (Figure drawn using Po Correlation)

The 3 disagreements in Table 5 need to be analysed using the conflict diagram before the final conclusion is possible, see Fig. 10. The conflict diagram shows separated pairs of rankings, where no pesticide is connected to more than one single pesticide. This indicates that there is no single responsible pesticide for the conflicts. The inclusion of other variables like physico-chemical parameters and the analytical detection limit is discussed by Sørensen et al. (2003) in relation to South Sweden monitoring data. However, any further addition of ranking variables will increase the number of discordant rankings in *Set* 2 and thus tend to damage the completeness of the correlation analysis. In this way there is a trade off between the number of variables to be included and the completeness of the correlation analysis. So, it seems not relevant to consider any additional variables and the simple information about usage seems to be rather powerful for describing the occurrence of current used pesticides in streams. This is not a trivial conclusion because the main paradigm for pesticides exposure behaviour is based on the hypothesis that basic physico-chemical properties including degradation and adsorption are governing factors for the quantification of differences in exposure for pesticides.

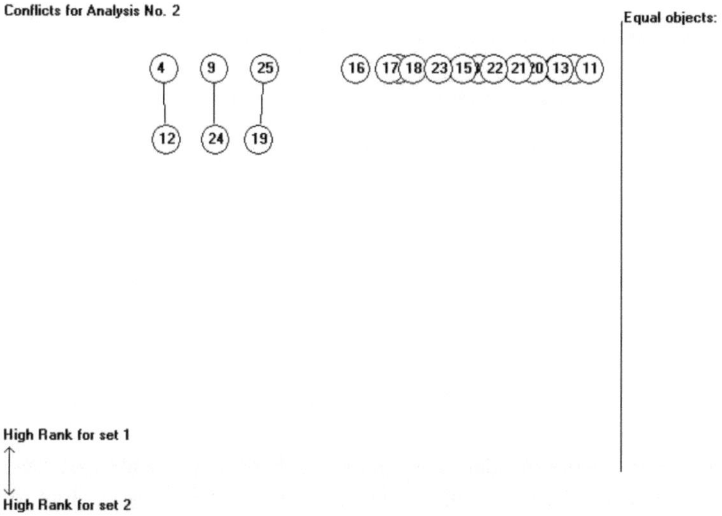

Fig. 10. Conflict diagram for the data set where the banned pesticides are neglected (1, 2, 6, 7, 8). (Figure drawn using Po Correlation)

The correlation profile of the single predictive variables and their interaction will be investigated in the following. This is easy to do using Po Correlation by repeating the correlation analysis only for respectively *Dose* and *SpArea* as single variable in *Set* 2. Inspection of the conflict diagrams shows how the two variables are acting as single variables and in relation to each other. In Fig. 11 the conflict diagram is displayed, where only the *Dose* is included as single variable in *Set* 2. The three pesticides (Ids. 5, 16 and 24) are ranked strongly downward having many concordant rankings. This tells that the *Dose* variable tends to rank these pesticides upward while downward rankings are more likely to happen for the variables *DetFreq* and *MedMax* together. The values in Table 1 also show that these pesticides are characterised by having a relatively high *Dose* value and a low *SpArea* value. Hence the occurrence seems limited due to limited propagation of use (limited sprayed area) event though the dose level is rather high when they are applied in the field. The Id. 13 is concordant ranked in the conflict diagram above 6 other pesticides and also associated with a very high *SpArea* value, which is not reflected in a high *Dose* value.

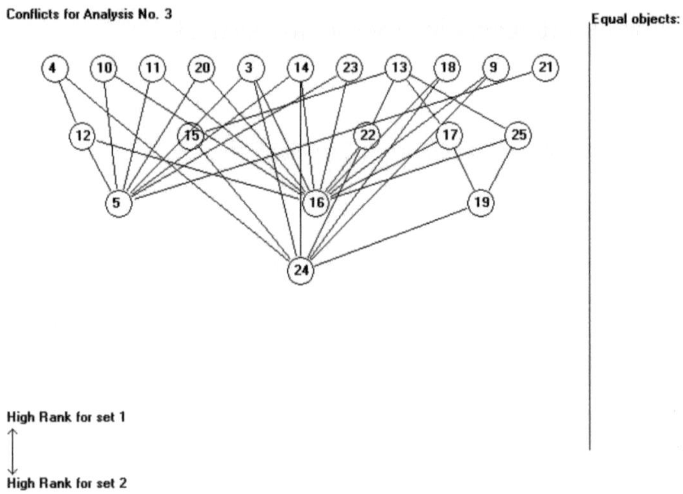

Fig. 11. Conflict diagram for the data set where the banned pesticides are neglected (1, 2, 6, 7, 8) and where the only variable used in *Set 2* is *Dose*. (Figure drawn using Po Correlation)

The conflict diagram using only the *SpArea* variable as *Set* 2 is shown in Fig. 12. Two strongly top ranked pesticides are identified (Ids. 15 and 9) having many concordant rankings downward. They both have a high *Dose* value and small *SpArea* value in Table 1 and thus cases where a low rank due to *SpArea* is in conflicts with the occurrence because of a high dose level. The complementary situation is also seen in Fig. 13 for Id. 21. This pesticide is ranked strongly downward in Fig. 13 and also a pesticide, which has low dose level and large sprayed area in according to Table 1. The Id. 21 is a very low dosage pesticide and this low dose level seems to prevent the pesticide to occur in the stream water even though the sprayed area is rather large.

The common set of rankings for Fig.'s 11 and 12 is displayed in Fig. 11, so only three rankings are in common between the Fig.'s 12 and 13. This shows that the two variables *Dose* and *SpArea* are working together by describing different parts of the information captured by *DetFreq* and *MedMax*. This is supported by the negative correlation between *SpArea* and *DetFreq* in Table 4.

In Fig. 13 a graphical display is made for a series of different variable combinations. The x and y axis is similar the axis in the significance plot (Fig.'s 8 and 9). Six different correlation analyses are performed for the pesticides, which have not been banned in year 2000.

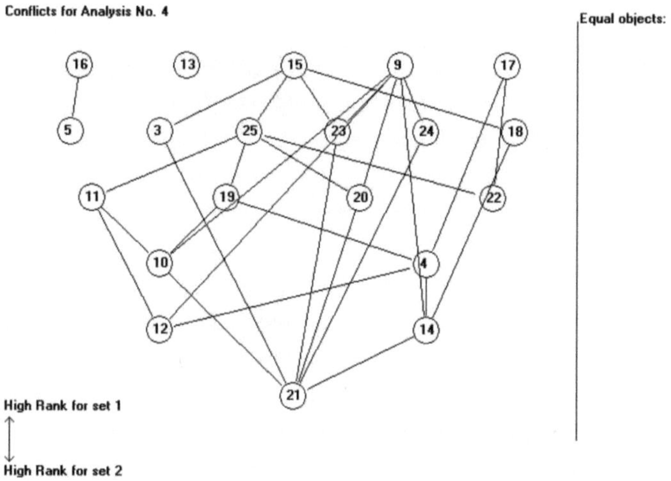

Fig. 12. Conflict diagram for the data set where the banned pesticides are excluded (Ids. 1, 2, 6, 7, 8) and where the only variable used in *Set 2* is *SpArea*. (Figure drawn using Po Correlation)

The numbered circles show the correlation results for a series of different variable combinations used for respectively *Set 1* and *Set 2*, which also refer to the columns in the Table to the right in Fig. 13. The small circles in this table identify the included variables. So *e.g.* for analysis no. 1, *Set 1* was made by the two variables *DetFreq* and *MedMax* and *Set 2* was made by only *Dose*. The variable denoted PRODUCT is the product between *Dose* and *SpArea* as presented by Table 4. The analysis 1, 2 and 3 is a test of every single variable including PRODUCT in relation to *Dose* and *SpArea* as *Set* 1. The product is best as single variable followed by respectively *Dose* and *SpArea* and this follows naturally the τ values in Table 4. However, the analysis no. 4 shows that the partial order of *Dose* and *SpArea* together performs better than the product if only the T (0,0) value is considered. The value drop on the y-axis from analysis no. 3 to analysis no. 4 is due to increased discordance in the ranking introduced when two variables are included in *Set* 2 instead of only a single one. The use of respectively *DetFreq* and *MedMax* as single variable in *Set* 1 is also tested as respectively no. 5 and no. 6. Neither analysis no. 5 nor 6 can make the same good correlation measured by T (0,0) as analysis no. 4.

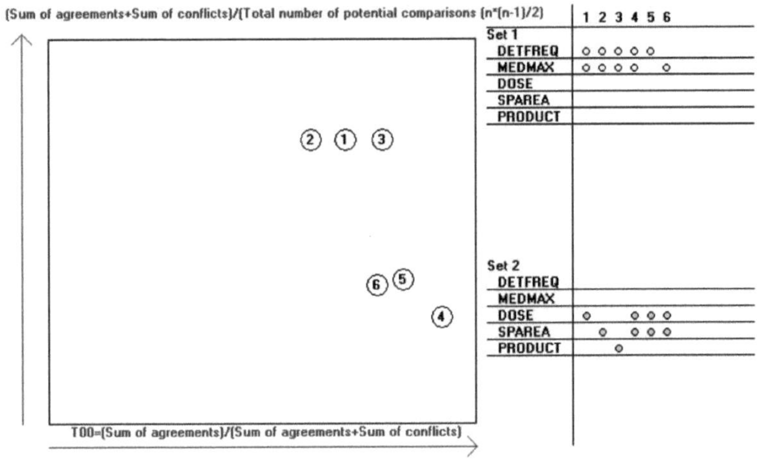

Fig. 13. A graphical display of the correlation result for a series of different variable combinations. The numbered circles refer to the numbering in the top row in the table to the right. Each number is a correlation analysis and the small circles in the tables indicate which variables that have been used in respectively *Set 1* and *Set 2*. The x and y axis is similar to the axis in the significance plot as explained for Fig. 8. (Figure drawn using Po Correlation)

References

Brüggemann R, Halfon E, Welzl G, Voigt K, Steinberg C (2001) Applying the Concept of Partially Ordered Sets on the Ranking of Near-Shore Sediments by a Battery of Tests. J Chem Inf Comp Sc 41:918-925

Brüggemann R, Bücherl C, Pudenz S, Steinberg CEW (1999) Application of the concept of partial order on comparative evaluation of environmental chemicals. Acta Hydrochim Hydrobiol Vol 23:170-178

Brüggemann R, Bartel HG (1999) A theoretical concept to rank environmentally significant chemicals, Journal of Chemical Information and Computer Sciences Vol 39(2):211-217

Conover WJ (1999) Practical nonparametric statistics, John Wiley & Sons, ISBN 0-471-16068-7

Gibbons JD (1993) Nonparametric measures of association, SAGE University papers. series: quantitative applications in the social sciences, ISBN 0-8039-4664-3

Goodmann LA and Kruskal WH (1963) Measures of association for cross-classifications III: Approximate sample theory. Journal of American Statistical Association Vol 58:310-364

Kendall MG (1938) A new measure of rank correlation. Biometrika Vol 30:81-93

Pavan M (2003) Total and partial ranking methods in chemical sciences. PhD thesis in Chemical Sciences, Cycle XVI, University of Milano – Bicocca

Pudenz S (2004) Stefan Pudenz, Criterion - Evaluation and Information Management. Mariannenstr. 33, D-10999 Berlin, Germany (web: www.criteri-on.de)

Pudenz S, Brüggemann R, Komossa D, Kreimes K (1998) An algebraic/ graphical tool to compare ecosystems with respect to their pollution by Pb, Cd III: comparative regional analysis by applying a similarity index. Chemosphere Vol 36:441-450

Sørensen PB, Brüggemann R, Thomsen M, Lerche DB (2005) Application of multidimensional rank-correlation, MATCH Communications in Mathematical and in Computer Chemistry 54:643-670

Sørensen PB, Brüggemann R, Carlsen L, Mogensen BB, Kreuger J and Pudenz S (2003) Analysis of monitoring data of pesticide residues in surface waters using partial order ranking theory - Data interpretation and model development. Environmental Toxicology and Chemistry Vol 22(3):661-670

Evaluation of Biomonitoring Data

Dieter Helm

Robert Koch-Institute, Seestraße 10, D-13353 Berlin, Germany

e-mail: helmd@rki.de

Abstract

The construction of posets or Hasse diagrams is a profitable means for the evaluation of biomonitoring data. In contrast to other statistical approaches the Hasse Diagram Technique enables the consideration of multiple attributes at the same time and will result in at least partially ordered data sets. Moreover, the calculation of averaged ranks allows the construction of a total order for a given data set. For the evaluation of biomonitoring data, as obtained for the German Environmental Specimen Bank, the Hasse diagram technique was applied to achieve partially or totally ordered data. The following scheme was applied: i) Careful rounding of the original data to increase the number of comparabilities; ii) splitting of the data in smaller sub-sets; iii) construction of the posets for each sub-set; iv) construction of the total order for each sub-set (by means of averaged ranks) and, v) synopsis of the sub-sets.

Introduction

At about the same time when Hasse published his famous book about algebraic topics and made Hasse diagrams popular (Hasse 1967), the environmental pollution reached its first depressing climax in Middle Europe. DDT, for example, which was invented during World War II, being the most powerful insecticide the world had ever known, was used with gay abundance in such large quantities, that this abuse caused the death of countless song birds and, as it accumulated in the food chain, prevented breeding success of the birds of pray, thus motivating Rachel Carson to write her famous book The Silent Spring (Carson 1962) – actual data on

DDT are summarised in Sørensen et al. 2004. DDT was so popular in these times that even the plant enthusiast treated ornamental pot plants in the home with this pesticide, often in too large a quantity and in a negligent fashion (Buxbaum 1958). In Germany the banks of the River Rhine were regularly covered with froth and the corpses of innumerable fish during the late 1960ies and the 1970ies. Acid rain killed trees in coniferous woodlands. Traffic-borne lead (from leaded gasoline) reached about 1970 an average air concentration of 125 ng/m^3 in the Hamburg area; human body burden was estimated retrospectively to be about 150 μg per litre blood at the same time (von Storch et al. 2003). (Actual values: approx. 20 ng lead per cubic metre air und 19 μg lead per litre blood for the student participants of the Environmental Specimen Bank (ESB). In due time, these alarming figures prompted the governments to action. In 1963 the Convention on the International Commission for the Protection of the Rhine against Pollution (Bern Convention) was signed by Switzerland, France, Luxembourg, Germany and the Netherlands. The European Economic Community joint in as contracting party in 1976. The DDT Act from 1972 prohibited production and use of DDT in Germany. The Petrol Lead Act of 1971 accomplished reduction of airborne pollution from lead in gasoline.

The German Environmental Specimen Bank

To monitor the effects of these and other activities of the legislative, the ESB was brought into being by the Federal Environment Ministry (FEM) of Germany in the 1970ies as a component of ecological environment surveillance (FEM 2000) and started sampling in 1981. After a series of development and trial phases, expansion of the ESB to full-scale operation (13 ecosystems and 4 human sampling sites) started on 1 January 1994. According to the concept (FEM 2000), the ESB, collects ecologically representative environmental and human specimens, which are analysed for environmentally relevant substances, and stored. Long-term storage is performed under conditions, which exclude any change in composition or chemical properties over a period of several decades. This archive retains specimens for retrospective analytical characterisation concerning unpredictable questions, which may arise in future. Although the specimens are analysed for a number of environmental substances prior to storage (monitoring), the genuine value of the ESB is the storage of samples (archive) to serve as records for the conservation of eco-toxicological and toxicological evidence. In order to attain a high level of quality assurance, all stages from sampling itself, to the transportation of specimens, the preparation and analysis of specimens, through to long-term storage, are set out in mandatory standard operating procedures (SOPs) for all types of environmental and human specimens. The environmental specimens are obtained from representative areas (ecosystems); representative ecosystems from the terrestrial, limnetic (riverine) and marine environment have been selected as examples. With regard to its entire composition each ecosystem is depicted by the biological, physical and chemical characterization of specimens taken from this system: producer, consumer, and destructor (FEM 2000). Table 1 gives an overview of the sampling programme of the ESB.

Table 1. Sampling programme of the ESB

Sampling area	Sampling sites	Ecosystem type	Specimens sampled
BR/NP Wadden Sea	3	Marine	1 - 4
Bodden NP of Western Pomerania	3	Marine	1 - 4
River Elbe	5	Riverine	5, 6, 15
River Rhine	4	Riverine	5, 6, 15
Bornhoeved Lake District	1	Agrarian	6 - 11, 14
Upper Bavarian Tertiary Uplands	1	Agrarian	7 - 9, 11, 14
Solling	1	Forestry	7 - 9, 11, 14
Palatinate Forest BR	1	Forestry	7 - 9, 11, 14
Saarland conurbation	3	Close to conurbation	5 - 7, 9 - 12, 14, 15
Dueben Heath	2	Close to conurbation	6, 9 - 13
Upper Harz NP	1	Nearly natural	7 - 9, 14
Berchtesgaden BR/NP	1	Nearly natural	7 - 9, 14
Bavarian Forest BR/NP	1	Nearly natural	7 - 9, 14
Human	4 Universities	Urban	Blood, urine, hair, saliva

Abbreviations: BR = UNESCO Biosphere Reserve (MAB programme); NP = National Park; 1 = common bladder wrack (*Fucus vesiculosus*); 2 = common mussel (*Mytilus edulis*); 3 = eelpout (*Zoarces viviparus*); 4 = herring gull's egg (*Larus argentatus*); 5 = zebra mussel (*Dreissena polymorpha*); 6 = bream (*Abramis brama*); 7 = spruce (*Picea abies*); 8 = beech (*Fagus sylvatica*); 9 = roe deer (*Capreolus capreolus*); 10 = domestic pigeon's egg (*Columba livia f. domestica*); 11 = earth worms (*Lumbricus terrestris* or *Aporrectodea longa*); 12 = Lombardy poplar (*Populus nigra* ‚Italica'); 13 = pine (*Pinus sylvestris*); 14 = soil; 15 = sediment

Most specimens are sampled in yearly intervals. Sampling is regulated by prescribed sampling timetables and by standard operating procedures (FEM 2000). Individual samples from each sampling site are pooled to give a pooled sample of approx. 5 kg wet weight. Further processing of the pooled samples includes grinding and freeze drying, resulting in a fine homogeneous powder, which is packaged as 10 g portions. These are stored for the long term over liquid nitrogen in the gas phase at temperatures below -150 °C. Results of the measurements prior to the storage are published in comprehensive reports (e.g. FEA 1999).

Evaluation Methods

The univariate approach

Since the temporal changes are of interest, investigations of trends are performed applying the Mann-Kendall-Test and the estimator of Theil (Conquest 2000). Fig. 1 shows a simple example of a trend analysis. Thallium values from 1994 to 2003 from bream musculature (*Abramis brama*) were obtained from pooled samples from the two Saarland sampling sites Rehlingen barrage weir und Güdingen barrage weir.

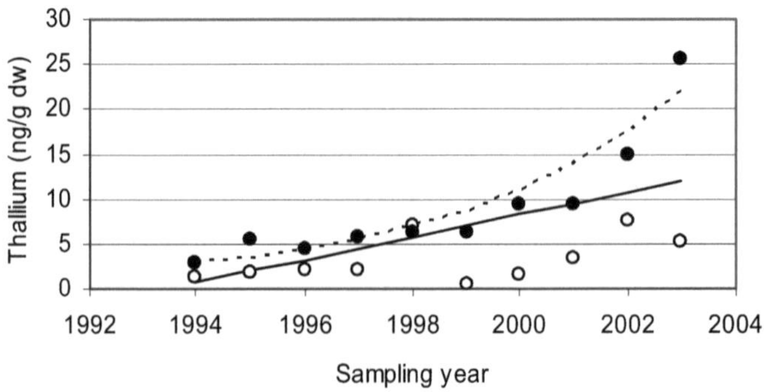

Fig. 1. Thallium measured in the musculature of the Bream (*Abramis brama*) from the Saarland conurbation. Closed circles: Rehlingen barrage weir; open circles: Güdingen barrage weir. The solid line shows the linear component of the trend as estimated with the Theil estimator and the dotted line depicts the trend estimator for data linearised prior to the analysis

Whilst the Mann-Kendall-Test found no trend in the Güdingen data ($p = 0.054$), the testing resulted in a positive trend for the Rehlingen data ($p < 0.001$) due to an input of the industrial zone of Dillingen located ca. 1 km up-stream from the Rehlingen weir. Theil estimator calculated the linear component of the trend to be an annual increase of 1.26 ng Thallium per g dry weight. Linearisation of the measurement values (by calculation of log10 values) prior to the Mann-Kendall-Test yielded a better fit of the trend line (dotted line in Fig. 1). What is much needed for the evaluation of biomonitoring data is a procedure that can evaluate the samples in terms of their contamination with environmental substances. In the above example, Rehlingen is always more strongly contaminated than Güdingen with the sole exception of 1998. For all other years one would order the both sites as Rehlingen > Güdingen when only Thallium is considered. (See also the chapter by Pudenz and Heininger p.111, where the results of ecotoxicological tests in river sediments are presented and analyzed.)

The multivariate approach

Like many other statistical methods for the evaluation of biomonitoring data, the above-depicted example of a trend analysis considers only a single variable. Although the multivariate procedures consider several measured variables at the same time, their results are often only limited meaningful. Cluster analyses can reveal structures in a given data set; principal component analyses concentrate the information contents of many variables in a set of a few latent variables, which are difficult to interpret correctly.

Fig. 2 depicts the result of a cluster analysis performed on the data in Table 2. Sampling sites are combined to clusters according to the similarity of their contamination pattern rather than the magnitude of this contamination. Consequently, four clusters are formed: cluster 1 consisted of the sites Weil, Koblenz, Cumlosen, Blankenese and Barbay; cluster 2 of Saale, Güdingen and Mulde; cluster 3 comprised the sites Iffezheim, Bimmen and Rehlingen; and the fourth cluster, more isolated, was composed by Prossen and Zehren. Even if this partition could reveal interesting structures, it was not possible to identify the most severely contaminated sites, let alone to order them according to their contamination. Similar results were obtained by principle component analysis (PCA; see Fig. 3). PCA yielded four principal components, which explained 82 % of the entire information (formerly distributed in 11 measured variables). The first component was dominated by DDE, OCS, HCB, Hg, and Pb. The second summarised the information of PCB, Cu and As. The third component was dominated by Tl and HCH, whereas the fourth component represented Se. The clusters differed somehow from those detected by cluster analysis (Fig. 3). Additionally, the sampling sites could be ordered along the component axes. In respect to component 1, Zehren was mostly contaminated whereas Güdingen and Rehlingen appeared to be the cleanest sites. In contrast to this, Rehlingen was the most contaminated site when component 2 was considered. Thus, no unequivocal order could be achieved.

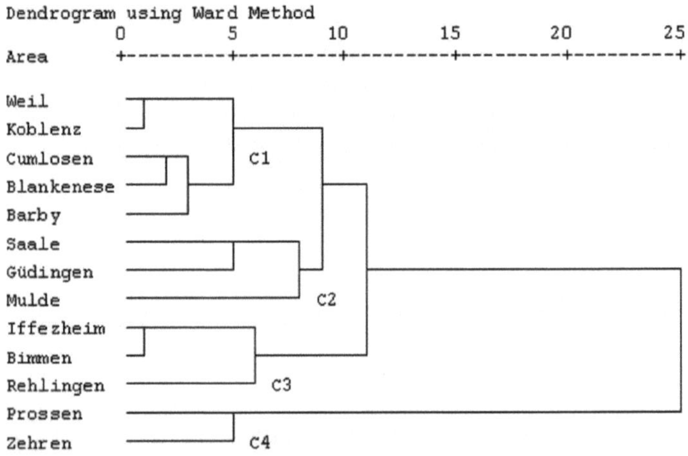

Fig. 2. Result of a hierarchical cluster analysis of the data in Table 2

The partial order approach: Evaluation of ESB data with the Hasse diagram technique

Neither cluster analysis nor PCA could order the sampling sites considering the complete information. One approach to judge more than one measured variable at the same time is the calculation of an index, which is supposed to summarise the information of all single variables. This approach, however, arises new problems, which are difficult to solve. First, the calculation of an index would require correct weighting factors for all single variables and secondly, before adding up the values all variables must be transformed to the same numeric scale. Consider, for example, the third row in Table 2. The index for Blankenese would be approx. 3 µg/g for the organic compounds when the weighting factors for all substances would be set to 1. This would mean that neither HCB nor the very much more toxic DDE will contribute equal amounts (this is, each approx. 20 %) to the index's value.

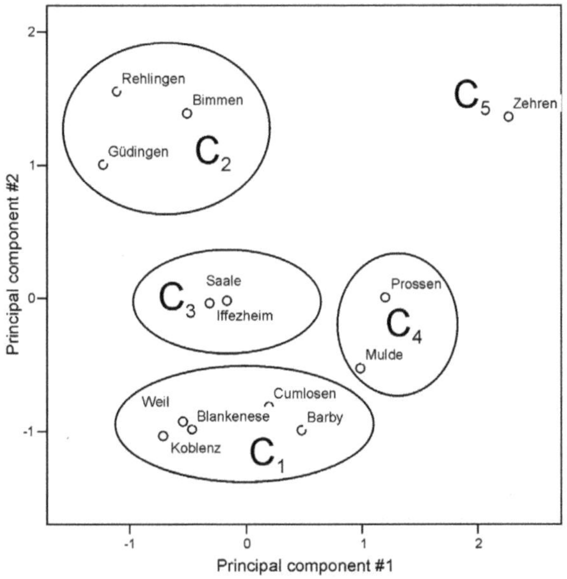

Fig. 3. Result of PCA performed on the data in Table 2

Since this is obviously unjustified we need to apply weighting factors accounting for the ecotoxicity of each compounds, but the problem is that we do not know them. An additional disadvantage of the index is the loss of information since all data are lumped together to give a stew.

The Hasse Diagram Technique (HDT) is a method out of the scale of the mathematical order theory, which was already applied productively to ecological and ecotoxicological problems (e.g. Brüggemann et al. 1999; Brüggemann 2001a; Brüggemann and Halfon 1997; Brüggemann and Steinberg 2000). More recent applications of the HDT in the environmental sciences can be seen elsewhere in this book. The principle of the technology is based on the fact that 'objects' (e.g. ecological systems, habitat diversity, field data, chemicals, databases, sampling sites of the environmental sample bank) are compared each with one another and ordered on the basis of their attributes (e.g. measured values of xenobiotics). Each comparison of objects considers all attributes at the same time. An object is classified to be 'worse' than another, if it is 'worse' or at least 'equal' in respect to each of the attributes than the other object (or, in other words, if the product order relation is fulfilled). The designation of the relation is then: '>'. In ecology the term 'better' can be replaced by 'more contaminated' since we consider higher concentrations to be 'worse'. If two objects are equal for all attributes, the symbol for their relation is '='; they are equivalent. Controversial or inconsistent results of a comparison will result when object a is 'better' than object b in respect of one attribute but at the same time 'worse' in respect to another. This will be labelled with the symbol '||'. In terms of the HDT the two objects are 'not comparable' and can, thus, not put into any order.

In short,

a > b	means that object a is more contaminated than b
a = b	means that object a and b are contaminated in the same way
a || b	means that objects a and b cannot be compared with another

Table 2. Original data for bream from 2002 (inorganic compounds in terms of dry weight; organic compounds related to the fat contents of the musculature)

Sampling site	HCB	γ-HCH	OCS	ΣPCB	DDE	As	Pb	Cu	Hg	Se	Tl
			[μg/g]			[μg/g]	[ng/g]	[μg/g]	[ng/g]	[μg/g]	[ng/g]
Barby	1.60	0.027	0.262	2.38	2.10	0.38	86.4	1.40	1050	3.40	2.7
Bimmen	0.24	0.016	0.089	7.40	0.57	0.72	128.8	1.15	927	2.30	8.8
Blankenese	0.57	0.038	0.193	1.58	0.58	0.54	62.3	2.04	524	2.45	0.5
Cumlosen	0.98	0.025	0.301	2.17	1.48	0.43	59.0	1.93	1270	1.91	6.4
Güdingen	0.10	0.033	0.002	7.25	0.61	0.24	91.8	1.48	507	1.47	7.5
Iffezheim	0.56	0.017	0.054	4.22	0.41	0.66	87.8	1.08	1155	2.53	1.8
Koblenz	0.17	0.012	0.012	1.36	0.23	0.43	62.4	1.00	426	2.82	0.8
Mulde	1.16	0.033	0.392	2.71	5.19	0.42	98.2	2.72	1520	2.49	12.3
Prossen	1.47	0.017	0.620	6.54	5.14	0.37	150.0	1.44	1585	1.83	1.5
Rehlingen	0.06	0.043	0.003	7.16	0.34	0.56	66.9	1.31	603	3.17	14.8
Saale	0.20	0.042	0.047	2.29	2.54	0.34	149.0	1.24	1638	2.28	2.0
Weil	0.12	0.013	0.058	2.64	0.46	0.45	51.1	1.51	841	2.27	1.4
Zehren	3.15	0.022	0.711	8.54	9.28	0.47	151.5	1.29	1973	2.25	16.6

For all objects, which meet the order relation, a (partial) order can be constructed. The position of any given object within this order enables direct reading of the relative ecological or ecotoxicological load of this object. Additionally, the applied software, WHASSE (Hasse for Windows, producer: GetSynapsed GmbH, München) allows presenting the pollutant profile as bar chart, increasing the information content of the Hasse diagrams (Bücherl et al. 1995). The differences made visible in this way allow conclusions to contamination processes and pathways (Brüggemann 2001a). The most crucial advantage of the HDT is the synoptic view of several attributes, without the need of an index or a quality function (Brüggemann 2001b). This synoptically view enables an ecosystemical evaluation. The helpfulness of the HDT to the comparison of ecological systems in respect to their environmental loading had been already demonstrated (Brüggemann et al. 1994).

Some terms often used in connection with the HDT must be explained since they are necessary to understand and interpret a Hasse diagram. Please note, that Hasse diagrams are read top-down or exclusively bottom-up. In this chapter the top-down view is preferred.

- Anti-chain: An alignment of objects, which are not comparable with one another. Elements of the same level (see chapter by Brüggemann and Carlsen, p. 61) are incomparable. They can be considered to be similarly polluted but with different pollution patterns. As sometimes the construction of levels cannot be done uniquely, their interpretation needs some care.
- Chain: An alignment of objects, which are all comparable with one another. The elements of a chain are connected with a line. Often a common mechanism is responsible for the formation of the chain, which leads to the synchronous, at least weakly monotonous increase of some attributes.
- Equivalence: Two objects are equivalent, if all attributes have equal values.
- Incomparability: Two objects are not comparably with one another if the first object is, in respect to at least one attribute, 'worse' than the other and, simultaneously, in respect to another attribute, 'better'. In the Hasse diagram incomparable objects are not connected with lines.
- Isolated object: Objects, which are comparable with none of the other objects in the data set. Because of conservatism isolated objects are always assigned to the highest diagram level. This corresponds with the assumption that high pollutant concentrations indicate a high endangerment. The software Hasse for Windows automatically adopts the correct arrangement.
- Maximal objects: Objects, for which no other objects exist in the data set, which can be classified 'worse' are called maximal objects. In WHASSE maximal objects are assigned to the uppermost level in the diagram.
- Minimal objects: Objects, for which no other objects exist in the data set, which can be classified 'better' are called minimal objects.

- Product order relation: The product order relation is fulfilled if i) an object is, in regard to at least one attribute, 'worse' than another and at the same time 'equal' in respect to all remaining attributes, or if ii) an object is, in regard to at least one attribute, 'better' than another and at the same time 'equal' in respect to all remaining attributes. The existence of the product order relation is the prerequisite for the construction of chains.
- Predecessor: An object ranked above a given object x. In our case the predecessor is more contaminated than x.
- Sensitivity: A measure for the importance of a particular attribute. How strongly will the omittance of a given attribute change the resulting Hasse diagram? Most often the matrix W, calculated in WHASSE, is used to perform sensitivity studies.
- Stability: An estimator for changes in the diagram to be anticipated, if any attributes will be added or omitted. Symbol: P(*IB*). Since P(*IB*) is normalised and can only take values between 0 and 1 it can easily be interpreted. If P(*IB*) = 1, then all objects are arranged in an anti-chain – the inclusion of additional attributes will not change the structure. If P(*IB*) = 0, then all objects are arranged in a chain or they are equivalent to each other - the chain (and/or the equivalence) remains, if attributes are omitted.
- Successor: An object ranked below a given object.

The Data Set

Specimens

The bream (*Abramis brama* Linné 1758, Cyprinidae) is a predominantly carnivorous teleost fish of the carp family inhabiting lakes and quiet parts of slowly running rivers and is widespread throughout Europe (Bond 1979). It feeds mainly on small molluscs (e.g. *Pisidium*, *Anisus*), tubifex and insect larvae (e.g. *Chironomus*), sucking in sediment with the aid of its protrucible mouth while foraging and extracting the food from the sediment, but will also take plant material occasionally.

Data were obtained for the bream's swimming musculature as collected for the ESB in 2002. Different numbers of 4 to 19 years old bream (n = 16 – 28) have been captured after the breeding season in each of the 13 different sampling areas. HCB, γ-HCH, octachlorostyrole (OCS), PCB 101,

PCB 118, PCB 138, PCB 153, PCB 180, and 4,4'-DDE were measured via GC-MS. The 5 PCB congeners were added up to give the sum parameter ΣPCB. With the exception of mercury, which was measured via DMA, all metal analyses were carried out using ICP-MS.

Fig. 4. The bream (*Abramis brama*), a specimen for the ESB (Photograph by Dr. Roland Klein, University of Trier)

Sampling areas

Bream were collected in 13 sampling areas belonging to the three riverine zones of Elbe, Rhine and Saar. ESB samples of bream were collected at 5 different sampling sites along the river Elbe; these are – in down-stream direction: Prossen (river km 13), Zehren (km 93), Barby (km 296), Cumlosen (km 470), and Blankenese near the Port of Hamburg (km 633). Additionally, bream from the Elbe tributaries Saale and Mulde (the latter near the mouth), were sampled too. In 2002, the year of the Elbe flood, at all sites (with the exception of Blankenese) bream were collected during the first days of the announcing flood. The Blankenese specimens were collected after the flood (in September). River Rhine was represented by sampling areas Weil (river km 174), Iffezheim (km 334), Koblenz (km 590,3), and Bimmen near the German-Dutch border (km 865). Bream from the river Saar stemmed from the barrage weirs of Güdingen (km 93) and Rehlingen (km 54).

Hasse Diagrams or POSETs

Hasse diagram for the entire data set

The entire raw data set, as listed in Table 2 (13 objects, 11 attributes), was used to construct a Hasse diagram. The result (Fig. 5A) is a single antichain, due to very individual contaminant patterns of the 13 sampling areas. None of the sampling areas is comparable to any other. This result, being somewhat unpromising, leads us to the development of a double fold stratagem (Helm 2002). Firstly, since the incomparableness of the objects is, in many cases, due to only minor differences of the measured values, many of them being in the magnitude of the measurement uncertainty, an appropriate, carefully performed rounding of the data will better the diagram. Take, for example, the As values for the Elbe sampling areas Cumlosen (0.43 [µg/g dw]) and tributary Mulde (0.42 [µg/g dw]). Since 0.43 is greater than 0.42, Cumlosen can be considered to me more contaminated than Mulde. This difference, however, lies within the measurement uncertainty of approx. 10% and is, therefore, not justified. (Note the similar ideas of "smoothing" of data (noise deficient QSAR) in chapter by Carlsen, p. 163).

Secondly, splitting the entire data set into two or more sub-sets, each with a lower number of attributes, will yield more comparabilities since a greater number of attributes will reduce the probability that the product order relation is fulfilled: $P((a_n > b_n) \cap (a_m > b_m)) < P(a_n > b_n)$.

Let us first focus on the rounding procedure. Desirable as it may be, rounding of the data according to the overall data error is not possible, due to the simple fact that we do not know it. The complete data error is composed by i) the simple (or multiple) measurement uncertainty, ii) the long-term lab error and iii) the sampling error. The simple measurement uncertainty is – depending on the contaminant and its concentration – about 10 to 20 %. The long-term lab error is even more important because the series of measurement in biomonitoring can easily span many years (like in ESB). Unfortunately the latter is unknown and the same is true for the sampling error. Thus we better adjust the rounding of the data to the scale of the measured values.

Pre-processing: Rounding of the raw data

A good measure for the scale (or magnitude) of data is the median. Other than the arithmetic mean the median is insensitive to the distribution of the data. After calculation of the median for every data column of Table 2, data are to be rounded to the *k*. decimal place.

$$k = \mathrm{trunc}(log10(\mathrm{median})) - 1 \qquad \text{if median} < 1 \tag{1}$$

$$k = \mathrm{trunc}(log10(\mathrm{median})) + 1 \qquad \text{if median} \geq 1 \tag{2}$$

If k is negative, the data will be rounded to k^{th} position after the decimal point; if k is positive, the data will be rounded to k^{th} position prior to the decimal point. Two examples illustrate the method (HCB and Hg from sampling area of Prossen, river Elbe).

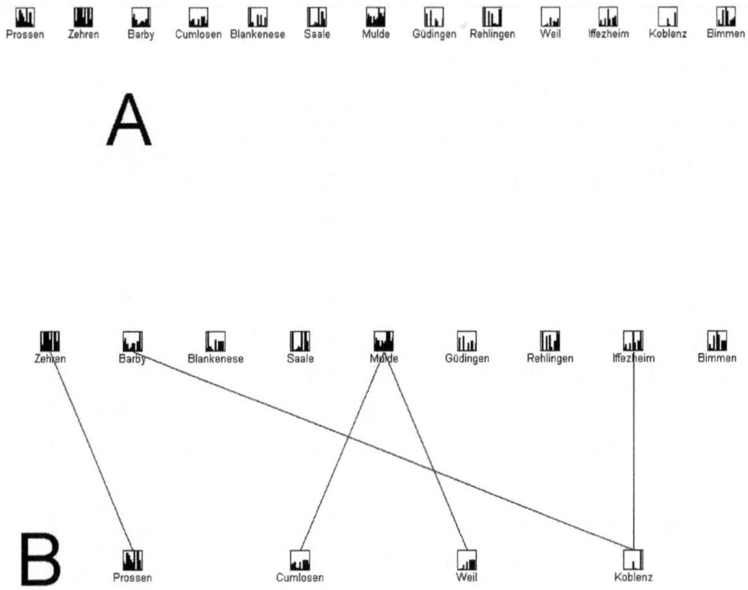

Fig. 5. Hasse diagram for the raw data (A) and for the rounded data (B)

Table 3. Stepwise implementation of the rounding process

Step #	Example 1	Example 2
1. Measured value	1.47	1585
2. Median of the data column	0.559	1050.00
3. The logarithm of the median	-0.253	3.021
3. Truncation of the decimal	0	3
4. decrement, if median < 1, or,	-1	
increment, if median ≥1		4
5. k	-1	4
6. Resulting rounded value	1.5	2000

In example 1 k is -1, which means that the HCB values will be rounded to the first digit after the decimal point, and the mercury values will be rounded to the fourth digit before the decimal point, because k is 4 in this case. The rounded data are listed in Table 4.

Table 4. Rounded data for bream from 2002 (inorganic compounds in terms of dry weight; organic compounds related to the fat contents of the musculature)

Sampling site	HCB	γ-HCH	OCS	ΣPCB	DDE	As	Pb	Cu	Hg	Se	Tl
			[μg/g]			[μg/g]	[ng/g]	[μg/g]	[ng/g]	[μg/g]	[ng/g]
Barby	1.6	0.03	0.26	2	2.1	0.4	90	1	1000	3	3
Bimmen	0.2	0.02	0.09	7	0.6	0.7	130	1	1000	2	9
Blankenese	0.6	0.04	0.19	2	0.6	0.5	60	2	1000	2	1
Cumlosen	1.0	0.02	0.30	2	1.5	0.4	60	2	1000	2	6
Güdingen	0.1	0.03	0.00	7	0.6	0.2	90	1	1000	1	7
Iffezheim	0.6	0.02	0.05	4	0.4	0.7	90	1	1000	3	2
Koblenz	0.2	0.01	0.01	1	0.2	0.4	60	1	0	3	1
Mulde	1.2	0.03	0.39	3	5.2	0.4	100	3	2000	2	12
Prossen	1.5	0.02	0.62	7	5.1	0.4	150	1	2000	2	2
Rehlingen	0.1	0.04	0.00	7	0.3	0.6	70	1	1000	3	15
Saale	0.2	0.04	0.05	2	2.5	0.3	150	1	2000	2	2
Weil	0.1	0.01	0.06	3	0.5	0.4	50	2	1000	2	1
Zehren	3.1	0.02	0.71	9	9.3	0.5	150	1	2000	2	17

The Hasse diagram constructed for the rounded data is depicted in Fig. 5B. As a consequence of the rounding, the number of levels increased from 1 to 2, the number of incomparabilities was reduced from 156 to 146.

Five chains have been formed, from which three do represent intra-river segments, indicating an increase of pollution for the river Elbe from Prossen towards Zehren, a decrease from tributary Mulde towards Cumlosen, and, for the river Rhine, a decrease of the pollution from Iffezheim towards Koblenz. However, the chains are very short (each comprising only two elements) and there are still 5 isolated objects (Blankenese, Saale, Güdingen, Rehlingen, Bimmen). The stability, P(*IB*), is 0.94, indicating, that the diagram is very near an anti-chain and that the inclusion of additional attributes will not change the structure. On the other hand, the exclusion of attributes may very well change the structure. This leads to the second stratagem, the splitting of the data set. The data set will be splitted into two sub-sets; one containing the inorganic components and the other containing the organic substances.

Hasse diagram for the inorganic compounds

The resulting Hasse diagram or Partial Order for the six inorganic compounds was similarly poorly structured as the Hasse diagram for all 11 pollutants when the original raw data were used (Fig. 6A). The diagram consisted of only two levels; 9 objects were assigned to an anti-chain. Only two short chains were formed, each of which consisted of two objects and none of these short chains represented intra-river segments. Application of rounded data produced an additional level and the number of incomparabilities was reduced from 152 to 130. Two three-link chains and 6 two-link chains were formed; some of them corresponded to river segments, thus allowing deriving a partial order for these streams with an enriched degree of comparabilities (Fig. 6B). Within the first chain for river Elbe the sampling site of Zehren appeared to be most contaminated, followed by Prossen and tributary Saale. The second Elbe chain showed site Mulde to be stronger contaminated than Cumlosen and within the Rhine segment site Koblenz was cleaner than site Iffezheim. Most important for the order achieved are Cu ($W = 14$) and Se ($W = 10$); of no importance is Hg ($W = 0$). The stability value (P(*IB*)) was reduced from 0.97 to 0.83 by the rounding of the data, but is still near that of an anti-chain.

Hasse diagram for the organic compounds

When the same scheme was applied to the sub-set of the five organic compounds, the improvement achieved by the rounding of the data was far less pronounced. The number of incomparabilities was reduced from 116 to 114; the number of levels remained 4 and that of the isolated objects remained 2 for both, the raw data and the rounded data. This remarkably lesser improvement can be attributed to the fact that the organic compounds are stronger correlated with each other than the inorganic compounds with each other and thus are more often increased simultaneously. The maximum chain length was 4.

In terms of inter-fluvial segments, both Bimmen and Iffezheim were recognised to be more contaminated than Koblenz (River Rhine), additionally Bimmen was more strongly contaminated than Weil. Within the Elbe segment the order Zehren > Prossen > Cumlosen was found. Elbe tributary Mulde was more strongly contaminated than Cumlosen. Güdingen and Rehlingen from river Mosel proved to be incomparable.

In general, the Elbe sites and tributaries appeared in the upper part of the diagram, whereas the river Rhine sites appeared in the lower half, allowing a tentative assessment of the contamination for both streams. The most important substances for the partial order shown in Fig. 7 are HCH and PCB-sum (for each: $W = 17$). The stability P(*IB*) is 0.73, indicating that the order achieved is still near an anti-chain.

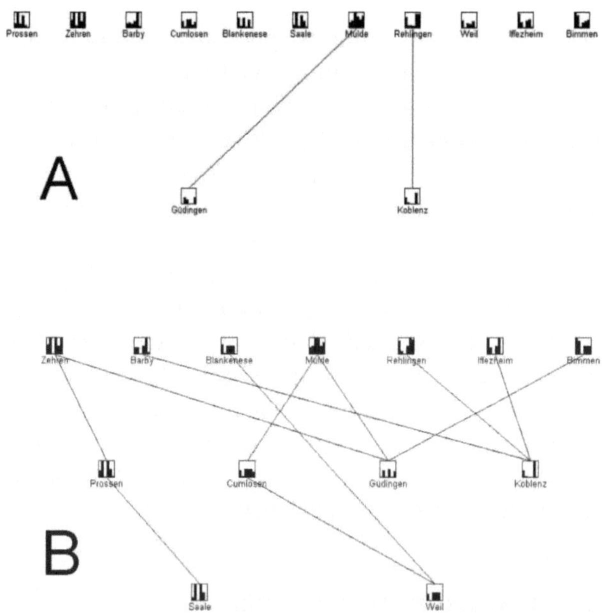

Fig. 6. Hasse diagram for 6 inorganic compounds; A: raw data; B: rounded data

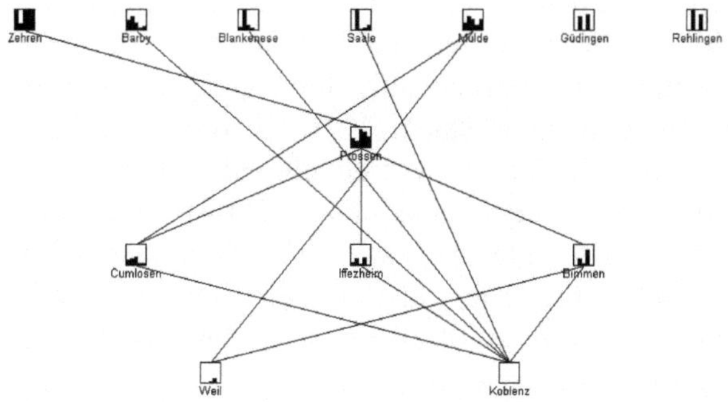

Fig. 7. Hasse diagram for 5 organic compounds; rounded data

From partial to total order: Calculation of averaged ranks

Though some improvement have been achieved, the Hasse diagrams or partially ordered sets shown in Fig.'s 6B and 7 are not entirely satisfactory. To overcome this, linear extensions can be constructed for each partially ordered set of objects. A linear extension is a total order where all comparabilities will be conserved (Brüggemann et al. 2004; see also chapters by Brüggemann and Carlsen, p. 61 and Carlsen, p. 163). This will yield a number of linear extensions for a given partially ordered set, taking all possible locations of the incomparable objects into account. Take e.g. the Hasse diagram for the inorganic compounds (Fig. 6B). To simplify the matter we will only consider the subgraph consisting of Zehren, Prossen, Saale and Güdingen. The possible linear extensions for this subgraph, when all comparabilities are preserved, are as follows.

Zehren > Güdingen > Prossen > Saale
Zehren > Prossen > Güdingen > Saale
Zehren > Prossen > Saale > Güdingen

Please note, that in all cases the orders Zehren > Prossen > Saale and Zehren > Güdingen, respectively, have been preserved. In a likewise manner all possible linear extensions for any partially ordered set can be constructed. Most unfortunately the number of linear extensions will increase exceedingly with the number of objects and incomparabilities in the given set. In case of Fig. 6B not less than 13,414,830 linear extensions can be constructed. When all possible linear extensions are known the most probable rank for each object can be calculated, thus yielding a totally ordered set (Brüggemann et al. 2004) without the use of weights or assuming a rank index function. Since, however, the calculation of all possible linear extensions will require an impracticably long computing time this approach cannot be taken in most cases. To solve this obstacle, a method based on Monte Carlo calculations was developed (Sørensen et al. 2001, Lerche et al. 2003) and Brüggemann et al. (2004) gave a simple and straightforward equation for the calculation of averaged ranks which are very close estimations of the 'true' ranks obtained by the examination the cumbersomely calculated full set of all linear extensions (Brüggemann et al. 2004), counting from bottom to top.

$$R_{kav(x)} = (S_{(x)} + 1) \cdot (N + 1) / (N + 1 - U_{(x)}) \qquad (3)$$

where $R_{kav(x)}$ is the averaged rank of object x, $S_{(x)}$ is the number of successors of x (objects ranked below x), $U_{(x)}$ is the number of objects incomparable to x and N is the total number of objects in the set. It must be mentioned that an object with many successors will tend to get a higher averaged rank than one, which has instead many predecessors. Therefore the partial order must be well justified, and only with significant comparabilities and incomparabilities the averaged rank approach is a reasonable one (Brüggemann et al. 2004).

Application of the averaged ranks approach to the partially ordered set of inorganic compounds (Fig. 6B) yielded the totally ordered set depicted in Fig. 8A. In words, the order is:

{Zehren, Mulde} > {Barby, Blankenese, Rehlingen, Iffezheim, Bimmen} > Prossen > Cumlosen > Saale > {Güdingen, Koblenz, Weil};

corresponding to the averaged ranks 2.8, 4.6, 7.0, 8.4, 10.5 and 11.2, respectively. (Please note that in Fig. 8 equivalent objects are 'hidden beneath' the first object of each equivalence group.)

Likewise a totally ordered set can be obtained for the organic data (Fig. 8C). The resulting order is:

Mulde > {Zehren, Prossen} > {Saale, Barby, Blankenese} > {Rehlingen, Bimmen, Güdingen} > Iffezheim > Cumlosen > Weil > Koblenz;

corresponding to the ranks of 2.8, 3.5, 4.6, 7.0, 8.4, 9.3, 11.6 and 12.7, respectively.

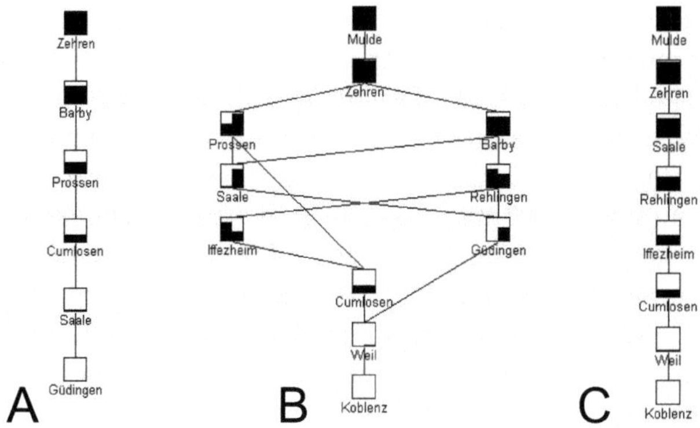

Fig. 8. Complete order sets for the inorganic (A), the organic compounds (C), and a synopsis of both (B), achieved by the calculation of averaged ranks. Note that there are equivalent objects (see text)

Synopsis of inorganic and organic data

We have now two different totally ordered sets; each one for the organic and the inorganic compounds which parallel in same respects. In both cases, the sampling sites of Mulde, Zehren, Barby and Blankenese appear at or near the top and can thus be considered to represent the mostly contaminated sites. In contrast, Koblenz, Weil and Cumlosen are assigned to the bottom end of the order. When defining the averaged ranks of both groups of compounds as new attributes the order as shown in Fig. 8B can be obtained. The stability (P(*IB*)) of this poset is 0.14, indicating nearness to a chain. This poset has 8 levels, only one minimal and one maximal object, and the number of comparabilities is 69 (the maximum number). The corresponding total order, achieved by the calculation of averaged ranks, is Mulde > Zehren > {Barby, Blankenese} > Prossen > {Rehlingen, Bimmen} > Saale > Iffezheim > {Güdingen, Cumlosen} > Weil > Koblenz.

This means that the Elbe sites are stronger contaminated than those of river Rhine, if all 11 compounds are considered. Within river Elbe and its tributaries we found the order: Mulde > Zehren > {Barby, Blankenese} > Prossen > Saale > Cumlosen. The order for the river Rhine sites is Bimmen > Iffezheim > Weil > Koblenz. Since none of these orders corresponds to the course of the streams we must assume a complex interplay of contamination and subsequent dilution and degradation processes. Coming from the Czech Republic the Elbe is medium strongly contaminated. Up to the mouth of tributary Mulde this contamination is strongly increased, but tributary Saale, bringing cleaner water, will dilute this contamination.

Barby, located at the mouth of tributary Saale, and Cumlosen have decreasingly less contamination, whereas Blankenese (located in close proximity and down-stream from the Port of Hamburg) shows a new increase of contamination. Whether this is due to local inlet of contaminants or the sampling time (after the flood) cannot be decided here. The river Rhine, too, starts with medium contamination, which is increased between Weil and Iffezheim, whereas the contamination is diluted (or otherwise reduced) in down-stream direction until to Koblenz. But the industrial regions located at the Lower Rhine seem to increase the contamination anew, so that location Bimmen, near the German-Dutch border, is the most strongly contaminated site of the river Rhine. River Saar, represented by only two sampling sites, shows the order Rehlingen > Güdingen; an order which reverses the direction of the stream. As already mentioned above, this is due to an input of the industrial zone of Dillingen located ca. 1 km up-stream from the Rehlingen weir.

Conclusions and Outlook

The construction of a partial or even total order is a valuable means for the evaluation of biomonitoring data. A critical point is the number of incomparabilities. To reduce this number the data should be pre-processed. For this contribution pre-processing was done by rounding the data according to the magnitude of the median. A somewhat more sophisticated approach could be a rounding procedure, which considers the measurement uncertainty, thus expressing the data as multiples of the measurement uncertainty, taking into account different uncertainties for different chemicals. Another possibility for pre-processing is the cluster analysis (Luther et al. 2000), resulting in a smaller number of objects, which can be used as input for the Hasse Diagram Technique. Disadvantage of this approach is, however, the loss of information.

It should be stressed that there are more possible applications for posets in biomonitoring. One promising application for the Hasse Diagram Technique is the ranking of chemicals. When chemicals are expressed as fractions of the corresponding guide values or limit value, a partial order as shown in Fig. 9 will result. Prior to the ranking, the data of six chemicals obtained from the sampling sites of river Elbe, including the two tributaries Mulde and Saale, have been transformed according to guide values of the EC. Then the data table was transposed so that the rows became columns and vice versa, thus making the chemicals to objects and the sampling sites to their attributes. The resulting partial order reveals that mercury is the most relevant chemical, whereas γ-HCH is of least importance.

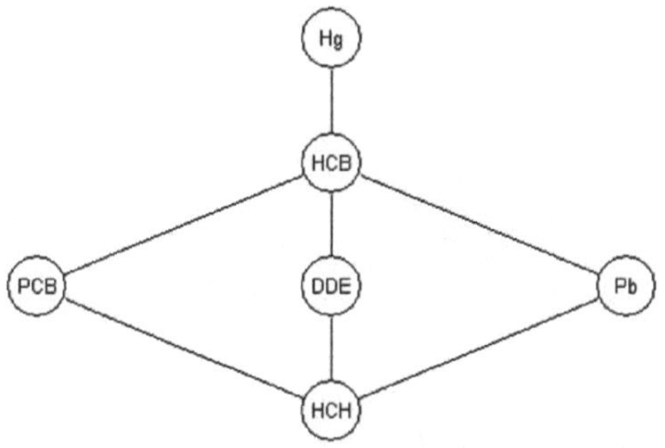

Fig. 9. Ranking of chemicals for the sampling sites of river Elbe

Analysis of the temporal course of contaminations is another example for further applications (Fig. 10). Based on essentially the same chemicals it seems that human beings experience an increasingly better protection than wild live. For bream sampled from 1997 to 2003 at the barriage weir of Rehlingen the improvement of the environment did not parallel the time course (Fig. 9A), since the order of contamination is, obviously, 1999 > {1997, 2001} > 2000 > {2002, 2003} > 1998. In contrast, samples taken from students of the Münster University show a clear parallel between time and bettering of the body burden. This would mean, that man is better protected against the chemical released into the environment by him than wild life is. But since man is an inseparable part of the environment contaminations will always backfire on the mankind.

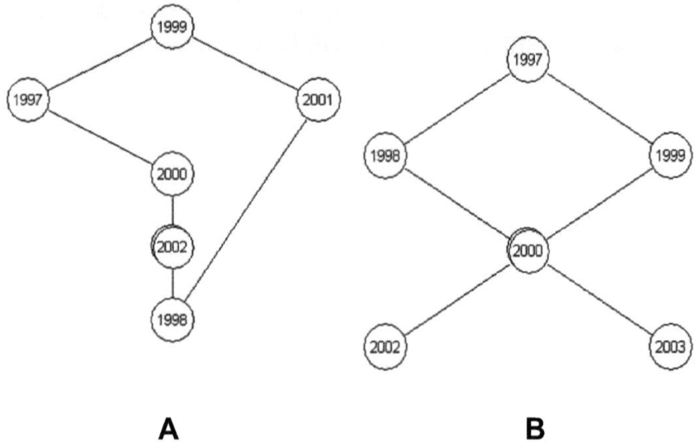

Fig. 10. Temporal course of contamination for environmental samples (A: bream from sampling site Rehlingen) and human samples (B: students of the University of Münster)

References

Bond CE (1979) Biology of fishes. Philadelphia, Saunders

Brüggemann R (2001a) Einsatzmöglichkeiten des Konzepts partiell geordneter Mengen für Bewertungen in ökosystemarem Kontext. In. Steinberg CEW, Brüggemann R, Kümmerer K, Ließ M, Pflugmacher S, Zauke GP (Hrsg.). Streß in limnischen Ökosystemen. Neue Ansätze in der ökotoxikologischen Bewertung von Binnengewässern 173-185

Brüggemann R (2001b) Ansätze zur vergleichenden Bewertung unter spezieller Berücksichtigung der Hassediagrammtechnik. In. Neumann-Hensel H, Ahlf W, Wachendörfer V (Hrsg.): Nachweis von Umweltchemikalien. Auswerte- und Interpretationsmethoden für Toxizitätsdaten aus einer ökotoxikologischen Testkombination. Initiativen zum Umweltschutz 29:27-46

Brüggemann R, Sørensen PB, Lerche D, Carlsen L (2004) Estimation of averaged ranks by a local partial order model. J Chem Inf Comput Sci 44:618-625

Brüggemann R, Halfon E (1997) Comparative analysis of nearshore contaminated sites in Lake Ontario: Ranking for environmenatl hazard. J Environ Sci Health A32:277-292

Brüggemann R, Steinberg C (2000) Einsatz der Hassediagrammtechnik zur vergleichenden Bewertung von Analysendaten – am Beispiel von Umweltuntersuchungen in den Regionen Baden-Württembergs. Analytiker Taschenbuch 21:3-33

Brüggemann R, Münzer B, Halfon E (1994) An algebraic/graphical tool to compare ecosystems with respect to their pollution – the German river 'Elbe' as an example – 1: Hasse-diagrams. Chemosphere 28:863-872

Brüggemann R, Bücherl C, Pudenz S, Steinberg CEW (1999) Application of the Concept of Partial Order on Comparative Evaluation of Environmental Chemicals. Acta hydrochimica et hydrobiologica 27:170-178

Bücherl C, Brüggemann R, Halfon E (1995) Hasse, Ein Programm zur Analyse von Hasse-Diagrammen, Projektgruppe Umweltgefärdungspotentiale von Chemikalien, ISSN 0721-1694, GSF-Bericht 19/95

Buxbaum F (1958) Cactus Culture: Based on Biology, Blandford Press, London

Carson R (1962) The Silent Spring, Greenwich, Connecticut, 1962

Conquest L (2000) Environmental monitoring: Investigating associations and trends. In: Statistics in Ecotoxicology (Edt. T. Sparks), Chichester, New York

FEA – Federal Environmental Agency (1999) Umweltprobenbank des Bundes – Ausgabe 1999. Ergebnisse aus den Jahren 1996 und 1997, Texte 61/99, Werbung & Vertrieb, Berlin

FEM – Federal Environment Ministry (2000) Environmental Policy: Environmental Specimen Bank. Concept, Federal Environmental Agency, Berlin

Hasse H (1967) Vorlesungen über Klassenkörpertheorie, Universität Marburg. Physica-Verlag

Helm D (2003) Bewertung von Monitoringdaten der Umweltprobenbank des Bundes mit der Hasse-Diagramm-Technik. UWSF – Z. Umweltchem Ökotox 15:85-94

Lerche D, Sørensen PB (2003) Evaluation of the ranking prob-abilities for partial orders based on random linear extensions. Chemosphere 53:981-992

Lerche D, Sørensen PB, Brüggemann R (2003) Improved estimation of the ranking probabilities in partial orders using random linear extensions by approximation of the mutual ranking probability. J Chem Inf Comput Sci 43:1471-1480

Luther B, Brüggemann R, Pudenz S (2000) An approach to combine cluster analysis with order theoretical tools in problems of environmental pollution. MATCH 42:119-143

Sørensen PB, Lerche DB, Carlsen L, Brüggemann R (2001) Statistically Approach for Estimating the Total Set of Linear Orders. Pudenz S, Brüggemann R and Lühr HP 14/2001, 87-97. 2001. Berlin, IGB Berlin. Berichte des IGB

Sørensen PB, Vorkamp K, Thomsen M, Falk K, Møller S (2004) Persistent organic Pollutants (POPs) in Greenland environment - Long-term temporal changes and effects on eggs of a bird of prey. National Environmental Research Institute - NERI Technical Report 509 pp 1-126

von Storch H, Costa-Cabral M, Hagner C, Feser F, Pacyna J, Pacyna E, Kolb S (2003) Four decades of gasoline lead emissions and control policies in Europe: A retrospective assessment. The Science of the Total Environment (STOTEN) 311:151-176

Exploring Patterns of Habitat Diversity Across Landscapes Using Partial Ordering

Wayne L. Myers[1]*, G. P. Patil[2] and Yun Cai[2]

[1] 124 Land & Water Research Bldg., The Pennsylvania State University, Univ. Park, PA 16802 USA

[2] Department of Statistics, The Pennsylvania State University, Univ. Park, PA 16802 USA

*Corresponding author: wlm@psu.edu

Abstract

Potential habitat suitability was assessed for species groupings of vertebrate fauna in the State of Pennsylvania, USA as part of a nationally coordinated GAP Analysis Program to find gaps in provision for conservation of important habitats. Diversity values were compiled spatially at a resolution of one square kilometre from species models developed at 30-meter resolution. Diversity patterns differ in varying degrees among species groups for mammals, birds, amphibians, snakes/lizards, turtles, and fishes. Comparing the patterns for partial ordering on watershed extents using statistical indices of ranking can facilitate determination of inter-group commonality and contrast. This helps to designate watersheds as having importance from multi-group and particular group perspectives. Partial ordering on the basis of rank-range runs is particularly informative when combined with levels of counter-indication corresponding to levels in a Hasse diagram. This serves to segregate sets having combinatorial clarity of condition relative to conservation from settings where disparate conditions may offer opportunities for targeted restoration. Disparity of conditions on multiple bio-indicators may arise from habitat heterogeneity as well as differential degradation. Broadening the spectrum of indicators will usually increase the apparent complexity of the conservation context.

Introduction

Patil and Taillie (2004) consider partially ordered sets (posets) in environmental contexts from the perspective of political contention whereby there is need for inferential extension of the observed data on multiple indicators in order to obtain a single induced ordering that resolves contentious issues (see for example Simon et al. 2004). There are, however, numerous environmental contexts in which incomplete orderings become directly useful from management perspectives without forcing a single induced ordering by inferential extension. In particular, incomplete orderings can answer four important questions pertaining to conservation and potential for remediation.

(1) The first question is which ones among a disparate population of n cases (landscape units) have consistency of expression (concordance) relative to a suite of p indicators. Subsets of the cases having consistent expression are subject to direct comparative ordering to address further questions.

(2) How to sort out superior cases for priority attention in conservation and protection and/or to serve as reference standards for comparative assessment.

(3) How can cases (landscape units) be recognized that are severely degraded in all relevant respects to the degree that preservation and protection concerns are effectively absent.

(4) Among the remaining cases that lack concordance in varying degree, are there cases of landscape units that could be elevated to superior status by remedial attention in some particular regard. These are the better cases for which there is consistency of expression among p-1 of the p indicators. The degree to which consistency is improved by deleting the most discordant indicator shows both the benefit that would accrue to targeted remediation and the level of effort that remediation would entail.

Our primary focus here is on these questions where partial or incomplete orderings are directly informative to conservation, remediation, or allocation issues of environmental management. Biodiversity and ecosystem health are multifaceted concerns that are most readily and objectively approached through suites of indicators. Prioritization of landscape units in these regards is also intrinsically complex. If a subset of indicators is essentially concordant across all cases of observation, then the indicators in the (sub)set are also substantially redundant. On the other hand, indicators representing largely independent dimensions of biodiversity or ecosystem health will almost necessarily complicate prioritization processes because there will be less consistency in how the respective landscape units reflect the indicators. There are also issues of scale related to the well known 'species-area effect' and its extensions whereby greater diversity is encountered as the extent of area under consideration is expanded. For practical purposes of conservation and ecosystem management, multiple indicators need to express related but at least somewhat different senses of the issues under consideration and also have a common polarity with regard to superior versus inferior. Substantially different indicator dimensions are going to present substantially different problems of prioritization that will then have to be reconciled to some degree in allocating scarce resources for management. The latter question will essentially be one of how to divide resources in dealing with different types of problems and conflicting priorities, which is beyond our present scope. The scale of landscape units is expected to span some divergence in level of spatial heterogeneity.

Pennsylvania Biodiversity Context

Our conservation context arises from model-based assessment of vertebrate biodiversity in the State of Pennsylvania (Myers et al., 2000) as part of a larger biodiversity assessment program known as GAP Analysis covering the entire United States, whereby 'GAP' refers to a gap in provisions for conservation. GAP Analysis is a coarse filter geographic approach to biodiversity assessment. Habitat factors for each of the vertebrate species that breed regularly in the state were mapped as layers in geographic information systems (GIS), and then combined into a potential habitat model for the species using GIS overlay analysis of the map layers (Scott et al., 1993). Land cover information as a component of models for all species was derived from remotely sensed data obtained from Landsat TM (Thematic Mapper) at a spatial resolution of 30-meter pixels. Therefore, the potential habitat maps were also prepared initially at a 30-meter scale. The habitat suitability mapping for each species was then generalized to a coarser scale resolution of 1-km^2 by considering a species to be present in the 1-km^2 cells if it is present in any 30-meter component pixel of the 1-km^2 cells. This provides for a richness value related to an ith 1-km^2, $R(1000m,i) = \sum b(i,j), j=1,\ldots, q,\ b(i) \in \{(0,1)q\}$, b(i,j) the jth component of the tuple b(i). The digits 0 or 1 indicate absence or presence of the jth species, and *q* is the number of species being considered.

Species (habitat) richness was thus compiled on a 1-km^2 basis for six taxonomic and life history groupings of species along with regional habitat importance ratings (Myers et al., 2001). The six species groupings were mammals, birds, amphibians, snakes and lizards, turtles, and fishes as $p =$

6 indicators of biodiversity. For the analyses that we present here, the mean of the kilometre-level species richness values was tabulated for each of these six species groups across each of 184 watersheds. For the ith watershed the group richness is $R_g(\text{watershed},i)= [\sum R_g(1000m,j), j=1,\ldots, t]/t$ where t is the total number of 1-km^2 cells in the ith watershed.

Fig. 1a. Location of State of Pennsylvania in North-eastern USA

Fig. 1b. Location of watersheds in Piedmont Plateau and South Mountain regions of Pennsylvania

Pennsylvania is situated in North-eastern USA, and the watersheds are located in the Piedmont and South Mountain physiographic regions in South-eastern Pennsylvania (Fig. 1). Other physiographic regions of the state could equally well have been included in the analysis, but this geographic extent was chosen for reasons of both presentational clarity and land use history. The Piedmont Plateau has moderate topography and the geologic materials weather into deep and fertile soils that have been conducive to extensive agricultural and urban development. Therefore, the natural forest cover has become highly fragmented in many areas and vertebrate habitat degradation is substantial. Thus, the areas where habitat degradation is less pervasive become high priority areas for cooperative conservation efforts among conservancies and landowners through reservation incentive programs such as easements that preclude development.

Mean species richness values in a watershed for the six groups provide us six indicators of biodiversity for the watershed as a case or landscape unit. Thus the objects (cases) to be analyzed are the watersheds as landscape units – or briefly 'units', the characterization of these units being done in terms of the six indicators. For comparative purposes, these six indicators can also be condensed to two indicators by adding together the amphibians, turtles, and fishes as an indicator of lowland habitats; and doing likewise for the mammals, birds, and snakes/lizards as an indicator for upland habitats. Henceforth, the latter two super-groups will be referred to as upland and lowland indicators. Hence two data matrices are to be analyzed: the first one consisting of 184 watersheds (the 'units') and six indicators, the second one consisting of 184 such units and two indicators, lowland and upland indicators, respectively.

Progression of partial and incomplete orderings

Conflicts in rankings can be viewed from two major perspectives. One perspective is that any conflict of rankings makes the units intrinsically incomparable. A second perspective attempts to resolve some of the conflicts on the basis of more liberal criteria.

Adopting the first perspective allows us to segregate subsets of landscape units whereby there is intrinsic ordering between the subsets but not within a subset among its members. These subsets are partially ordered sets (posets) corresponding to the levels that would be depicted on Hasse diagrams (Neggers and Kim 1998). If the positive direction for each indicator is better, then the primary (number 1) subset consists of units that are not dominated by any other unit. There is domination if some unit is equal to or better than another on all indicators. The secondary (number 2) subset is found by removing the primary subset and then finding the non-dominated subset of the remaining units. The tertiary (number 3) and subsequent subsets are found by applying the process recursively (levels and their construction, see chapter by Brüggemann and Carlsen, p. 61). Multiple units with identical values on all indicators are precluded if the subsets are to be posets in the mathematical sense because of the anti-symmetry condition (Patil and Tallie 2004). One may also define a quasi order; see Brüggemann and Bartel (1999). For practical purposes, identical units can be re-introduced retrospectively by placing identically. The conventional computational formalities are typically presented in terms of matrices. Computational efficiency for large numbers of units is better served, however, by recursively marking non-dominated units among the unmarked residual units. Identical units are easily accommodated from the outset in the latter approach to reach the same retrospective result as the matrix methods. We refer to these orderings of subsets as counter-indication (CI) levels because of the interpretive information that they convey. One superior indicator is sufficient to place a unit in the primary set, regardless of how inferior other indicators may be. However, a unit in the highest numbered subset will not have any superior indicators. In a general sense, the level of agreement regarding inferior status increases with increasing subset number.

Under the strict perspective of intrinsic ordering with inadmissibility of inconsistent evidence, the ranking of units would remain incomplete except for appeal to cover relations in depiction as a Hasse diagram. Since Hasse diagrams lose their utility rapidly with increasing numbers of units, we do not pursue that route here. This is because we may be faced with need to consider several thousand landscape units such as watersheds. Further progress depends on supplementing the poset CI level results through less restrictive criteria.

Patil and Taillie (2004) seek to induce a full ordering by considering the cumulative distribution of ranks that are not directly expressed in the data matrix, but which exploit the observed consistencies (linear extensions) in the data matrix. The concept of linear extensions is broadly used (Brüggemann et al. 2001; Brüggemann et al. 2004; Carlsen et al. 2002; Lerche et al. 2003; Lerche and Sørensen 2003; Sørensen et al. 2001); see the chapters by Brüggemann and Carlsen, p. 61; Carlsen, p. 163 and Helm, p. 285. The number of initial rankings to be considered becomes very large rapidly as the number of indicators increases, causing need for recourse to Monte Carlo Markov Chain (MCMC) methods. To avoid the somewhat nebulous nature of indirect evidence, we restrict ourselves here to supplemental views based entirely on direct evidence that help to clarify the prioritization picture while allowing the orderings to be incomplete.

Although CI (poset) levels can be determined directly from the indicator data matrix without conversion to ranks, we proceed immediately to convert the data for each indicator to rank numbers. To be consistent with the interpretive sense of CI levels, we adopt the 'place' convention for ranks whereby the first place (rank 1) is superior. We then liberalize the criteria for comparison so that only the best rank and the worst rank among any of the indicators are considered for each unit along with the rank range. If a unit is equal or superior to another in its best rank and also better in its worst rank, then there is superiority in a limited sense. Likewise, there is superiority if a unit is better than another in the best rank and better than or equal in the worst rank. Additionally, units with a narrow range of ranks reflect consistently on the indicators, whereas a wide range of ranks shows divergent expression among the indicators. These properties provide the basis for a series of internally ranked subsets that speak to the management questions that were raised earlier, and can be readily coupled with intrinsic partial orderings for enhanced insights.

As a simple example, consider the data in Table 1 on five hypothetical units. This table shows the best and worst ranks for each unit on an unspecified number of indicators. There are two series of progressive superiority based on extreme ranks. One series is E > B > C > D, where the > symbol implies superiority of the left member to the right member. The other series is E > A > C > D. These two series differ only in exchanging A for B. The wider range of ranks for A relative to B gives the second series more dissonance or discrepancy (less consensus) than in the first.

Table 1. Example of five hypothetical units

Unit	Best rank	Worst Rank
A	1	4
B	2	3
C	3	5
D	4	5
E	1	3

Rank range runs

Fig. 2 shows a series of watershed sequences that comprise ordered subsets according to rank range relations. Of course, rank range relations only have meaning when there are two or more indicators. Conversion of indicators to ranks is a preparatory operation for sequencing of rank ranges. In converting an indicator to ranks, we treat the best value as having rank number 1 and tied units are assigned the average of the tied ranks. Ranking is, of course, a form of rescaling that removes all statistical differences in the distribution of indicators.

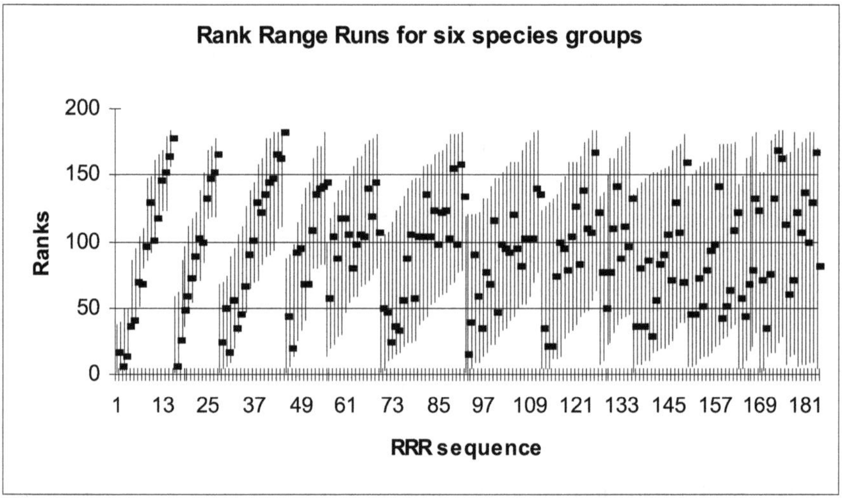

Fig. 2. Rank range runs for 184 watersheds according to species richness of six taxonomic groups as indicators. The dots along the lines show the position of the median rank for the watershed

Therefore, the focus shifts from indicator to landscape unit (watershed in this case). Statistics are computed on the set of rank values for each landscape unit (watershed). Statistics needed for the present operation are minimum rank, maximum rank, and median rank. These become three derived observational variables for each landscape unit (watershed). Given these preparatory computations, the operations for rank range sequencing are as follows:

a) First, order the landscape units according to increasing value of minimum rank.
b) Suborder any tied units according to increasing value of maximum rank.
c) Suborder any tied units according to increasing value of median rank.
d) Create an empty stack for shifting (landscape/watershed) units, and set series number to 1.
e) Proceeding from end of ordered list of units toward beginning of list, remove any unit to the top of the stack if a unit having lesser maximum rank follows it in the remaining list.

f) Upon reaching the head of the list, assign the current series number to all units remaining in the list.
g) Increment the series number.
h) Apply steps d) through g) to units in the stack working from bottom of stack toward top of stack.
i) Append this series to the previous series and then return to h) if the stack is not empty.
j) Assign sequential numbering (rank range run sequence) to the entire list.

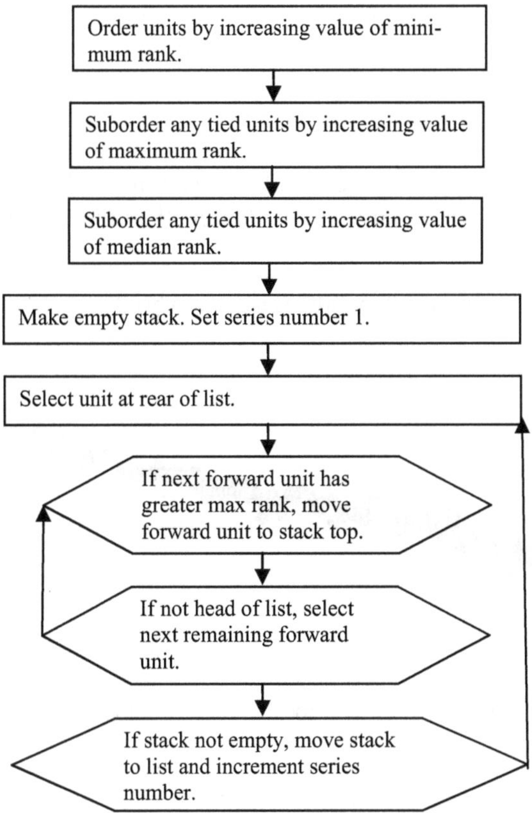

Fig. 3. Flowchart for obtaining rank range runs

Fig. 3 gives a flowchart of the process except for the final sequence numbering across the entire set of series. This algorithm allows the computer scientist ample opportunity for exercising strategic skills of their craft in designing the implementation. For example, a single data array can be handled in three portions that are dynamically repartitioned. The leading portion contains the rank range series that have been processed, the middle portion contains the units that are currently being processed, and the tail portion contains the stack of residuals. Large numbers of units can be accommodated relatively easily.

The graph in Fig. 2 is termed a "stock chart" in Microsoft Excel spreadsheet software that is intended for depicting the activity of a stock market during a day. Before making the chart, the landscape units must be sorted in the spreadsheet according to ascending value of the rank range run sequence number assigned in the j) step above.

Hereafter, we refer to each series in Fig. 2 as a rank range run (RRR). Each unit appears as a member of only one run. It will be observed that the ranges of the units generally increase for each successive rank range run, as shown in Fig. 4. This means that the lower numbered rank range runs contain landscape units that have greater consistency with respect to the indicators than do higher numbered rank range runs. Additionally, each successive member of a rank range run is 'better' than the unit that follows it.

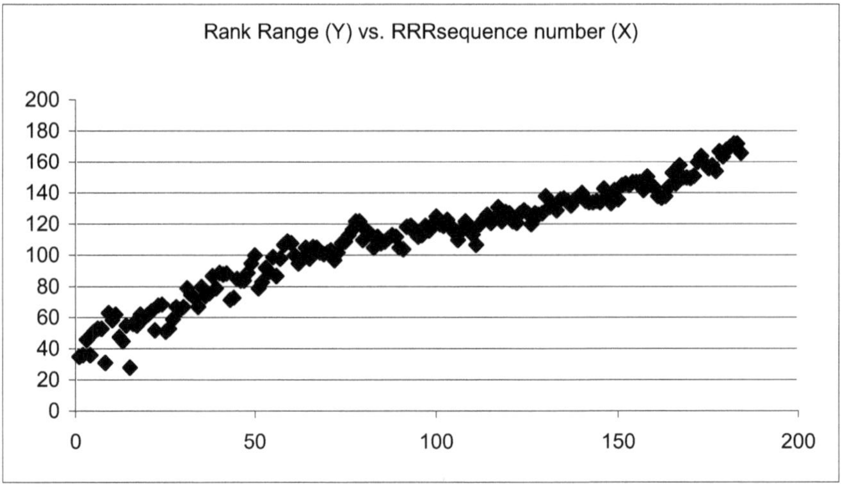

Fig. 4. Trend of rank range with progression along rank range run sequence for six species richness indicators on watersheds

It may be possible to 'graft' part of one rank range run into another rank range run as a replacement for some of its members. An example of this can be found in the illustrative data of Table 1 referenced earlier. The RRR process would yield a first run of E > B > C > D, with A trailing as a unitary second run. As noted earlier, the A unit can be grafted into the first sequence as a replacement for the B unit. Part of an RRR series can be grafted into another if (1) both minimum and maximum rank of the leading unit in graft are greater than or equal to those of the unit (if any) that it follows, and (2) both minimum and maximum rank of the trailing unit in the graft are less than or equal to those of the unit (if any) that it precedes. The rank range runs and their possible inter-run grafts define (incomplete) ordering relations among the units according to the rank range criteria.

The position of the median mark relative to the midpoint of a range line adds information on the distribution of ranks for that unit. Having the median mark off-centre toward the lower end implies a tight grouping of better ranks and more dispersion of poorer ranks. Having the median mark off-centre toward the higher end implies the reverse. This is a reflection of skewness in the distribution of ranks for the particular unit. A symmetric distribution of ranks for a unit would place the median at the midrange. When the median is below the midrange, there is a concentration of lower numbered (better) ranks below the midrange with a tail of higher numbered (poorer) ranks above.

An apparent tendency toward cyclic variation in rank range within a series can also be noted for Fig. 4. The middle members of a series tend to have greater rank range than the end members. This is an empirical observation that bears study for other contexts.

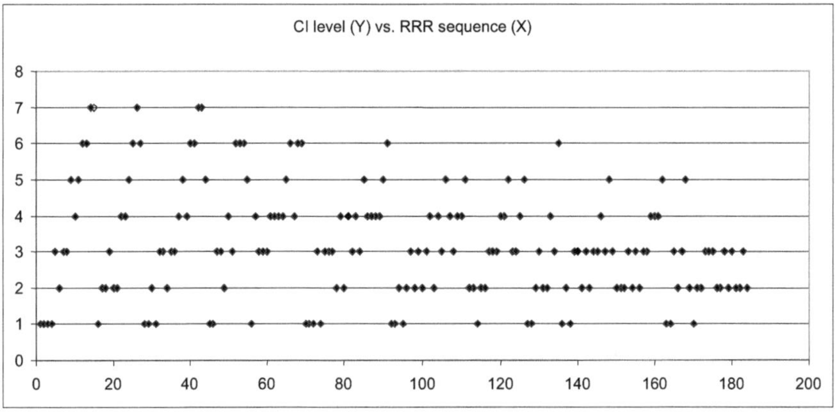

Fig. 5. Poset CI level in relation to rank range run sequence for watersheds with six species richness indicators

Fig. 5 shows graphical combination of intrinsic ordering by poset CI level with rank run sequencing. It is notable that the RRR series toward the left-hand side have a strong correspondence of CI level with position in the rank range run, but this relationship weakens progressively with shift toward the right. Answers to the first three managerial prioritization questions posed at the outset are to be found in the rank range runs toward the left side of a graph like that of Fig. 5. The challenge is how to portray this information in a manner that is easily interpreted.

When working with mapped information, this challenge can be met by strategic use of red-green-blue (RGB) coloration for the map units. Each unit has a colour component of red proportional to its CI level and a colour component of green inversely proportional to its CI level. A colour component of blue is added in proportion to the RRR series number. In this colour regime, inferior units in low numbered (low dissonance) RRR series will appear strongly red, whereas superior units will appear strongly green and intermediate units will appear yellowish. As RRR series number and dissonance increase, the increasing blue component will grade red into magenta and green into cyan. Likewise, yellow will grade into grey and then further into blue-grey. Unfortunately, direct illustration of this regime for visualization would require prohibitively expensive colour figures. However, Fig. 6 shows the watershed units that appear strongly black in such a display and thus reflect consensus of indicators regarding superior characteristics. These watersheds primarily occupy highland areas having relatively less impact of development and agriculture.

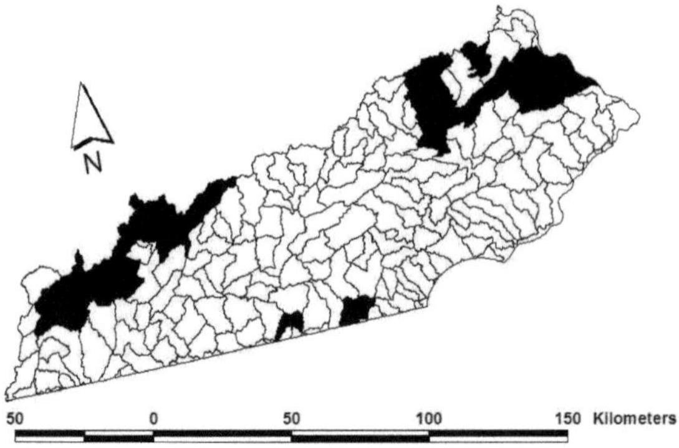

Fig. 6. Watersheds indicated as having superior status

An important advantage of this approach is that it facilitates working with large numbers of units that would have Hasse diagrams so complex as to be completely lacking in interpretability. Even if units do not lend them to portrayal in map form, the RRR series can be sorted with a spreadsheet according to series number and extracted one or a few a time for progressive graphical display and interpretation. The RRR series are logical sets for comparison with regard to the management questions posed at the beginning. Furthermore, adjacent RRR series are more readily comparable than series that are farther apart in their numbering. It should also be noted that sorting the units according to the final composite RRR sequence number also serves to place the RRR series in order by number.

It is interesting to explore how the structure changes for the watershed context when the six species richness indices are combined into only two indices that represent predominantly upland species versus predominantly lowland species. Of course, the segregation is neither absolute nor necessarily in opposition relative to watersheds. Complex topography will imbed both uplands and lowlands in the same watershed, thus creating units having greater habitat diversity. Such complex topographic units tend to be of special interest for conservation both because the rugged terrain discourages development and because there is opportunity to capture a spectrum of habitats in an area of relatively modest size.

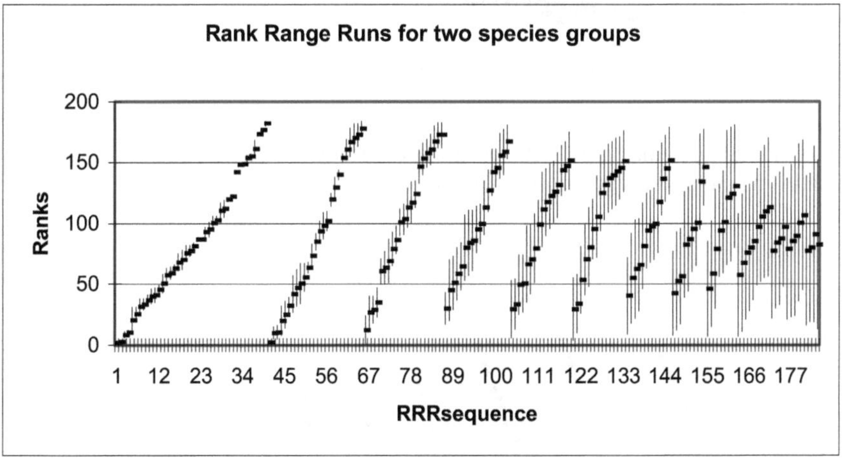

Fig. 7. Watershed rank range runs for upland and lowland species richness groups as indicators

Fig.'s 7, 8, and 9 are the two-indicator counterparts of Fig.'s 2, 4 and 5 for six indicators. There are similarities and differences that invite speculation regarding both the general utility and propensities for the approach. The lower numbered rank range run series (to the left) in Fig. 7 are even more distinctive than those in Fig. 2 where there are more indicators. Fig. 8 is like Fig. 4 in exhibiting an increase of rank range with RRR sequence and a propensity of the rank range to increase in the middle of each series. However, there are more units with narrow ranges in Fig. 8 that give the trend a concave (upward) shape instead of a somewhat convex shape. Fig.

9 exhibits considerably more posets CI levels than Fig. 5, and there is substantial linearity between CI level and position within the rank range run series. However, it is to be noted that moderately good classes of intrinsic ordering (posets) occur even in substantially conflicted cases (rightward rank run series).

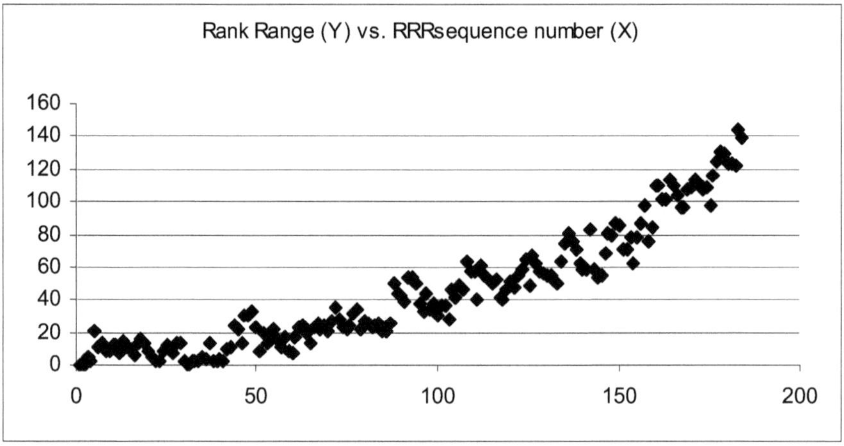

Fig. 8. Trend of rank range with progression along rank range run sequence for upland/lowland species groups on watersheds

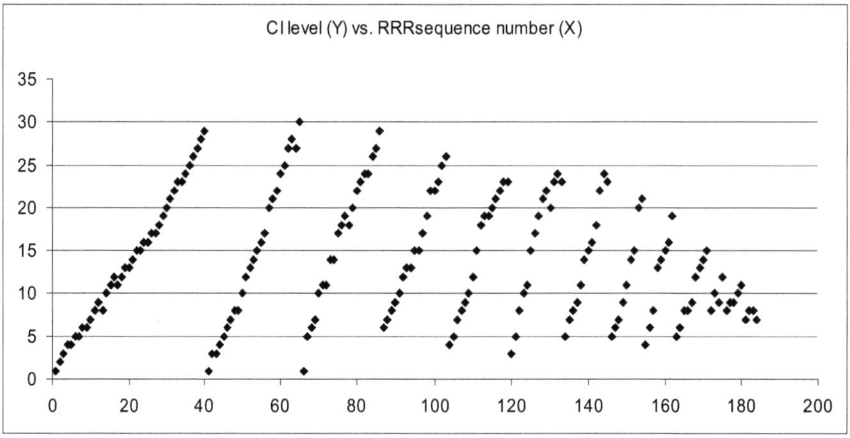

Fig. 9. Poset CI level in relation to rank range run sequence for watersheds with upland/lowland species richness indicators

These observations suggest robustness in the nature of the patterns, and a sharpening of the patterns with increasing concordance of indicators. The rank correlation (after Spearman) is 0.411 between upland and lowland indicators, and rank correlations among the six indicators are highly varied as shown in Table 2. Among the six, the maximum rank correlation is 0.778 between birds and mammals. The rank correlation is 0.517 between amphibians and turtles, and 0.530 between amphibians and snakes/lizards. All other rank correlations are less than 0.400 among the six. Four of the rank correlations are essentially indicative of independence, having magnitude less than 0.01 and even carrying a negative sign. Reversing polarities is not appropriate for any of the indicators since it would induce negativity of a larger magnitude in other pairs of rank correlations. It should also be kept in view that these are correlations between the ranks rather than between the observed species richness data. For the present biodiversity context, there is some compensatory interaction among the indicators as evidenced by the result of aggregation to upland versus lowland perspective.

Table 2. Rank correlations between six species richness indicators of biodiversity

	Mammals	Fishes	Amphibians	Turtles	Snake/Lizard
Birds	0.778	0.383	0.343	-0.001	0.329
Mammals		0.292	0.278	-0.092	0.187
Fishes			0.215	-0.071	-0.007
Amphibians				0.517	0.530
Turtles					0.357

Performing principal component analysis on the ranks can help to assess the dimensionality of the ordering context. Since the marginal distributions of the ranks are the same except for ties, the difference between covariance matrix and correlation matrix is not critical. If there are subsets of indicators that segregate strongly in their loadings, then complexity is confirmed and it may be prudent to consider partitioning of the prioritization process.

Rank range end-member elimination effect

The management question regarding opportunities for remediation and restoration has still not been addressed. This question can be approached through rank ranges by considering the reduction in rank range for a unit that would be obtained by deleting one of the end-members that determine the range. We call this the 'End-member Elimination Effect' (EEE). If the greatest rank range reduction is obtained by eliminating the lowest numbered rank, then the EEE is given a negative sign. If a greater reduction of rank range for the unit is obtained by eliminating the highest numbered rank, then the EEE is given a positive sign. Units having a large positive EEE value offer the opportunity to improve the rank range standing considerably by ameliorating a single indicator factor. Conversely, units having large negative EEE values indicate that the overall standing can worsen considerably by deterioration in a single indicator factor.

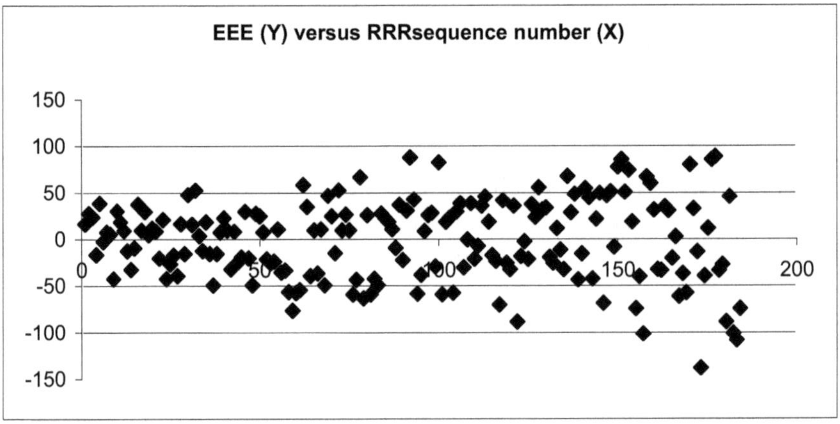

Fig. 10. End-member elimination effect versus rank range run sequence number for six indicators on watersheds

The end-member effects for six watershed indicators are plotted against RRR sequence number in Fig. 10. As would be expected the effects are small toward the left where rank ranges are small, with larger effects appearing more frequently toward the right where rank ranges are larger. Rank range reduction is not applicable to the case of only two indicators, since eliminating either of them would also eliminate the range.

Second order analysis

Managers may also be interested in the comparative differences between perspectives provided by rank summary statistics such as median rank, mean rank, and midrange of ranks. The rank range approach can also provide an interesting alternative for conducting such comparative analysis. The statistics of interest can be treated as second-order multiple indicators for the units instead of using the indicator data from which the statistics were derived. Poset CI level and rank range analyses are then conducted in a parallel manner to the procedures described above. It would not be informative, however, to consider the end-member elimination effects in this manner.

References

Brüggemann R and HG Bartel (1999) A theoretical concept to rank environmentally significant chemicals. J Chem Inf Comp Sci 39:211-217

Brüggemann R, Halfon E, Welzl G, Voigt K and Steinberg C (2001) Applying the concept of partially ordered sets on the ranking of near-shore sediments by a battery of tests. J Chem Inf Comp Sci 41:918-925

Brüggemann R, Sørensen PB, Lerche D and Carlsen L (2004) Estimation of averaged ranks by a local partial order model. J Chem Inf Comp Sci 44:618-625

Carlsen L, Lerche DB and Sørensen PB (2002) Improving the predicting power of partial order based QSARs through linear extensions. J Chem Inf Comp Sci 42:806-811

Lerche D, Sørensen PB and Brüggemann R (2003) Improved estimation of the ranking probabilities in partial orders using random linear extensions by approximation of the mutual probability. J Chem Inf Comp Sci 53:1471-1480

Lerche D and Sørensen PB (2003) Evaluation of the ranking probabilities for partial orders based on random linear extensions. Chemosphere 53:981-992

Neggers J and Kim HS (1998) Basic Posets. World Scientific: Singapore

Myers W, Bishop J, Brooks R, O'Connell T, Argent D, Storm G and Stauffer J Jr (2000) The Pennsylvania GAP Analysis final report. The Pennsylvania State University, Univ. Park, PA 16802

Myers W, Bishop J, Brooks R and Patil GP (2001) Composite spatial indexing of regional habitat importance. Community Ecology 2(2):213-220

Patil GP and Taillie C (2004) Multiple indicators, partially ordered sets, and linear extensions: Multi-criterion ranking and prioritization. Environmental and Ecological Statistics 11:199-228

Scott JM, Davis F, Custi B, Noss R, Butterfield B, Groves C, Anderson H, Caicco S, D'Erchia F, Edwards TC Jr, Ulliman J and Wright RG (1993) GAP Analysis: a geographic approach to protection of biological diversity. Wildlife Monographs No 123

Simon U, Brüggemann R and Pudenz S (2004) Aspects of decision support in water management – example Berlin and Potsdam (Germany) I – spatially differentiated evaluation. Wat Res 38:1809-1816

Simon U, Brüggemann R and Pudenz S (2004) Aspects of decision support in water management – example Berlin and Potsdam (Germany) II – improvement of management strategies. Wat Res 38:4085-4092

Sørensen PB, Lerche DB, Carlsen L and Brüggemann R (2001) Statistical approach for estimating the total set of linear orders. In: Pudenz S, Brüggemann R and Luhr HP 14/2001, Berlin, IGB Berlin, Berichte des IGB (Report)

Information Systems and Databases

Kristina Voigt[1]*, Rainer Brüggemann[2]

[1] GSF-Research Centre for Environment and Health, Institute for Biomathematics and Biometry, Germany

[2] Leibniz-Institute of Freshwater Ecology and Inland Fisheries Department: Ecohydrology, Germany

* e-mail: kvoigt@gsf.de

Abstract

The main objective of the European Commission's White Paper on a future chemicals strategy (EEC 2001) is to facilitate the risk assessment of chemicals leading to, where necessary, risk reduction. Important roles play the chemical and environmental databases, which can be regarded as an information turnover. In this paper the emphasis lies on the evaluation of 12 numerical databases available on the free Internet, which focus on environmental fate and ecotoxicity as well as on high production volume chemicals. Hence we analyse a 12x27 data-matrix in the first place. Two multi-criteria evaluation and decision support instruments are applied: The Hasse Diagram Technique (HDT), a method derived from discrete mathematics, and the Method of Evaluation by Order Theory (METEOR). The original data-matrix of 12 databases (objects) and 27 parameters (attributes) will be subject to several logical aggregation steps. The aim of the aggregation procedure that can be performed by applying iteratively the Hasse Diagram Technique (HDT) is to get a unique prioritisation scheme. Significant data gaps even on the chosen well-known high production volume chemicals as well as on ecotoxicity and environmental fate parameters are identified by the chosen methods and weighting procedures. This indicates an alarming signal concerning the new existing chemicals policy of the EEC.

Environmental Data, Databases and Information Systems

Future Chemicals' Policy in the EEC

The current number of existing substances marketed in volumes above 1 ton is estimated at 30.000 (EEC 2001). These substances amount to more than 99 % of the total volume of all substances on the market. In the so-called White Paper, the paper on the Strategy for a future Chemicals Policy of the Commission of the European Communities, the testing and evaluation of a large number of existing substances in the coming 10 years is envisaged. Initiative was taken to collect data on chemicals for their risk assessment leading to, where necessary, risk reduction (Heidorn et al. 2003).

The gap in knowledge about intrinsic properties of existing substances should be closed to ensure that equivalent information to that on new substances is available. The available information on existing substances should be thoroughly examined and best use made of it in order to waive testing, wherever appropriate. Studies show significant gaps in publicly available knowledge of existing chemicals especially in environmental fate and pathways as well as in ecotoxicity parameters (Allanou et al. 2003).

Environmental Chemicals' Databases and Information Systems

It is evident that the topic "environmental chemicals' data" is strongly related to the subject of structuring and archiving them in environmental and chemical databases (Page and Voigt 2003). These databases are not only found in every medium, that is to say online, CD-ROM, and on the Internet, but are quite varied in their type and contents. The interested community urgently needs support in finding relevant information and data about environmental chemicals. Recently a very valuable and unique guide for searching scientific and technical information online, CD-ROM and Internet was published (Poetzsch 2004). Furthermore it is of utmost importance to give precise indications of the importance and quality of the databases. This means performing a comparative evaluation of databases with respect to several different evaluation criteria. Several approaches exist. Comparisons of eight large chemical structural databases are performed (Voigt et al. 2001). Another evaluation approach for structural and reaction databases is given by Cooke and Schofield (2001). Approaches using the mathematical method of partial orders are also used to evaluate environmental and chemical databases. A comparative evaluation of data-sources of online databases and databases on CD-ROM based on research results gained in the years 1996/1997 is given by the first author (Voigt 1997), (Voigt et al. 2000). An overview on evaluation approaches for chemical databases was also published recently (Voigt and Welzl 2002a).

Selection of Databases (Objects) and Parameters and Chemicals (Attributes)

Selection of Environmental Chemical Databases

A huge amount of environmental databases is available worldwide. Most of these databases can be accessed via the Internet. However one must distinguish between fee-based and non-fee based databases. In our approach we only take databases on the free Internet into account. The second focus we have in mind is that we only chose numerical and full-text databases. In these databases the data and information can be retrieved immediately. For further reading on the types of environmental databases the article of the first author in the Handbook of Chemoinformatics, edited by Gasteiger and Engel is recommended (Voigt 2003). All the chosen 12 non-fee databases are incorporated in the DAIN - Metadatabase of Internet Resources for Environmental Chemicals, which can be found under http://www.wiz.uni-kassel.de/dain (Voigt 2000). As in this database more than 100 numerical databases can be found, we selected those ones which are commonly known and recommended in several publications (Wexler 2004), (Russom 2002), (Felsot 2002). Besides the US databases we added two German data-sources, one Japanese database and one Australian information source. Furthermore we selected the international database Screening Information Data Sets that is described in recent publications (Gelbke et al. 2004), (CERHR 2004).

The entire study was performed in late 2002. The authors are aware that the selection is arbitrarily and that many more interesting sources for information on environmental chemicals exist worldwide. The possibility of the move or remove of sites exists. The chosen databases which are listed together with their later used abbreviations and their Internet address (URL = Uniform Resource Locator) are given in Table 1. Three different types of numerical databases can be distinguished:

- Single databases which cover only one data collection (BID, CIV, HSD, UMW)
- Multi-database databases which encompass several databases under the same name and search interface (ECO, ENV, EFD, EXT)
- Monograph databases, which cover extensive reviews on very few chemicals (EHC, NRA, PES, SID).

Table 1. List of Chosen Numerical Databases Focusing on Environmental Chemicals

Name off Database	Abb.	URL
Biocatalysis/Biodegradation Database	BID	http://umbbd.ahc.umn.edu/
Chemicals Information System for Consumer-relevant Substan ces (CIVS)	CIV	http://www.bgvv.de/cms/detail.php?te mplate=internet_en_index_js
ECOTOX	ECO	http://www.epa.gov/ecotox/
Envirofacts	ENV	http://www.epa.gov/enviro/html/emci/ chemref/
Environmental Fate Database	EFD	http://esc.syrres.com/efdb.htm
Environmental Health Criteria Monographs (EHCs)	EHC	http://www.inchem.org/pages/ehc.html
EXTOXNET	EXT	http://ace.ace.orst.edu/info/extoxnet/
HSDB	HSD	http://toxnet.nlm.nih.gov/cgi-bin/sis/htmlgen?HSDB
NRA Chemical Review Program	NRA	http://www.nra.gov.au/chemrev/chemr ev.shtml
Pesticide Database, Japan	PES	http://chrom.tutms.tut.ac.jp/JINNO/PE SDATA/00alphabet.html
SIDS	SID	http://www.chem.unep.ch/irptc/sids/sid spub.html
UmweltInfo	UMW	http://www.umweltinfo.de/ui-such/ui-such.htm

Selection of Environmental Parameters

The environmental fate and pathways and the ecotoxicity parameters implemented in the IUCLID database (Allanou et al. 1999) will be looked upon. These are:

Environmental fate and pathways: photodegradation, stability in water, stability in soil, monitoring data (environment), transport between environmental compartments, distribution, mode of degradation in actual use, biodegradation, BOD5, COD or BOD5/COD ratio, bioaccumulation.

Ecotoxicity: acute/prolonged toxicity to fish, acute toxicity to aquatic invertebrates, toxicity to aquatic plants e.g. algae, toxicity to microorganisms, e.g. bacteria, chronic toxicity to fish, chronic toxicity to aquatic invertebrates, toxicity to soil dwelling organisms, toxicity to terrestrial plants, toxicity to other non-mammalian terrestrial species, biological effects monitoring, biotransformation and kinetics.

In Table 2 a list of these attributes with their abbreviations and their membership to a class of super-attributes is given (see text).

Table 2. Environmental Parameters

Parameter	Abbre-viation	Super-attribute[a]
photodegradation	PHO	FATE
stability in water	SWA	FATE
stability in soil	SSO	FATE
biodegradation	BDE	FATE
BOD5, COD or BOD5/COD ratio	BOD	FATE
bioaccumulation	BAC	FATE
acute/prolonged toxicity to fish	ATF	ETOX
acute toxicity to aquatic invertebrates	ATD	ETOX
toxicity to aquatic plants e.g. algae	ATP	ETOX
toxicity to microorganisms, e.g. bacteria	ATB	ETOX
chronic toxicity to fish	CTF	ETOX
chronic toxicity to aquatic invertebrates	CTD	ETOX
toxicity to soil dwelling organisms	TSO	ETOX
toxicity to terrestrial plants	TTP	ETOX
toxicity to other non-mammalian terrestrial species	TNT	ETOX

[a] see eqn. 3

Selection of Environmental Chemicals

The databases are not only looked upon with respect to their parameters but also with respect to some selected chemicals. The selection of a pragmatic number of existing chemical substances, which are not only relevant in one aspect, is difficult and here solved rather pragmatically. The following 12 high production volume chemicals are chosen which are listed in Table 3.

Table 3. List of Chosen Chemicals for the Evaluation of Environmental Chemicals' Databases

CAS Number	Chemical Name	Remarks	Super-attribute
100-00-5	1-chloro-4-nitrobenzene	HPVC	CHID
100-01-6	4-nitroaniline	HPVC	
100-02-7	4-nitrophenol	HPVC	
1912-24-9	Atrazine	HPVC, ED	
999-81-5	Chlormequat chloride	HPVC	
333-41-5	Diazinon	HPVC	
60-51-5	Dimethoate	HPVC	
26761-40-0	Ethofumesate	HPVC	
1071-83-6	Glyphosate	HPVC	
34123-59-6	Isoproturon	HPVC	
121-75-5	Malathion	HPVC, ED	
137-26-8	Thiram	HPVC	

HPVC= High Production Volume Chemical, ED = endocrine disruptor, CHID, see eqn. 3

The chemicals were taking of a ranking approach for chemical substances performed by Lerche et al. (2002). Note that we are aware that pesticides are not covered by REACH; anyway: here the chemicals serve as an example how to analyze data availability.

The 12 databases (objects) will be evaluated by 27 attributes (environmental parameters and chemicals). Recently this 12x27 data-matrix was analysed putting the emphasis to different order theoretical issues (Voigt et al. 2004a).

Environmetrical and Chemometrical Methods Used

For analyzing environmental chemicals' data well-known chemometrical and environmetrical methods are used. A good overview of established methods of chemometrics and environmetrics is given in the literature (Massart et al. 1997), (Einax et al. 1997), (Stoyan et al. 1997), (El-Shaarawi and Hunter 2002), (Welzl et al. 2004). As the data situation in environmental sciences in combination with chemical substances becomes more and more complex, this poses a great challenge for establishing new data-analysis methods. Einax summarizes in a recent publication important new chemometrical methods (Einax 2003). One of these challenging new chemometrical and environmetrical method is the method of evaluation by order theory, based on the theory of partially ordered sets, and its specific application, known in literature as the Hasse Diagram Technique (HDT).

Hasse Diagram Technique (HDT)

Introduction of the History of HDT

The Hasse Diagram Technique is well explained in a variety of different environmental and chemical as well as statistical journals. Brüggemann, Carlsen in the chapter p. 61, explains the scientific background. Further comprehensive descriptions can be found in Brüggemann et al. (2001) and Brüggemann and Welzl (2002). A comparison of the Hasse Diagram Technique with multi-variate statistical methods is given by Voigt et al. (2004b). Therefore only some few aspects will be explained, which will be useful in the subsequent application. Hasse diagrams visualize the order relations within objects: Two objects, also called elements (if the aspect of belonging to sets is important) x, y of an object set are considered as being ordered, e.g. $x \leq y$, if all attribute values (often called scores) of x are less or equal than those of y. Hasse diagrams are acyclic digraphs and objects are drawn as small circles together with an appropriate identifier. The edges of this graph are the cover-relations; that means, edges, which express simply the transitivity, are omitted, as they bear redundant information. In our applications the circles near the top of the page (of the Hasse diagram) indicate objects that are the "better" objects according to the criteria used to rank them: The objects not "covered" by other objects are called maximal objects. Objects, which do not cover other objects, are called minimal objects. In some diagrams there exist also isolated objects, which can be considered as maximal and minimal objects at the same time. Sometimes it is useful to call those elements as 'proper', which are not at the same time both, maximal and minimal elements. When there is exactly one maximal and one minimal element respectively, then these unique objects are called greatest and least element, respectively.

The WHASSE program is developed, improved and updated by Brüggemann (a brief technical information about the WHASSE-program, written in DELPHI, can be found in Brüggemann et al. (1999) and is available for non-commercial use from the second author. For commercial applications it is recommended to contact the company Criterion – Evaluation and Information Management (Criterion 2004). Further theoretical developments concerning order theoretical tools in environmental sciences and their applications are discussed in regularly held workshops.

Notation and some theoretical background of HDT

Attributes are -in the case of the object "x" denoted as q(1,x), q(2,x),...,q(m,x) and often written as a tuple q(x). We avoid the term vector, because the properties of a linear space are not needed in the HDT. Often the properties are gathered to a set without reference to actual values realized by the objects. This set of properties is called an information base *IB*. If METEOR is to be applied, often subsets of *IB* are needed.

The main frame of HDT is therefore (the four-point-program):

1. Selecting a set of elements of interest which are to be compared, *E*. The so-called ground set.
2. Selecting a set of properties, by which the comparison is performed, called the information basis *IB*.
3. Find a common orientation for all properties, according to the criteria they are assigned.
4. Analysing $x,y \in E$ whether one of the following relations is valid:
   $x \sim y$ (equivalence, we call the corresponding equivalence relation, the equality of two tuples q(x), q(y))
   $x \leq y$ or $x \geq y$ (comparability)

$x \parallel y$ (incomparability, there is a "contradiction in the data of x and y", see also the chapter by Simon, p. 221 where -based on incomparabilities- the concept of antagonistic attributes (syn. indicators, descriptors) is outlined.).

The relation defined above among all objects is indeed an order relation, because it fulfils the axioms of order, namely

- reflexivity (one can compare each object with itself)
- anti-symmetry (if x is preferred to y then the reverse is only true, if the two objects are equal (or equivalent)
- transitivity (if x is better than y, and y is better than z, then x is better than z).

A set *E* equipped with an order relation $\leq$ is said to be an ordered set (or partially ordered set) or briefly "poset" and is denoted as (E $\leq$). Because the $\leq$-comparison depends on the selection of the information basis (and of the data representation (classified or not, rounded, etc.) we also write (*E*, *IB*) to denote this important influence of the *IB* for any rankings (Brüggemann and Welzl 2002).

Sometimes it is useful to refer to the quotient set, which is induced by the equivalence relation of equality, *R* (see for details: Brüggemann and Bartel 1999). As usual we write *E/R* for the quotient set, and (*E/R, IB*) for the partially ordered quotient set.

If empirical posets are to be examined it is important to establish orientation rules, i.e. which value of attributes are considered to contribute to badness and which values to goodness. Concerning the evaluation of the ecotoxicity of environmental chemicals by lethal concentrations i.e. LC_{50} values for example, the orientation is the other way round. Here the following situation arises; see Brüggemann and Carlsen, p. 70:

- small values: "good", relatively unhazardous, objects are drawn in the lower part of the Hasse diagram.
- large values: "bad", relatively hazardous, objects are drawn in the upper part of the Hasse diagram.

This consideration is very important and relates to point 3 of the four-point program: One has clearly to state, which orientation is selected and what is related with the bottom-top gradient of a Hasse diagram.

Concerning the evaluation of environmental chemical databases with respect to environmental parameters and chemicals, the value 1 means available information, hence "good", the value 0 means information unavailable, hence "bad".

The total number of comparabilities V and incomparabilities U and their local analogues (i.e. the no of comparabilities V(x) and incomparabilities U(x) of a certain element x are useful quantities for the documentation of the Hasse diagram and for the estimation of ranking uncertainties (Brüggemann and Welzl 2002).

W-Matrix (Dissimilarity Matrix)

The theoretical background of this dissimilarity matrix, which describes the influence of the attributes on the Hasse diagram, is given by Brüggemann and Carlsen, p. 61.

The entries of the W-matrix are a measure for the metric distance among posets, based on the same ground set of objects, but induced by different subsets of *IB* of m-1 attributes, i.e. subset generated by *IB* – $\{q_i\}$, i= 1,..., m. The definitions of the entries of the W-matrix depend on the actual selected subset of elements of *E*. Mostly the full ground set *E* is used. More details can be found in Brüggemann et al. (2001). For further reading we refer to background publications by Brüggemann and Welzl (2002).

METEOR - Method of Evaluation by Order Theory

Aggregation procedures of the data-matrix will be performed by applying METEOR (Method of Evaluation by Order Theory). The basic idea is that subsets of the *IB* can be combined by weighted sums; see Brüggemann and Pudenz (2001). In order to combine attributes freely, a common scaling level must be assumed. Each positive monotonous combination of -say- two attributes, leading to a "superattribute" corresponds order theoretically to an order-preserving map. Therefore the role of weighting can be traced back, when the final result, a linear order, is found by a stepwise aggregation. Note that an aggregation of an ordinal quantity with a metric one or the aggregation of ordinal attributes must be carefully examined. Often it is better to stop the aggregation than to mix attributes of different scaling level.

Furthermore, checking the local incomparability of any element, it is possible to identify weight-sensible and weight-insensible elements of the ground set *E* and *E/R*, respectively, see Brüggemann et al. (2001). Further explanations concerning the mathematical background are given in a recent publication (Voigt et al. 2004b).

Averaged Ranking

The application of weighting schemes as performed by METEOR is not the only way to get linear orders. Another possibility was found by (Winkler 1982) and worked out by Lerche and Sørensen (2003), Brüggemann et al. (2004). The principle to get a linear order is first to find all order preserving maps of an empirical poset. Hereby a set of linear extensions, LE is found, where each element of this set is a single linear order preserving all ≤-relations of the empirical poset. The set LE can be very large. A very crude upper estimation of LT, the number of all linear extensions of an empirical poset is n!, with n the number of all elements of the quotient set. The set LE of all linear extensions can be interpreted as probability space: Let us assume that the rank of an object x, found for one specific linear extension, rk(x) has a certain value, Rk. Then the probability of x to get this value Rk is the number of linear extensions where rk(x) = Rk, L(rk(x) = Rk), divided by LT (see chapter by Brüggemann and Carlsen, p. 86). We write

$$\text{prob}(\text{rk}(x) = \text{Rk}) = L(\text{rk}(x) = \text{Rk})/\text{LT} \quad (1)$$

and then the averaged rank Rkav(x) is defined by:

$$\text{Rkav}(x) = \sum_{\text{Rk}} \text{prob}(\text{rk}(x) = \text{Rk}) \cdot \text{Rk} \qquad (2)$$

Application of Evaluation Methods to Data-Matrix

Hasse Diagram Technique

As described in detail in section 'Selection of Databases (Objects) and Parameters and Chemicals (Attributes)' we have to cope with 12 databases (objects), which we want to analyse by 27 criteria (attributes). These criteria are divided into environmental parameters and chemicals. Hence we will evaluate a 12x27 data-matrix in a first step. We calculate a Hasse diagram for this data-matrix (see Fig. 1). It can be seen that the databases ECO - ECOTOX, EXT – EXTONNET, EFD – Environmental Fate Database show best results as well as HSD – Hazardous Substances Database. These objects are proper maximal objects. There are no other databases, which are better in all aspects than these proper maximal objects. UMW – UmweltInfo, ENV – EnviroFacts Databases, BID – Biocatalysis / Biodegradation Database and PES – Pesticide Database give bad results in comparison to most other databases. These are the proper minimal objects. The databases NRA – NRA Chemical Review Programme and SID – Screening Information Datasets from the OECD are so-called isolated objects. They cannot be compared to any other object. Hence four proper maximal, four proper minimal and two isolated objects are found in the diagram.

For example the successor set of the maximal object ECO is composed of the objects CIV, EHC, PES and UMV. The maximal object EFD has only two successors, namely ENV and BID. There are four levels:

Level 1: {ENV, BID, PES, UMW}
Level 2: {EHC}
Level 3: {CIV}
Level 4: {EFD, HSD, EXT, ECO, NRA, SID}
Level 1 < Level 2 <Level 3 < Level 4

"bad" "good"

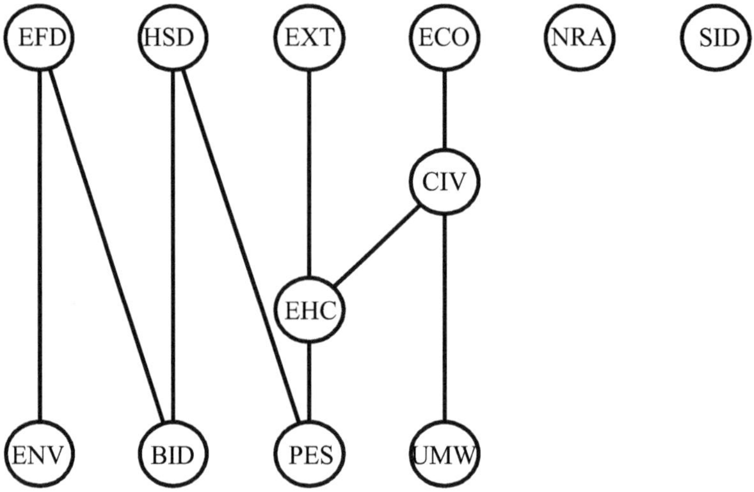

Fig. 1. Hasse diagram of 12x27 Data-Matrix

This Hasse diagram displays many incomparabilities and only few comparabilities. The condensed information on this diagram is listed in Table 4.

As we have to cope with a broad data-matrix, that is to say more attributes than objects, data reduction procedures on the attributes' side seem to be appropriate.

Application of METEOR

Weighting Schemes (Overview)

The original data-matrix of 12 databases (objects) and 27 parameters (attributes) will be subject to several logical aggregation steps. The aim of the aggregation procedure, which can be performed by applying the Hasse Diagram Technique Program WHASSE (Brüggemann and Welzl 2002) is to get a unique prioritisation scheme or at least a greatest or least element.

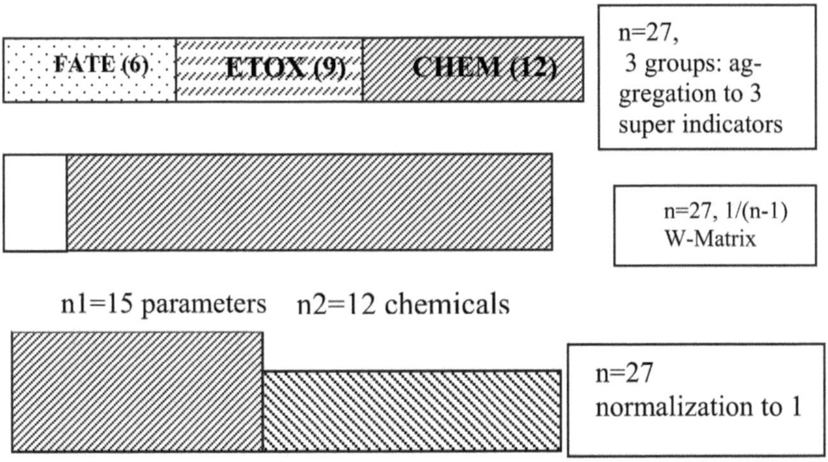

Fig. 2. Different Weighting Schemes

Several different weighting procedures are considered and performed (Fig. 2).

1. Aggregation to get three super indicators concerning environmental fate (FATE), ecotoxicity (ETOX), and chemical information (CHID) (Section 'Aggregation equal weight of environmental Parameters and Chemicals')
2. W-Matrix: 1 attribute is left out; all other 26 attributes are kept (Section 'W-Matrix: Leaving out the most important attribute'). Based on this finding, the remaining 26 attributes are equally weighted, whereas the weight of the omitted attribute is formally considered as 0. This is an example of an extreme case of weighting
3. Two different weighting, normalization to 1, n=27 (Section 'Different weighting schemes')

Aggregation Equal Weight of Environmental Parameters and Chemicals

The aggregation of the data-matrix will be performed and the results presented by the application of METEOR (Method of Evaluation by Order Theory). The criteria (attributes) encompass ecotoxicity as well as environmental fate. As all the chemical substances are high production volume chemicals and used as pesticides, we aggregate them into one group. Hence we cope with three aggregation groups; where each of the following super indicators "FATE", "ETOX", and "CHID" are calculated by a sum with equal weights. For example see eqn. 3:

$$\text{Fate}=\sum_{i=1}^{i=6} w_i \cdot q_i \qquad q_i \in IB_{environmental\ Fate} = \{\text{PHO, SWA,...,BAC}\}$$

$$\{\text{PHO, SWA,...,BAC}\} \subset IB \tag{3}$$

$$w_i = 1/6$$

Aggregation of 6 environmental fate attributes: FATE
Similarly: Aggregation of 9 ecotoxicity parameters: ETOX
Similarly: Aggregation of 12 chemicals: CHID.

By this aggregation the aspects of fate, ecotoxicity and chemicals are maintained and kept separately for further analysis. This "thematic" aggregation can be interpreted as taking a more abstract level of consideration: Not which specific fate parameter is important, but the whole concept of fate in comparison to other criteria.

Some visible changes took place comparing the original Hasse diagram of the 12x27 data-matrix (Fig. 1) with the reduced 12x3 data-matrix (Fig. 3). Nevertheless all ≤ - relations found in the Hasse diagram of Fig. 1 are reproduced in this "enriched" Hasse diagram. For examples, see Table 4:

Table 4. Comparison between Hasse diagrams based on 27 attributes (HD_{27}) and that on 3 attributes (HD_3)

HD_{27}	HD_3	Remarks
ENV < EFD	ENV < EFD	cover relation in HD_{27}, cover relation in HD_3
BID < EFD	BID < EFD	cover relation in HD_{27}
EHC < ECO	EHC< ECO	in both Hasse diagrams no cover relation
UMW || NRA	UMW < NRA	An incomparability is vanishing due to compensation and a < - relation appears.

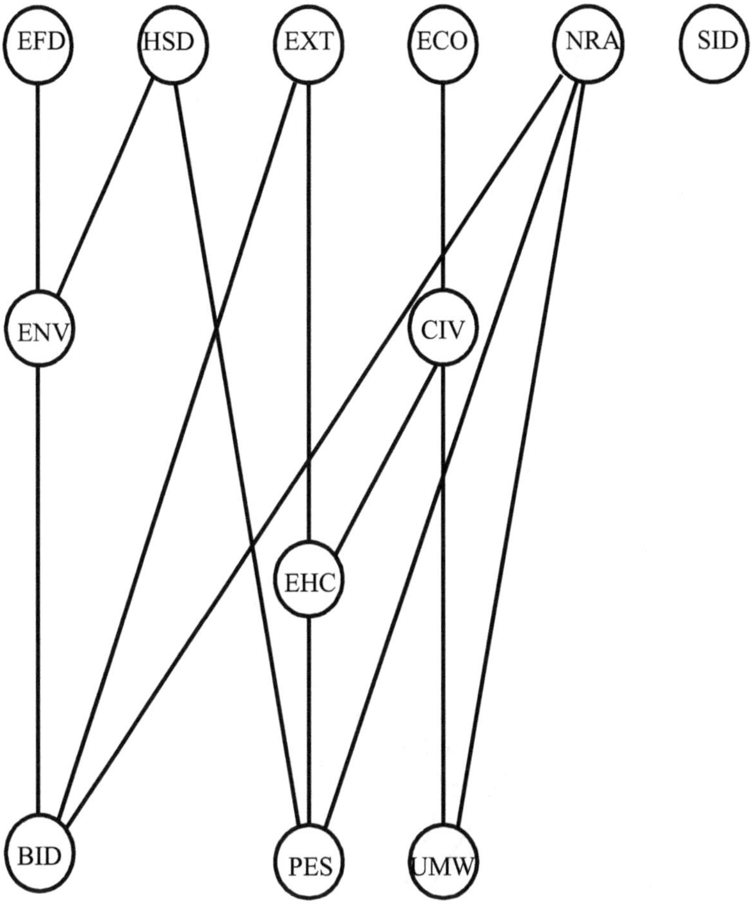

Fig. 3. Hasse diagram of *IB*= {FATE, ETOX, CHID}, 12x3 Data-Matrix

Any ≤-relation of the original poset will be maintained by this kind of aggregation. This is generally true if as aggregation function positive weak monotonous functions with respect to the attributes are applied.

NRA is now a maximal object and no longer an isolated object. The only isolated object in this approach is SID. ENV is no longer a minimal object but one level above. Further differences are found in Table 4 where three diagrams are compared.

W-Matrix: Leaving out the most important attribute

As explained earlier, the W-matrix describes the influence of the attributes on the Hasse diagram. The W-matrix is calculated for all objects given in the original diagram in Fig. 1. It reveals that the criterion acute fish toxicity is the most important attribute in this approach. Three changes take place leaving out this criterion acute fish toxicity. The Hasse diagram of this described case is given in the following diagram (Fig. 4).

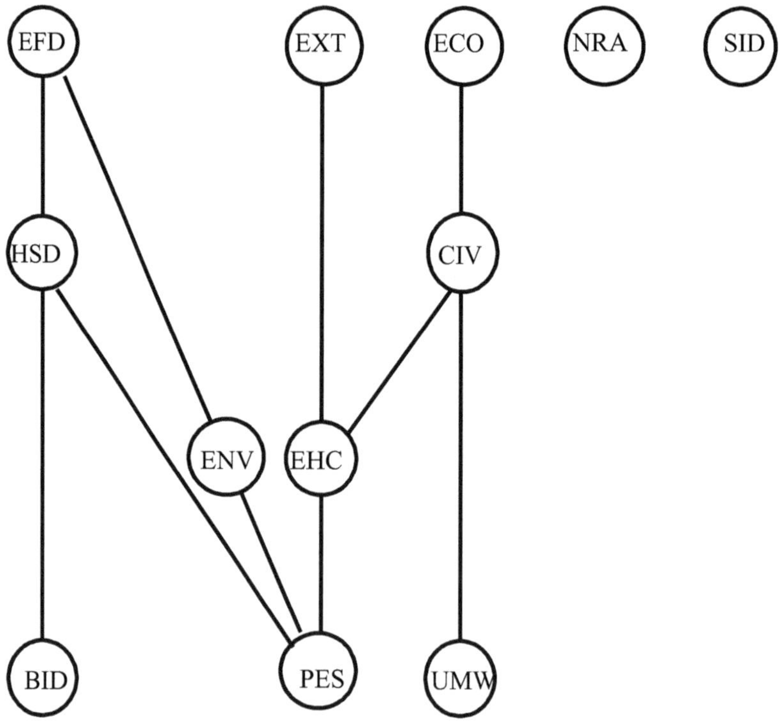

Fig. 4. Hasse diagram of Case 7 (leaving out the attribute acute fish toxicity)

Visible changes took place, e.g. HSD is no longer a maximal object but found in the second highest level. ENV is no longer a minimal object. Several differences can be detected while reducing the initial 12x27 data-matrix applying the METEOR method and the W-Matrix. In our examples we reduced the data-matrix to a matrix of 12x3 (equal weighting) and to 12x26 (W-Matrix). The results are listed in Table 5.

Table 5. Comparison of Three Hasse diagrams

Data-matrix	12x27	12x3	12x26
Number of Levels	4	4	4
Objects in largest level	6	6	5
Equivalent objects	0	0	0
Maximal objects	4	5	3
Minimal objects	4	3	3
Isolated Objects	SID, NRA	SID	SID, NRA
Comparabilities V(N)	14	21	17
Contradictions U(N)	104	90	98

All three diagrams also show some similarities. They all have 4 levels and no equivalent objects. The number of comparabilities decreases with the number of attributes. For the number of incomparabilities the situation is the other way round. In other words: The more attributes the data-matrix encompasses, the less comparabilities are found.

The next logical step is the aggregation of all attributes 1/26 = 0,03846 and the setting of acute fish toxicity ATF = 0. The linear order Hasse diagram is given in Fig. 5, left hand side.

Different weighting schemes

The next step is to weight in a logical and subject-oriented way selected attribute-groups differently. For example: One might be more interested in parameters than in chemicals or verse visa.

For the first attempt we weight the parameters double with respect to the chemicals.

Let n1 and n2 be the numbers of attributes in two attribute groups (here: parameters and chemicals, respectively), w1 and w2 the weights by which the attributes of the one group and those of the other group are combined, and m the relative weight. Then, maintaining the normalization to 1 the weights can be calculated as follows in eqn. 4 and 5:

$$w_2 = \frac{1}{(n1 \cdot m) + n2} \tag{4}$$

$$w_1 = m \cdot w_2 \tag{5}$$

Equivalent Objects: {NRA;SID}, {PES;UMW}(left), {BID;PES} {EFD;EXT} {HSD;NRA}(middle), {BID;PES}(right)

ECO		ECO		ECO
EFD		CIV		EFD
CIV		EFD		CIV
HSD		SID		HSD
EXT		HSD		ENV
ENV		ENV		EXT
NRA		EHC		NRA
EHC		UMW		EHC
BID		BID		SID
PES				BID
				UMW

Fig. 5. Linear orders represented in tabular form (ECO > EFD > CIV >...). First column: ATF=0 and all other attributes: weights 1/26 equivalence classes: {NRA, SID}, {PES, UMW}, Second and third column: Parameters and Chemicals aggregated with different weights (see text); equivalence classes (2nd column) {BID, PES}, {EFD, EXT}, {HSD, NRA}; (3rd column) {BID, PES}

As the parameters are considered to be more important than the chemicals, m must be selected > 1. Here for m, arbitrarily the value 2 is given:

w_1 = (weight of parameters) = 0,0238
w_2 = (weight of chemicals) = 0,0476

The corresponding Hasse diagram is given in the Fig. 5 in the middle section.

Clearly this Hasse diagram comprises a linear order, not a partial order. Three equivalence classes are shown: {BID;PES} {EFD;EXT} {HSD;NRA}. The database ECO comes best whereas the databases BID and PES are the least important ones considering the parameters to be of higher importance than the chemicals. UMW is also not recommendable as it is found on the second lowest level.

Alternatively we weight the chemicals double with respect to the environmental parameters, i.e. m=0.5.

w_2 = (weight parameters) = 0.0256
w_1 = (weight chemicals) = 0.0513

the corresponding Hasse diagram being given in the Fig. 5, right hand side.

In this diagram only one equivalence class {BID;PES} is given. The maximal object is still ECO like in the diagram where the emphasis is put on the parameters. This means that ECO is a very comprehensive database as well for chemicals as for parameters. In this diagram UMW is the minimal object and BID / PES can be found on the second lowest level. These three databases are neither recommendable for chemicals nor for parameters. The location of SID in both diagrams is worth discussing. Whereas it can be found on a rather good position in the diagram where the parameters are important (weighted higher than chemicals), it is situated at a comparably low position in the Hasse diagram in which the chemicals are of more importance. This means that SID comprises a huge range of parameters but only on a small amount of chemicals. The variation of the ranking position of SID is expected: SID is an isolated element, therefore any aggregation can locate SID everywhere, whereas objects, which are related to others by the ≤-relation must preserve this relation independent which kind of weighting one is applying. Note however that different weighting schemes will not necessarily lead to very different ranking positions, even if the object is isolated. See for example the object NRA. This object is isolated in HD27, however its ranking variation is not large -at least in comparison to the object SID.

Discussion of Linear Orders

All diagrams Fig. 5 left (W-Matrix – AKF weighted 1/26), Fig. 5 middle (parameters more important than chemicals), Fig. 5 right (chemicals more important than parameters) show linear orders. These diagrams will be compared now. Six objects are arbitrarily selected, not to overburden the

Hasse diagram: ECO, CIV, SID, ENV, EHC, and UMV. The equivalence classes are left out (Table 5) (Fig. 6).

Table 6. Comparison of Linear Orders

Subset of objects /numbers	Fig. 7 (middle) parameters more important than chemicals	Fig. 7 (right) chemicals more important than parameters	Fig. 7 (left) Attribute fish toxicity left out, the other uniformly weighted
ECO	6	6	6
CIV	5	5	5
SID	4	2	3
ENV	3	4	4
EHC	2	3	2
UMV	1	1	1

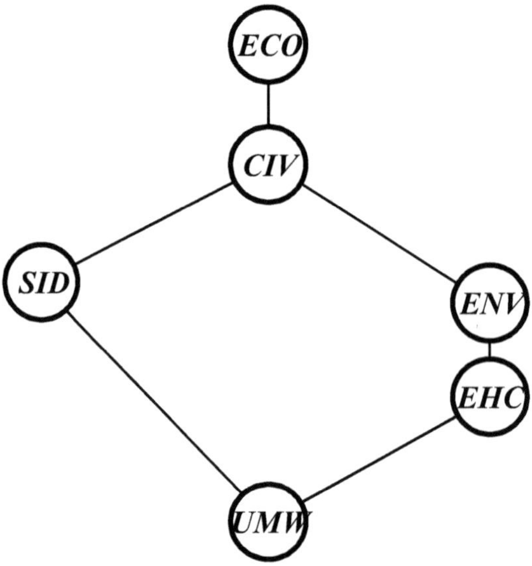

Fig. 6. Hasse diagram of the Rankings Induced by the Linear Orders

Obviously different weighting schemes even that for which the most important attribute is neglected lead to the more or less same ranking. The main uncertainty for this evaluation is found for the database SID. Its rank position will severely depend on how the weights are selected. Once more, the advantage of the inclusion of order theoretical tools is evident.

Averaged Ranking

By applying the average ranking procedure briefly explained above in section 'Averaged Ranking' and in more detail in chapter Brüggemann and Carlsen, p. 61 the following result is obtained: ECO > EFD > EXT > HSD > CIV > NRA > SID > ENV > EHC > UMW > BID > PES.

Performing the averaged rank, nearly the same ranking effect than that which is induced by a specific weighting scheme, namely parameters two times more important than chemicals, can be detected. This means that an averaged ranking is often a good policy for a first screening of priorities. If no knowledge is known about weightings, it might be a pragmatic way to estimate the linear order by calculating the averaged ranks (Brüggemann et al. 2004), (Lerche and Sørensen 2003). Additionally probability distribution, prob(rk(x) = Rk) as a function of Rk can be derived by which the uncertainty of any rank can be calculated. For example the probability distribution of three databases is shown in Fig. 7.

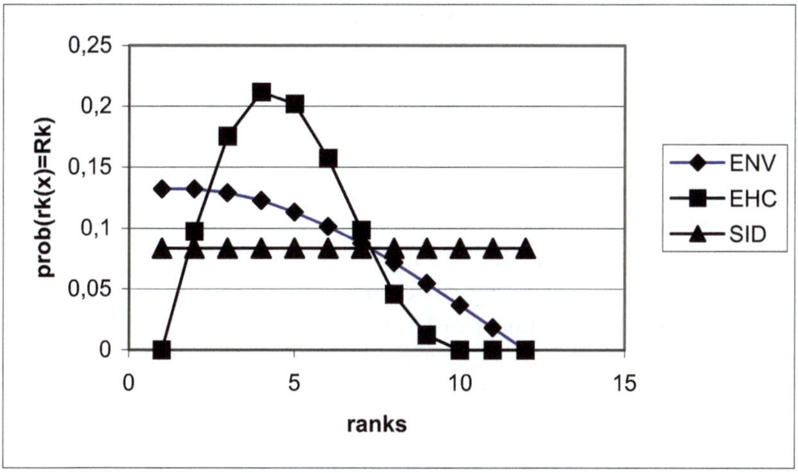

Fig. 7. Probability Distribution of Ranks of Three Databases

Thus a broad and nearly uniform distribution of ranks is typical for objects, which are isolated (SID) or nearly isolated (ENV). EHC has many connections and can be therefore more safely ranked. This is seen by the relatively sharp unimodal curve in Fig. 7. One can see that even if a single number is generated by the averaged ranking procedure, the same theoretical setting helps to identify uncertainties (Brüggemann et al. 2004).

Discussion and Conclusion

The availability of environmentally relevant parameters (environmental fate and ecotoxicity) as well as the availability of data on 12 high production volume chemicals in 12 well-known international databases on the free Internet was analysed. Different mathematical and statistical methods were taken into account. The emphasis is set on the METEOR – Method of Evaluation by Order Theory, a discrete mathematical method. All methods revealed significant shortcomings in the databases. The Hasse diagram of the complete data-matrix 12 objects (databases) x 27 attributes (parameters + chemicals) revealed that the databases ECO- ECOTOX, EFD – Environmental Fate Database and EXT Extoxnet, also called multi-database databases came best. One has to consider the fact that these databases comprise different sizes of data. Whereas EFD comprises approximately 20.000 chemicals, ECO encompasses data on approximately 8.000 chemicals, and EXT only on 400 chemicals. EXT has extended profiles on high production volume chemicals. HSD – Hazardous Substances Database has a broad data collection on 4.500 chemicals. Most single databases, which are specialised, are found in a minimal position in the Hasse diagram. These are BID- Biocatalysis/Biodegradation Database, PES – Pesticide Database, and UMW – UmweltInfo.

The aggregation of environmental parameters and chemicals (equal weight) leads to a slimmer data-matrix on the attribute side. However, no significant differences are found in the "best" and "worst" objects. The aggregation procedure, assigning different weights to either parameters or chemicals, which lead to linear order diagrams, also reveals that the multi-database databases get a high position. Differences can be detected in the other positions. The monograph databases EHC, ENV, EFD, EXT come in the low positions when the chemicals are weighted twice, they come in middle and low positions when the weighting of parameters is emphasised. This is a slight indication that monograph databases have a huge variety of parameters but comprise only few chemicals.

Performing the averaged rank, a very similar ranking effect than that which is induced by a specific weighting scheme, namely parameters twice as important as chemicals, can be detected. This means that getting an averaged ranking might be a good policy for a first screening of priorities. This approach will be followed in our future studies.

The whole approach indicates a rather bad situation on the data availability on existing chemicals and hence an alarming signal concerning the new existing chemicals policy of the EEC.

For future steps concerning the data availability of chemicals five ways should be taken into account:

- Foster many data-sources (timeliness)
- Foster new publications and enter the data into the numeric databases
- Estimate data by well-established methods (QSAR) and fill up data gaps indicating that the data are estimated ones.
- Test chemicals in the described way by the EEC according to the White Paper (EEC 2001)
- Evaluate dynamically the actually best databases

Only by improving the data situation extensive testing can be reduced to a pragmatic size.

References

Allanou R, Hansen BG, van der Bilt Y (1999) Public Availability of Data on EU High Production Volume Chemicals. EUR 18996EN, http://ecb.jrc.it/

Brüggemann R (ed.) (1998) Order Theoretical Tools in Environmental Sciences held on November 16th, 1998 in Berlin, Berichte des IGB 1998, Heft 6, Sonderheft I, Institut für Gewässerökologie und Binnenfischerei, Berlin

Brüggemann R, Bücherl C, Pudenz S, Steinberg C (1999) Application of the Concept of Partial Order on Comparative Evaluation of Environmental Chemicals. Acta Hydrochem Hydrobiol. 27:170-178

Brüggemann R, Bartel HG (1999) A Theoretical Concept to Rank Environmentally Significant Chemicals. J Chem Inf Comp Sci 39:211-217

Brüggemann R, Pudenz S (2001) Hassediagrammtechnik zur Bewertung wasserwirtschaftlicher Massnahmen. In: Wasserforschung e.V. (ed.) Nachhaltige Entwicklung in der Wasserwirtschaft - Konzepte, Planung und Entscheidungsfindung; Interdisziplinäre Fachtagung am 27. und 28. Juni in Berlin; Dokumentation, IFV Berlin pp 225-239

Brüggemann R, Halfon E, Welzl G, Voigt K, Steinberg C (2001) Applying the Concept of Partially Ordered Sets on the Ranking of Near- Shore Sediments by a Battery of Tests. J Chem Inf Comp Sci 41:918-925

Brüggemann R, Welzl G (2002) Order Theory Meets Statistics – Hasse Diagram Technique. In: Voigt K, Welzl G (eds.) Order Theoretical Tools in Environmental Sciences, Order Theory (Hasse Diagram Technique) Meets Multivariate Statistics. Shaker-Verlag Aachen pp 9-40

Brüggemann R, Sørensen PB, Lerche D, Carlsen L (2004) Estimation of Averaged Ranks by a Local Partial Order Model. B. Leibniz-Institut für Gewässerökologie und Binnenfischerei J Chem Inf Comp Sci 44:618-624

CERHR, Center for the Evaluation of Risks to Human Reproduction, (2004) NTP-CERHR Expert Panel Report on the Reproductive and Development Toxicity of Propylene Glycol. Reproductive Toxicology 18:533-579

Cooke F, Schofield H (2001) A Framework for the Evaluation of Chemical Structure Databases. J Chem Inf Comput Sci 41:1131-1140

Criterion (2004) WHasse Software, http://www.criteri-on.de/hdtws-site/index.html

EEC, Commission of the European Communities (2001) White Paper, Strategy for a Future Chemicals Policy, COM 88 final. http://www.europa.eu.int/comm/environment/chemicals/index.htm

Einax JW, Zwanzinger HW, Geiss S (1997) Chemometrics in Environmental Analysis. VCH, Weinheim

Einax JW (2003) Chemometrics in Environmental Sciences - Challenges and New Methods. In: Gnauck A, Heinrich R (eds.) The Information Society and the Enlargement of the European Union. Metropolis Verlag Marburg pp 51-57

El-Shaarawi AH, Hunter JS (2002) Environmetrics, Overview In: El-Shaarawi AH, Piegorsch WW (eds.) Encyclopedia of Environmetrics. John Wiley and Sons, Chichester pp 698-702

Felsot AS (2002) Web Resources for Pesticide Toxicology, Environmental Chemistry, and Policy: A Utilitarian Perspective. Toxicology 173:153-166

Gelbke HP, Fleig H, Meder M (2004) SIDS Reprotoxicity Screeening Test Update: Testing Strategies and Use. Regulatory Toxicology and Pharmacology 39:81-86

Halfon E, Reggiani MG (1986) On Ranking Chemicals for Environmental Hazard. Environ Sci Technol 20(11):1173-1179

Heidorn C, Rasmussen K, Hansen BG, Norager O, Allanou R (2003) An Information Management Tool for Existing Chemicals and Biocides. J Chem Inf Comp Sci 43(3):779-786

Lerche D, Brüggemann R, Sørensen PB, Carlsen L, Nielsen OJ (2002) A Comparison of Partial Order Techniques with Three Methods of Multi- Criteria Analysis for Ranking of Chemical Substances. J Chem Inf Comp Sci 42:1086-1098

Lerche D, Sørensen PB (2003) Evaluation of the Ranking Probabilities for Partial Orders based on Random Linear Extensions. Chemosphere 53:981-992

Lerche D, Sørensen PB, Brüggemann R (2003) Improved Estimation of the Ranking Probabilities in Partial Orders Using Random Linear Extensions by Approximation of the Mutual Ranking Probability. J Chem Inf Comp Sc 53:1471-1480

Massart DL, Vandeginste BGM, Buydens LMC, De Jong S, Lewi PJ, Smeyers-Verbeke J (1997) Handbook of Chemometrics and Qualimetrics. Elsevier, Amsterdam

Page B, Voigt K (2003) Recent History and Development of Environmental Information Systems and Databases in Germany. Online Information Review 27(1):37-50

Poetzsch E (2004) Naturwissenschaftlich-technische Information, Online, CD-ROM, Internet. Materialien zur Information und Dokumentation, Band 23. Verlag für Berlin-Brandenburg, Potsdam

Pudenz S, Brüggemann R, Lühr HP (Ed.) (2001) Order Theoretical Tools on Environmental Science and Decision Systems, 222, Heft 14, Sonderheft IV. Leibniz Institute of Freshwater Ecology and Inland Fisheries, Berlin

Russom CL (2002) Mining Environmental Toxicology Information: Web Resources. Toxicology 173:75-88

Sørensen PB, Carlsen, Mogensen BB, Brüggemann R, Luther B, Pudenz S, Simon U, Halfon E, Voigt K, Welzl G, Rediske G (Ed.) (2000) Order Theoretical Tools in Environmental Sciences, NERI – Technical Report No. 318, National Environmental Research Institute, Roskilde

Sørensen P, Brüggemann R, Lerche DB, Voigt K (2004) Order Theory in Environmental Sciences. NERI Technical Report No. 479, National Environmental Research Institute, Ministry of the Environment, Copenhagen

Stoyan D, Stoyan H, Jansen U (1997) Umweltstatistik, Statistische Verarbeitung und Analyse von Umweltdaten. B.G. Teubner Verlagsgesellschaft, Stuttgart, Leipzig

Voigt JH, Bienfait B, Wang S, Nicklaus MC (2001) Comparison of NCI Open Database with Seven Large Chemical Structural Databases. J Chem Inf Comput Sci 41:702-712

Voigt K (1997) Erstellung von Metadatenbanken zu Umweltchemikalien und vergleichende Bewertung von Online Datenbanken und CD-ROMs. http://vermeer.organik.uni–erlagen.de/dissertationen/data/dissertation/Kristina_Voigt/html/

Voigt K, Gasteiger J, Brüggemann R (2000) Comparative Evaluation of Chemical and Environmental Online and CD-ROM Databases. J Chem Inf Comp Sci 40:44-49

Voigt K, Welzl G (2002a) Chemical Databases: An Overview of Selected Databases and Evaluation Methods. Online Information Review 26:172-192

Voigt K, Welzl G (Ed.) (2002b) Order Theoretical Tools in Environmental Sciences, Order Theory (Hasse Diagram Technique) Meets Multivariate Statistics. Shaker-Verlag, Aachen

Voigt K (2003) Databases on Environmental Information, in: Gasteiger J, Engel T (Ed.) Chemoinformatics – From Data to Knowledge, Volume 2. Wiley-VCH, Weinheim, pp 722-741

Voigt K, Brüggemann R, Pudenz S (2004a) Chemical Databases Evaluated by Order Theoretical Tools. Analytical and Bioanalytical Chemistry 380:467-474

Voigt K, Welzl G, Brüggemann R (2004b) Data Analysis of Environmental Air Pollutant Monitoring Systems in Europe. Environmetrics 15:577-596

Welzl G, Faus-Kessler T, Scherb H, Voigt K (2004 online) Biostatistics. In: EOLSS Encyclopedia of Life Support Systems, Sydow A (Ed.) EOLSS Publishers Co. Ltd.: Oxford. http://www.eolss.net

Wexler P (2004) The U.S. National Library of Medicine's Toxicology and Environmental Health Information Program. Toxicology 198:161-168

Winkler P (1982) Average height in partially ordered sets. Discr Math 39:337-341

6 Rules and Complexity

Partial order has for its own a rich mathematical theory and there is a manifold of relations to combinatorics, graph theory and algebra. Even relations to experimental designs and variance analysis can be established. However, mathematicians like to find more structure for their objects to be studied. In that sense posets are poor, because there is only one binary operator, i.e., the $\leq$-relation. Comparing with the daily life example where we have addition and multiplication as binary operators the mathematical multitude in posets is somewhat restricted. Should this deficiency bother the applications? This question can be answered with "yes" if the chapters of Kerber and of Seitz are examined.

In the chapter by Kerber a posetic structure is explained, which obeys the axioms of order but fulfil additional properties, namely those of lattices. Similar to daily life calculations lattice theory combines the objects of interest by two operators. Kerber shows by a very simple chemical example (what is standard knowledge) how one can derive systematically a classification for chemical systems. He uses the Brønsted definition of acids and shows by a stepwise process how additional knowledge can be systematically introduced. In a second example Kerber shows how implications about the pollution status in regions of Baden-Württemberg, Germany, can be derived and how general characteristics, like density of forests, or of traffic roads could be related to the pollution status. Methodologically the reader gets an impression about the powerful Formal Concept Analysis, which is a variant of lattice theory.

Lattices can be obtained by several approaches. In the chapter by Seitz the lattice of Young diagrams is introduced and applied to study of complexity. Complexity as a research area is of high interest as this concept can be applied to almost all disciplines where interactions (in a very general sense; for example stock market analysis, chemical structures, social structures) are studied. Clearly the question arises, what is complexity, how can we measure it. Indeed Seitz states that there are "dozens of mutually inconsistent definitions". Nevertheless, similar to the concept of biodiversity the item 'complexity' has begun to have a life for its own. Seitz relates complexity with the degree of order in a Gaussian like fashion. Hence, a total order (a chain) and a complete disorder (an anti-chain) are examples of non-complexity, whereas in between the degree of complexity takes a maximum value. In his chapter, Seitz describes this concept and how it is quantified by means of the lattice of Young diagrams. For small systems one may derive a complexity measure just by optical inspection of the Hasse diagram of Young diagrams. In the general case a measure is needed which is computational tractable, which is derived in Seitz's chapter. Three examples clarify the arguments.

Contexts, Concepts, Implications and Hypotheses

Adalbert Kerber

Department of Mathematics, University of Bayreuth
D-95440 Bayreuth, Germany

e-mail: Adalbert.Kerber@uni-bayreuth.de

Abstract

We give a brief introduction to the notion of concept, a mathematical model of conceptual thinking. It serves very well in the organization of interviews, tests and evaluations, since it allows a systematic way of drawing conclusions and establishing hypotheses. Thus, it can be considered as an efficient tool for decision support, for example, in environmental risk management. In fact, it models a certain way of doing research by gathering examples and trying pattern recognition.

Contexts and their concepts

Suppose that we are given information on properties of certain objects, say on chemical substances and their environmental properties. We call this knowledge a context, and we should like to explore the information contained in it. For example, we should like to

- visualize its content
- to draw conclusions
- or to form hypotheses

based on the context and on nothing else. For this purpose, we can use concept analysis, introduced by R. Wille about twenty-five years ago, the standard reference is Ganter and Wille (1999). Its main tool is the notion of lattice of concepts, which is a particular partial order that beautifully reflects the information contained in the given context. Hence, the notion of concepts applies in particular to situations, where the information given is of the form

The object ω has (has not) the attribute α for a given set of objects ω ∈ Ω and a given set of attributes α ∈ A. Such a set of information can be gathered e.g. in the form of a table, a context. Fig. 1 shows the oldest context of environmental chemistry (Bartel 1995), it describes the classical "four elements":

ω / α	w	c	d	h
F	×		×	
E		×	×	
W		×		×
A	×			×

Fig. 1. The context of the "four elements"

Its objects are: Fire, Earth, Water, Air, and the attributes: warm, cold, dry, humid. The entry at the intersection of the row corresponding to an object ω and the column of the attribute α is put × if the object has this property, otherwise it is left empty.

The concepts reflect its information content. They are pairs (B, C), B ⊆ Ω, C ⊆ A, such that the corresponding entries form a rectangle of ×'s, not contained in a bigger rectangle full of ×'s. In this case, the set of objects B is called the extent, while the set of attributes C is the intent of the concept. For example, if $B = \{b_1, \dots, b_m\}$ is the extent of a concept and $C = \{c_1, \dots, c_n\}$ its intent, the situation looks as follows (after a suitable rearrangement of rows and columns, if necessary).

ω / α	…	c_1	…	c_n	…
⋮					
b_1		×	…	×	
⋮		⋮	…	⋮	
b_m		×	…	×	
⋮					

In formal terms: If we define the derivation of B by

$$B' := \{\alpha \in A \mid \text{each } \omega \in B \text{ has the attribute } \alpha\} \qquad (1)$$

and, analogously, the derivation of C:

$$C' := \{\omega \in \Omega \mid \text{each } \alpha \in C \text{ is an attribute of } \omega\} \quad (2)$$

then (B, C) is a concept, if and only if B = C′ and C = B′.

The concepts obtained from a given context form a partial order ≤, defined by

$$(B, C) \leq (D, E) \Leftrightarrow B \subseteq D \Leftrightarrow C \supseteq E \quad (3)$$

(B, C) being said to be a subconcept of (D, E), (D, E) a superconcept of (B, C). It is even a lattice, i.e. for two concepts there exists both an infimum (= the biggest subconcept of both of them) and a supremum (= the smallest superconcept of them). Hence we may call this partial order the lattice of concepts corresponding to the context in question. Fig. 2 shows, for example, the Hasse diagram of the lattice of concepts corresponding to the context on the "four elements".

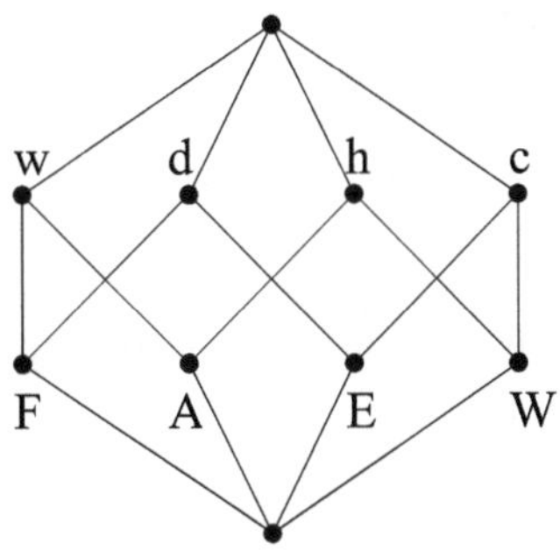

Fig. 2. The lattice of concepts of the context on the "four elements"

The letters F, A, E, W label the smallest concepts containing the objects F, A, E, W in their content. F for example, labels the concept (B, C) = ({F}″, {F}′). Correspondingly, the attribute ω say, labels the concept (B, C) = ({ω}′, {ω}″).

It is most important to note that the lattice of concepts given obviously allows reconstructing the context of Fig. 1. In fact, also in the general case, the lattice of concepts, better say the Hasse diagram of the lattice of concepts perfectly reflects and visualizes the information contained in the context in question! For example, we can easily read off from the above diagram, that fire is supposed to be warm and dry (and neither humid nor cold). This is why Hasse Diagram Technique (HDT) is so helpful and important.

Such contexts can be replaced by 0-1-matrices. For example, the above context about the "four elements", by Moreover, there are contexts that contain, besides 0, more than just one entry.

ω / α	w	c	d	h
F	1	0	1	0
E	0	1	1	0
W	0	1	0	1
A	1	0	0	1

Fig. 3. The context of the "four elements" as 0-1-matrix

For example, in order to express that water is "more humid" than air, we may establish the many-valued context of Fig. 4.

ω / α	w	c	d	h
F	1	0	1	0
E	0	1	1	0
W	0	1	0	2
A	1	0	0	2

Fig. 4. A multi-valued context on the "four elements"

Such many-valued contexts give, via scaling (see Ganter and Wille 1999), one-valued contexts, in our example by replacing the attribute h by two attributes $h_{\geq 1}$ and $h_{\geq 2}$, say.

Implications

For sake of simplicity, we start with a description of the conclusions that can be drawn from a context. For sets of attributes $A_1, A_2 \subseteq A$ we say that we have the implication

$$A_1 \rightarrow A_2 \tag{4}$$

if $A'_1 \subseteq A'_2$, i.e. if each object which has all the attributes $\alpha \in A_1$ has all the attributes $\beta \in A_2$ as well. The crucial point is that there exist bases of the set of all these implications! In order to describe a particular basis, we note that for $A_1, A_2, X \subseteq A$,

- X respects the implication $A_1 \rightarrow A_2$ iff it respects A_1, A_2) i.e. iff $A_1 \subseteq X \Rightarrow A_2 \subseteq X$.

- $A_1 \rightarrow A_2$ follows from a set $\mathcal{L}$ of implications, iff each X that respects every element of $\mathcal{L}$ respects $A_1 \rightarrow A_2$.
- $\mathcal{L}$ is complete iff every implication follows from $\mathcal{L}$.
- $\mathcal{L}$ is reduced iff no $(A_1 \rightarrow A_2) \in \mathcal{L}$ follows from $\mathcal{L} \backslash \{A_1 \rightarrow A_2\}$.

In order to describe such a complete and reduced set of implications, we define pseudo contents $P \subseteq A$ recursively as follows, supposing Ω and A to be finite:

$$P \neq P'' \text{ and [pseudo content } Q \subset P \text{ implies } Q'' \subseteq P] \quad (5)$$

The main result, which is very interesting for various applications, is

Theorem (Duquenne/Guigues)

$$\mathcal{I} := \{P \rightarrow P'' \mid P \text{ pseudo content}\} \quad (6)$$

The set $\mathcal{I}$ is a complete and reduced set of implications, the Duquenne/Guigues-basis.

Here are two important applications:

1. If you are given a context, and the Duquenne/Guigues-basis contains unacceptable implications, then you need further information, i.e. knowledge on further objects and their properties!
2. Otherwise, the implications in the Duquenne/Guigues-basis are either acceptable or you just don't know, in which case you may consider them as being hypotheses!

Here is a simple didactic example that I owe to Brüggemann (see e.g. Bell 1974). Suppose we want to find out what Brønsted's concept of acid might be, along examples. For this purpose we consider the following set of attributes:

- H: contains H,
- Cl: contains Cl,
- O: contains O,
- diss.: dissociates in water by protonating it,
- $AH \rightarrow A^- + H^+$, $H^+ + H_2O \rightarrow H_3O^+$.
- Br.: is an acid in the sense of Brønsted.

We are now going to establish a context by collecting a few examples of acids that we know. For example, we may start with HCl since we have heard, say, that this is an acid in Brønsted's sense, obtaining the context

	H	Cl	O	diss.	Br.
HCl	×	×		×	×

Using a suitable software package we evaluate its Duquenne/Guigues basis which turns out to be $\{\ \} \to \{H, Cl, diss., Br.\}$.

It says that every molecule contains H and Cl, dissociates and is an acid in the sense of Brønsted. The reason for that clearly unacceptable implication is the fact that we started with a single example. But, knowing the fact that CCl_4 is not an acid in Brønsted's sense, we obtain the following context, containing our information on CCl_4 in addition:

	H	Cl	O	diss.	Br.
HCl	×	×		×	×
CCl_4		×			

Its Duquenne/Guigues basis is

- $\{\ \} \to \{Cl\}$
- $\{Br.\} \to \{H, diss.\}$
- $\{diss.\} \to \{H, Br.\}$
- $\{O\} \to \{H, Cl, O, diss., Br.\}$
- $\{H\} \to \{diss., Br.\}$

The last one of these implications is obviously unacceptable, the presence of hydrogen does not imply that the substance is an acid in the sense of Brønsted. For this reason we extend the context by CH_4 obtaining

	H	Cl	O	diss.	Br.
HCl	×	×		×	×
CCl_4		×			
CH_4	×				

the implication basis of which is

- $\{Br.\} \to \{H, Cl, diss.\}$
- $\{diss.\} \to \{H, Cl, Br.\}$
- $\{O\} \to \{H, Cl, O, diss., Br.\}$
- $\{H, Cl\} \to \{diss., Br.\}$

As the presence of oxygen does not mean that the molecule in question is an acid we extend the context in the following way:

	H	Cl	O	diss.	Br.
HCl	×	×		×	×
CCl_4		×			
CH_4	×				
CH_3OCH_3	×		×		

with its implication basis

- {Br.} → {H, Cl, diss.}
- {diss.} → {H, Cl, Br.}
- {O} → {H}
- {H, Cl} → {diss., Br.}

The first element of the basis means that each acid in the sense of Brønsted contains chlorine. We know that this is not true, and so we add H_2SO_4 obtaining the context

	H	Cl	O	diss.	Br.
HCl	×	×		×	×
CCl_4		×			
CH_4	×				
CH_3OCH_3	×		×		
H_2SO_4	×		×	×	×

with its implication basis

- {Br.} → {H, diss.}
- {diss.} → {H, Br.}
- {O} → {H}
- {H, Cl} → {diss., Br.}

We see that the first two implications suggest that molecules containing hydrogen atoms are acids in the sense of Brønsted if and only if they dissociate water by protonating it. This is an interesting hypothesis! The third element in the basis indicates that the presence of oxygen implies the presence of hydrogen, which is obviously false, but we don't care, we could easily add counter examples. The fourth element in the basis suggests that the presence of hydrogen and chlorine means that the substance dissociates water and is an acid in the sense of Brønsted, but there are counterexamples, for example CH_3Cl which we add to the context, obtaining

	H	Cl	O	diss.	Br.
HCl	×	×		×	×
CCl_4		×			
CH_4	×				
CH_3OCH_3	×		×		
H_2SO_4	×		×	×	×
CH_3Cl	×	×			

with its implication basis

- {Br.} → {H, diss.}
- {diss.} → {H, Br.}
- {O} → {H}
- {Cl, O} → {H, Cl, O, diss., Br.}

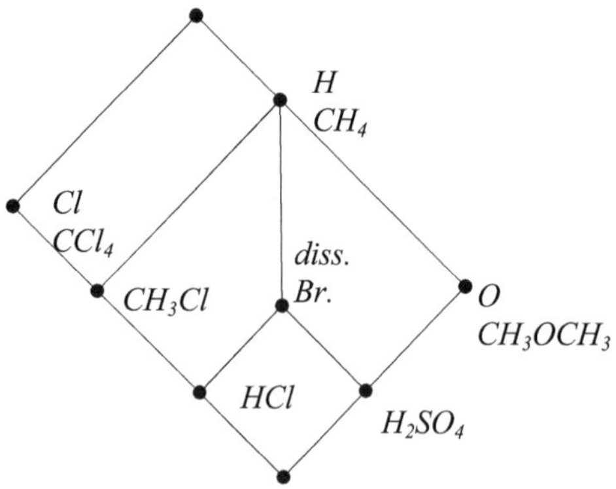

Fig. 5. The concept lattice connected with Brønsted's definition

The first two implications again suggest that molecules containing hydrogen atoms are acids in the sense of Brønsted if and only if they dissociate water by protonating it. Generalizations to other solvents can be done, however here the focus is to model the typical learning process by studying examples and counter examples. Any new example is narrowing the range of possible definitions. Although the learning process is not yet finished as the other two implications are obviously unacceptable, we stop here, since we arrived at two reasonable implications. In fact, if we look at the "Lexikon der Chemie", 1999 we find out that dissociation of water by protonating it is in fact Brønsted's definition of acid. The corresponding lattice of concepts is shown in Fig. 5.

Since we have to read Hasse diagrams from bottom to top, the concept lattice visualizes that HCl and H_2SO_4 are the only acids in Brønsted's sense, that they dissociate water in contrast to all the other molecules of the lattice, and that they contain hydrogen.

This example, although it is very small and trivial (but the reader may also think of doing the same with contexts of much larger size) demonstrates an interesting form of computer assisted learning or knowledge acquisition by successively collecting information on objects. Here is a brief description of the method that can be used:

Algorithm:

- Start from an empty context, the columns associated with the given elements of a set of attributes.
- Pick an object and fill the first line according to its properties.
- Evaluate the corresponding basis of implications and check if each of its elements is acceptable.
- If there are unacceptable elements in the basis, then add a further object (=line), until no unacceptable implications remain in the basis.
- If implications remain that are neither acceptable nor unacceptable, they can be considered as hypotheses and we can try to argue why they might be acceptable or unacceptable.

The last item shows that this method allows the automatic generation of hypotheses along the collection of examples and the evaluation of the Duquenne/Guigues–basis. Moreover, we note that this is a mathematical model of a popular scientific method: Collect examples and do pattern recognition!

Here is another example, a context on environmental situations at various places in Baden-Württemberg due to Brüggemann et al. 1998, 2003. The objects are regions in Baden-Württemberg, the attributes are

- F: forest
- A: agriculture
- T: traffic
- S_m: settlement
- I: industry
- L: lead
- Cd: cadmium
- Zn: zinc
- S: sulfur

We deduce from Brüggemann 1999 and Brüggemann et al. 1998 a multi-valued context, part of which is shown in Fig. 6. Its basis of implications contains – among many others – the following implications:

- $\{Zn \geq 1\} \rightarrow \{Cd \geq 1\}$
- $\{Cd \geq 1, S \geq 1\} \rightarrow \{T \geq 2\}$
- $\{Cd \geq 2, Zn \geq 1\} \rightarrow \{Zn \geq 2\}$
- $\{L \geq 1, S \geq 1\} \rightarrow \{A \geq 2\}$
- $\{L \geq 1, Cd \geq 1\} \rightarrow \{A \geq 2\}$
- $\{L \geq 2\} \rightarrow \{A \geq 2, Cd \geq 1\}$
- $\{T \geq 2, S_m \geq 3, I \geq 3\} \rightarrow \{L \geq 1\}$
- $\{T \geq 3\} \rightarrow \{F \geq 2, A \geq 2, I \geq 2\}$
- ⋮

For example, the first of these implications might be considered as the hypothesis "zinc comes with cadmium". The second implication relates the pollution of the herb layer by cadmium and sulfur with traffic. Thus, formal concept analysis helps in formulating valuable working hypotheses.

region	F	A	T	S_m	I	L	Cd	Zn	S
Lörrach	1	1	2	3	3	1	0	0	0
Bad Säckingen	2	2	1	1	2	0	0	0	0
Waldshut-Tiengen	2	1	2	2	3	0	0	0	0
Stühlingen	2	3	2	2	2	0	0	0	1
Immendingen	1	2	2	1	2	1	0	0	1
Engen	2	2	2	2	2	1	0	0	0
Salem	2	3	2	1	2	1	0	0	0
Überlingen	2	1	2	2	2	1	0	0	0
Wangen	1	2	1	1	1	0	1	2	0
Kandern	2	2	2	2	1	1	0	0	0
Schönau	2	1	2	1	1	1	0	0	0
Donaueschingen	1	1	1	2	1	0	1	2	0
⋮									

Fig. 6. Part of a context on regions in Baden-Württemberg

The problem with applications to contexts, which are based on field measurements, is that outliers may completely destroy an interesting implication. Hence the scaling process (see Ganter and Wille 1999) must be performed carefully and in fact it is more reasonable to apply concept analysis to contexts on Boolean attributes only.

Further interesting applications to chemistry can be found in particular in the cited literature by Bartel and Brüggemann.

References

Bartel HG (1995) Formale Begriffsanalyse und Materialkunde. Match Commun Math Comput Chem 32:27-46

Brüggemann R, Friedrich J, Kaune A, Komossa D, Pudenz S, Voigt K (1998) Vergleichende ökologische Bewertung von Regionen in Baden-Württemberg: GSF-Bericht 20/98. GSF–Forschungszentrum für Umwelt und Gesundheit Neuherberg, Germany pp 1-148

Brüggemann R (1999) A Theoretical Concept To Rank Environmentally Significant Chemicals. J Chem Inf Comp Sci 39:211-217

Brüggemann R, Welzl G, Voigt K (2003) Order Theoretical Tools for the Evaluation of Complex Regional Pollution Patterns. J Chem Inf Comp Sci 43:1771-1779

Ganter B, Wille R (1999) Formal Concept Analysis – Mathematical Foundations. Springer Verlag, Berlin

Bell RP (1974) Säuren und Basen und ihr quantitatives Verhalten. VCH - Verlag pp 1-111

Lexikon der Chemie in drei Bänden (1999) Spektrum-Verlag

Partial Orders and Complexity: The Young Diagram Lattice

William Seitz

Department of Marine Sciences
Texas A&M University at Galveston
P.O. Box 1675, Galveston, Texas 77539, USA

e-mail: seitzw@tamug.edu

Abstract

A partial order of longstanding interest to mathematicians and chemists, the Young Diagram Lattice (YDL) is discussed in the context of complexity. Ruch's (1975) identification of this partially ordered set with that appropriate to a general partial ordering for mixing is discussed. A mathematical quantity associated with each member of the set (the cardinality of maximal anti-chains for that member) is argued to provide a quantitative measure for complexity for members of the set. The measure has the desirable feature that low complexity is associated with both highly ordered and very random systems, while systems that have intermediate "structure" have larger complexity. Several quantitative examples based on the YDL are briefly discussed including statistical mechanics, diffusion, and biopolymeric complexity. Finally, a metaphor for complexity suggested by the YDL associates high complexity with posetic incomparability. Examples from sociology, ecology, and politics are discussed.

Introduction

It is intuitive that the study of partially ordered sets (posets) might be related to the currently popular study of complexity. Here we use a well-studied, mathematically beautiful, and physically fundamental poset – the Young diagram lattice (YDL) – to suggest an appropriate quantitative measure of complexity for a member of this poset, and possibly other posets as well. For the YDL, the measure is currently mathematically intractable for large systems, but an approximate computational approach can be applied. We will also propose a qualitative (and not intuitive) general complexity metaphor that is suggested by this partial order example.

Partial orders were first introduced in the late 19th century (Pierce 1880, Dedekind 1897). Shortly thereafter, in the early 1900's Young Alfred Young, a cleric interested in substitution expressions, introduced diagrams. (Young 1900, Young 1933, Rutherford 1947) His work is used by quantum theorists to generate wave functions that satisfy the Pauli exclusion principle (Matsen 1971). But it also has been discussed in the context of the general problem of mixing that arises naturally in statistical mechanics (Ruch 1975), as well as in diversity discussions in biology. For some more information about the historical development of the theory of posets, compare by Halfon, p. 385 Applications of the YDL in chemistry can also be found in another chapter of this book, see e.g. chapter by El-Basil, p. 3. For reference, the YDL (poset) for 10 objects is shown in Fig. 1.

Mixing

Statistical mixing is a concept that is not widely discussed by chemists. It is not the same as disorder, but clearly related. All chemists are familiar with the second law of thermodynamics that says that systems evolve to states with higher disorder – i.e. greater entropy. But how, exactly, does the evolution occur? Indeed, it might be true that different evolutionary paths exist. Each path proceeds from order to disorder in steps where each step results in higher entropy, but paths may differ. Since every path follows steps that increase entropy, there is no inconsistency with the second law. Referring to the YDL in Fig. 1, for example, one might go from top to bottom of the figure on many different paths that always proceed downward but along different routes. So we now proceed to describe in more detail the partial order represented in the YDL.

Suppose one has a set of objects each object of which may be characterized by some principle – e.g. a group of croquet balls with the characterization principle being colour. Imagine many sets (of croquet balls) with each set containing the same number of objects (balls). Sets in which each object is distinct from every other are maximally mixed, while sets where all objects are identical are minimally mixed. Other sets are intermediate in terms of mixing. (Note, with regard to mixing, it doesn't matter which colours are present, i.e. a set with 6 green balls has the same mixing character as one with 6 white balls, and a set with 3 green, 2 blue and 1 white ball would have identical mixing character as a set with 3 yellow, 2 green and 1 black ball.)

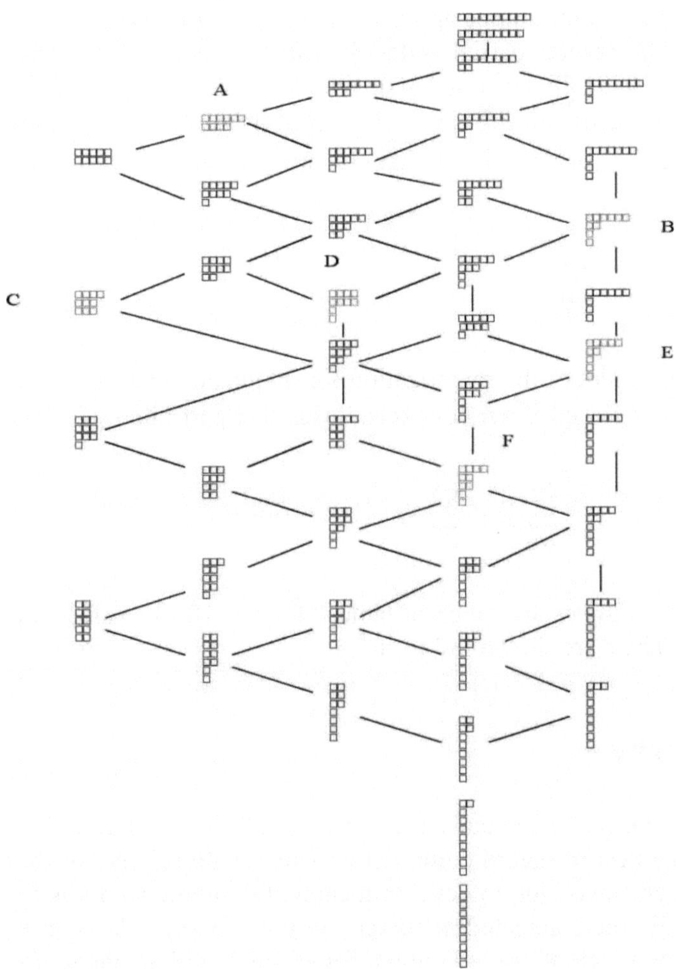

Fig. 1. Partitions of N=10

After some consideration, it is obvious that mixing character of a set is simply the partition of objects in the set according to their classification (in the example above [3,2,1]). If partitions represent mixing, one need only seek a partial order among these partitions that appropriately represents the concept of increased mixing character.

Ruch showed in his seminal 1975 paper (Ruch 1975) that a partial order relation for partitions already well known to mathematicians – namely the majorization partial order – is precisely the same partial order that corresponds to mixing. This fundamental result means that the mathematics of the Young Diagram Lattice applies to the physics of mixing.

The majorization partial order for integers can be stated as follows. (Marshall 1979) Given a set of n objects, one can represent its mixing character by a partition $\lambda=[\lambda_1,\lambda_2,\lambda_3, ...,\lambda_v]$ where λ_i is the number of objects of type i (or class i). In general we can take $\lambda_i \geq \lambda_j$ if i < j and clearly

$$\sum_{i=1} \lambda_i = n \tag{1}$$

The statement of the majorization (or dominance) partial order is that a partition λ is placed above (or exceeds) another partition μ if

$$\sum_{i=1}^{m} \lambda_i \geq \sum_{i=1}^{m} \mu_i \quad \forall m = 1 \text{ to } n \tag{2}$$

A few examples are given in table 1 for n = 10 (the selected partitions and their labels are shown in Fig. 1.[1]

Complexity

There are dozens of mutually inconsistent definitions of complexity. Thus it might be best to regard complexity theory as descriptive of the work of scientists endeavouring to develop meaningful, predictive tools for dealing with highly interconnected or mixed systems. Despite the apparent chaos in the field, much of use is being accomplished, with applications made to stock market analysis, chemical structure, military strategy, industrial processes, transportation, social structure and even baggage handling. Some applications of complexity theory to chemistry can be found in 'Complexity in Chemistry', edited by Bonchev and Rouvray (Bonchev 2003).

[1] A similar definition can be made for n-tuples of real numbers, but this doesn't add to the present argument.

Table 1. Selected partition of 10

Partition	Partition Label	Partial Sums in equation (2)	Incomparable with	Greater than	Less than
[6,4]	A	6,10,10,10, …		B,C,D,E,F	
[6,2,1,1]	B	6,8,9,10,10, …	C	D,E,F	A
[4,3,3]	C	4,7,10,10 …	B,D,E	F	A
[4,4,1,1]	D	4,8,8,10,10, …	C,E	F	A,B
[5,2,1,1,1]	E	5,7,8,9,10,10 …	C,D	F	A,B
[4,2,2,1,1]	F	4,6,8,9,10,10, …			A,B,C,D,E

It is sometimes helpful to distinguish between the complexity of objects and that of processes. If one seeks to compare different objects to one another in terms of their “complexity”; that is object complexity theory. The objects might be molecules, people, social structures, poems, toxins, etc… An enormous literature exists with much of the chemical focus being on molecular complexity indices that attempt to order molecular graphs according to some measure of their complexity. While not yet directly concerned with complexity, much environmental focus has been on the relative toxicity of pollutants partially ordered by their effects on different things.

The second type of complexity seeks to understand the behaviour of systems consisting of many components interacting in different ways, sometimes leading to novel collective properties – termed emergent properties. Again, an enormous literature exists, with much current chemical attention being addressed to self-assembly of nanostructures, and with much biology and physics attention focused on the behaviour of neural networks.

While there are definitions of complexity that essentially equate it to randomness, many scientists have come to regard complex systems as lying between fully ordered and fully disordered ones. A qualitative graph is shown Fig. 2 (Huberman 1986).

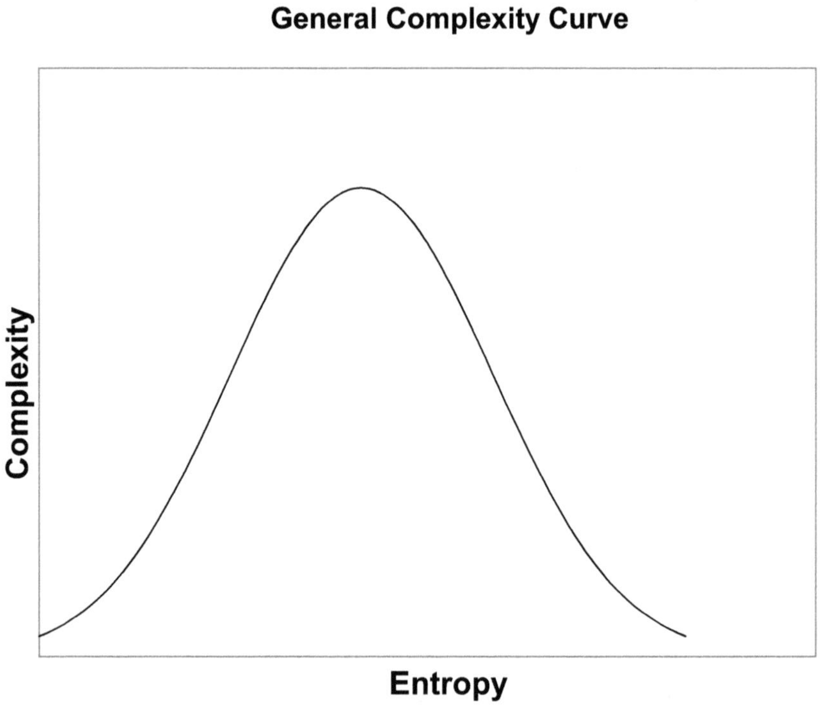

Fig. 2. Qualitative Complexity. The x-axis represents disorder or entropy while the y-axis represents complexity

According to this view of complexity, both highly ordered and highly disordered systems have low complexity values, with more complex systems being associated with intermediate levels of disorder. A simple example might be of use. Consider a pure crystalline substance (low entropy) and then heat it through its melting point and then continue to heat it until it vaporizes. If one assumes a complexity curve such as Fig. 2 applies, the implication is that solids and gases are relatively simple compared to liquids – a view quite consistent with the state of theory. We have used this idea elsewhere to generate a qualitative value for liquid complexity (Seitz 2003).

Complexity and the Young Diagram Lattice

By laying the YDL for 10 objects in Fig. 1 on its side, one may see a similarity to the complexity curve in Fig. 2. To partly quantify this idea we might simply relate complexity to the "breadth" of the diagram at any point ("width" in poset theory, see Brüggemann and Carlsen, p. 61). Now recall that the entropy for a partitioning (mixing) of objects is simply the logarithm of the number of ways that they can be rearranged, i.e., for a partition $\lambda=[\lambda_1,\lambda_2,\lambda_3, ...,\lambda_n]$ the entropy (applying the well known approximation of Stirling) is given as

$$S_\lambda = \ln\left(\frac{n!}{\prod_i \lambda_i!}\right) \cong n\ln(n) - \sum n_i \ln(n_i)\,, \tag{3}$$

so that low entropy diagrams lie at the top of Fig. 1 whereas high entropy diagrams lie at the bottom. But near the middle of the diagram, where the mixing and entropy are intermediate, there are, in general, a very large number of diagrams that cannot be compared – i.e. ones that are neither more nor less mixed than one another. Now we propose that complexity of any partition relates to the "breadth" of the YDL at the location of the partition. This we take as the extent to which a diagram is incomparable, in other words the number of diagrams to which it cannot be compared by the majorization partial order. A more explicit definition follows.

Mathematicians have termed a set of elements in a poset that are all mutually incomparable an anti-chain. (See chapter by Brüggemann and Carlsen, p. 61 and for more detailed mathematics and definitions see 'Combinatorics and Partially Ordered Sets: Dimension Theory' by Trotter (Trotter 1992)). If we consider all anti-chains that contain a partition [λ] as an element, the complexity of [λ] is the number of elements in those anti-chains (i.e. the cardinality or size of the anti-chains) that have the maximum number of elements, maximum anti-chains. Clearly, this concept can be generalized to any poset, though, as we have seen, the case of the YDL is of particular interest and relevance to physics and chemistry.

For small YDLs (and other posets), complexity values may be obtained from inspection of the poset diagram, but the difficulties in dealing with large YDL are great because the number of partitions grows exponentially. For example, for $n = 500$ there are 10^{21} partitions in the lattice. Also, there is no known algorithmic method to obtain the sizes of maximum anti-chains for arbitrary members of the YDL. It is therefore useful to consider alternative definitions for the breadth of the YDL that can be handled numerically and which yield complexities as a function of entropy. One such approach is discussed below.

We define a new, computationally accessible incompatibility (complexity) measure for the YDL that may be used efficiently for large systems. To motivate our proposed incompatibility measure, we again consider Fig. 1. The chain of partitions along the far right hand side of the figure contains diagrams that partition n into the maximum number of groups (i.e. types of objects, or number of rows in the Young Diagram) for a given level or entropy. We might call this chain the maximum diversity chain. It is straightforward to algorithmically generate this chain for large n. Similarly, the chain along the far left hand side of the figure contains diagrams that partition n into the minimum number of groups, and we might call this the minimum diversity chain. This chain can also be generated numerically via a more complicated, but still straightforward numerical algorithm.

Fig. 3 shows the entropies from equation (3) of the minimum and maximum diversity chains for n=500. In general, there are more minimum diversity partitions (for n=500 there are 495 maximum diversity partitions and 4387 minimum diversity partitions), but the number of partitions in both "extreme" chains grows linearly with n so large systems can be treated. It is important to recognize that the entropy range is the same for both chains since both begin and end at the same points.

Finally the numerical incompatibility measure for any element in the diagram lattice can be defined. Compute the entropy of the element of interest using equation 3 and identify partitions on the maximum and minimum diversity chains with entropies closest to it. The incompatibility measure is then defined to be the "distance" between these maximum and minimum diversity partitions, where this distance is defined as the number of boxes (particles) that must be moved to convert the minimum diversity partition to the corresponding maximum diversity partition, and vice-versa. Fig. 4 shows these (normalized) distances for n=500.

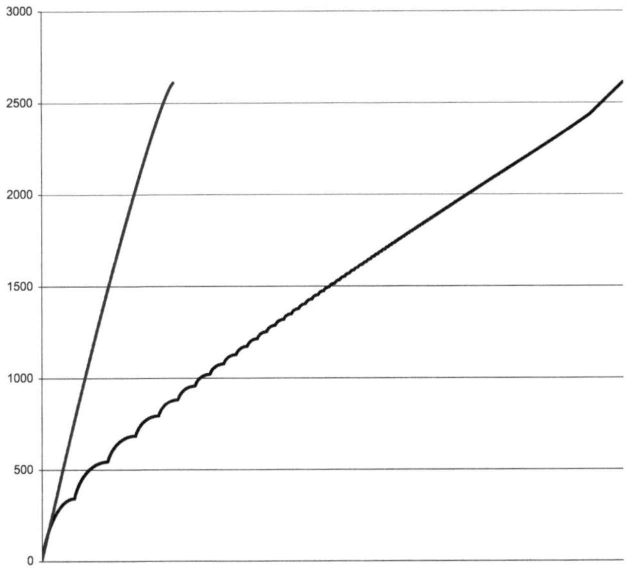

Fig. 3. Entropy of maximally and minimally diverse chains for n=500

Finally, it is possible to develop an asymptotic relationship for the numerical incomparability at medium entropy. The minimally diverse partitions (furthest to the left on the diagram lattice) are those that minimize the number of rows (for a given entropy – or vertical position on the lattice). This is accomplished if the rows have nearly the same number of objects (boxes). We let m be the number of rows in a partition in the left-hand chain: then there are approximately n/m boxes in each row for partitions on the left-hand chain to minimize m (or number of rows).

Next consider the right-hand chain. There the maximally diverse partitions (for given entropy) maximize the number of rows. If there were n-q elements (boxes) in the first row, then the corresponding maximally diverse partition would have q rows of one box each.

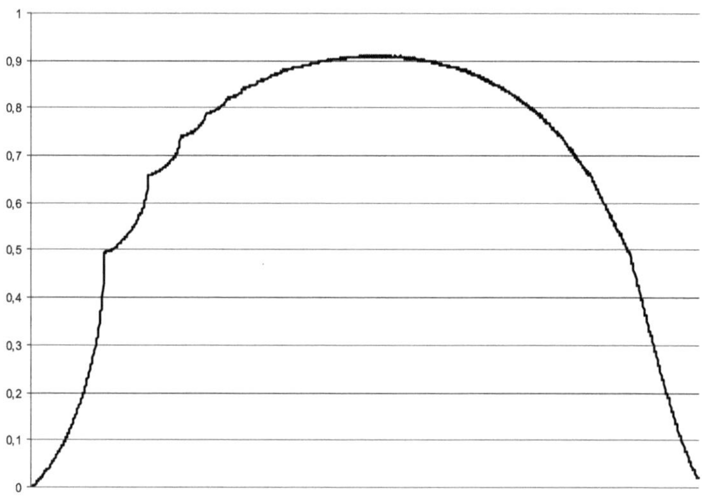

Fig. 4. Incompatibility curve for n=500. The horizontal axis is the Shannon entropy from equation (3) with maximum order to the left and maximum disorder on the right. The vertical axis is the (normalized) per-object "distance" from maximum to minimum diversity partitions at the same entropy. Note the similarity to Fig.2

(Note: partitions in the maximal diversity chain actually alternate between $[n\text{-}q, 1^q]$ and $[n\text{-}q, 2, 1^{q-2}]$ but this detail does not affect the asymptotes) At large n/m, q>>m and we see that the maximum and minimum diversity partitions have n/m+m-1 boxes in common so that the "distance" between them is n-(n/m)-m+1.

Thus the incomparability measure per site is

$$\frac{IM[at\, S = n\ln(m)]}{n} \approx 1 - \frac{1}{m} - \frac{m}{n} + \frac{1}{n} \qquad \text{for (n/m) large} \qquad (4)$$

where the entropy is that of the minimally diverse partition obtained from equation (3). Thus the asymptotic value for the maximum numerical incomparability (complexity) of a large YDL is *n*.

YDL Examples

We briefly consider some examples. The first is the microcanonical ensemble (alluded to earlier) in statistical physics where the Young diagrams represent the assignment of particles to states (Ruch 1975). The top partition assigns all members of the ensemble to a single accessible state, while the bottom diagram (partition) corresponds to the uniform distribution over all allowable states (the equilibrium distribution). It is this case that Ruch considered and proposed the generalized time development in terms of increasing mixing character discussed above. For the microcanonical ensemble, the complexity of a state is the number of different states incomparable to it through which evolution from order to disorder could occur without passing through the state itself.

As a second example we consider the problem of diffusion (or more specifically random walks in two dimensions). We attempt to develop visual representations that correspond to simple and complex distributions of walkers. We consider a collection of N walkers each beginning at the origin on a two dimensional square lattice and each executing a random walk of L steps with each step randomly chosen to be right, left, up or down. After 1 step, all walkers are a distance 1 from the origin (approximately 25 % moved up, 25 % down, etc…), while after two steps some are a distance 2 from the origin, (both steps in the same direction), some are at a distance $\sqrt{2}$ (e.g. left then up), and some are at a distance 0 (step 2 being the reverse of step 1). Let the integer portion of the distance of a walker classify the walker – then for 1-step walks, all are in the same class (distance 1) and for 2-step walks there are 3 classes, (distance 0,1,2). Thus it is straightforward to assign a partition (based on the integer portion of its distance from the origin) to each finite L-step walk, and, by applying eqn. 3, also obtain its entropy. Fig. 5 shows representative results for 300 walkers.

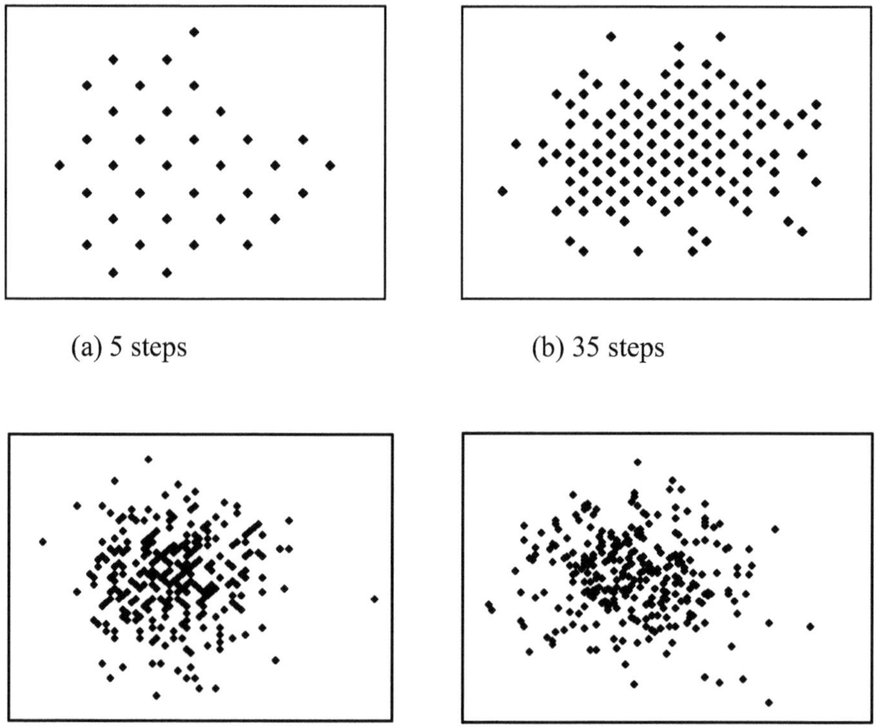

(a) 5 steps (b) 35 steps

(c) 271 steps (d) 811 steps

Fig. 5. Distributions of 300 random walkers after the indicated numbers of steps. (a) represents low entropy, low complexity (d) is high entropy low complexity, while (b) and (c) are high complexity patterns

Fig.'s. 5(a) and 5(d) both have low incomparability measures, while 5(b) and 5(c) have high incomparability measures. While qualitative, it is interesting to observe the existence of "complex" patterns in (b) and (c) as opposed to (d) where the distribution of walkers appears almost random and (a) where the distribution is "simple".

A third example may be the quantification of complexity for biological sequences. This area has, of course, been widely investigated by many laboratories. Most closely related to our work is that being done at the National Center for Biotechnology Innovation at NIH. There, Wan and Wootton developed a compositional complexity measure for biological sequences in proteins and nucleotide sequences in terms of longest paths on the YDL (Wan and Wooton 2000). It is interesting to note that there may be a relationship between maximum anti-chains and Wootton's maximum paths suggesting that the definitions of complexity are related. This relationship is the subject of ongoing work in our laboratory.

Complexity Metaphor – Other Posets and Beyond

The discussions so far have focused on the Young Diagram poset, but the general ideas may possibly be of use more widely. Clearly, since every (non-trivial) poset contains (non-trivial) anti-chains, the complexity measure of an element in any poset may be defined in the same way as was done here for the YDL poset. However, the utility of such a quantity will obviously depend on the system being described by the partial ordering.

However, the idea that incomparability is fundamental to complexity may be useful even where there is no clear partial ordering relation defined. If we take extent of incomparability as a metaphor for complexity, a number of interesting observations follow. There may be different ways to state the metaphor depending on context. Examples are:

- The complexity of a system is the number of incomparable things in it.
- The complexity of an object is the number of others with which it is incomparable.
- The complexity of an evolving system is the number of incomparable things in it at that point in its evolution.

Thus, if one is considering an ecosystem, one might use the first or third statement of the metaphor, while if one is considering a single component of the ecosystem, the second statement would apply.

The utility of any metaphor lies in its ability to expose subtleties so that new insights may be obtained. A few examples follow.

First consider the evolution (life histories) of a group of people (who all have lived to an old age, say 85, and now reside in a nursing home). As newborns, there were few distinctions among the members – all required similar food and care, and all appeared quite similar (except of course to the parents). At the end of their lives, they are again quite similar as they complete their life journey. Next consider them as pre-teens – now there are significant differences, but still there are a relatively small number of different groups depending on growth rates, progress in school, etc... Teenagers have a larger number of incomparable groups than pre-teens due to the beginnings of specialization into career paths, dating, choice of college vs. working, etc… Beginning as young adults and continuing through late middle age, the group has the largest number of incomparable individuals. Upon reaching retirement age, there is a reduction in the number of different types of people, with many retirees settling into areas with less and less diversity.

The analogy to the YDL is obvious where the "approximate" classification scheme is personal history. However it would be extremely difficult to make the relationship explicit. The metaphor suggests that (1) observed behaviour/lifestyle changes in humans is the natural consequence of evolution in a complex society, (2) the perception that the lives of mid-life individuals are more complicated than they were or will be is accurate and (3) there may be similarities between teen-agers and retirees yet to be explored (beyond their mutual fascination with Disney world). The term "second childhood" in the aged may reflect complexity more than mental acuity.

As a second example consider societal structures/governments. At the extremes are autocracy and anarchy – complete order and complete disorder. In autocracy, the individuals in the state are all the same in that they all obey the rulings of the leader. In anarchy, each individual makes his/her own rules and each is unique. As social structures go, however, both autocracy and anarchy can be regarded as simple – and indeed characterize less "civilized" societies. Intermediate structures such as democracy, communism, and socialism are more complex – entities within them are incomparable. Indeed one of the most significant aspects of democracy in the US is the separation of powers into judicial, executive and legislative – three "incomparable" branches of government.

Finally, as a qualitative quasi-scientific example, consider the classical evolution of a pond (Dashnowski 1912). A newly formed pond is simple with little life. Then algae appear followed by floating plants, submersed plants and finally emersed plants appear creating the mature pond where there is a maximum of incomparable plant-types present. However, the evolution of the pond does not stop there. As plants die, they sink to the bottom – most densely at the edge – reducing the size and complexity of the pond. This continues until the pond disappears entirely, returning to a "simpler" state.

Another possible use for the complexity/incomparability metaphor begins by determining which side of the complexity curve a system presently lays. Suppose we can determine that complexity is increasing with time under a set of conditions. Then simply allowing the system to evolve (to higher entropy) will result in increased complexity even if we do nothing. However, if complexity of a system is decreasing with time, energy must be provided in order to increase complexity.

Now consider two systems of the same complexity, one on either side of the complexity maximum, and suppose that the goal of each system is to maintain its level of complexity (or our goals as managers for the system). In both cases, we are far from equilibrium so that to maintain their state, energy must be expended to arrest evolution toward higher entropy. However, observation of the systems would show very different uses for the energy provided. On the left of the maximum, the natural evolution is toward increased complexity so that to maintain the state, energy is expended to reduce incomparability. I.e., the energy goes to make things more alike. On the right hand side, complexity is decreasing, so energy is expended to increase incomparability, i.e. to make things more different.

A possible example of this would be to consider the pond evolution example. Take the "left hand side" case to be where algae, floating and submersed plants exist, but emersed plants have not appeared. To maintain that state requires that as emersed plants do appear naturally, they are removed leaving the pond stable. The "right hand side" might be where the pond has begun to fill in and some plants are dieing. To maintain the pond requires that dieing plants be replaced with new plants to maintain the system in a steady state. Both strategies require the introduction of energy, but in one case plants are removed; while in the other plants are added. The example may appear trivial, but there may be cases where one may be guided by the complexity metaphor into the choice of appropriate action to maintain the status quo.

Conclusion

Complexity continues to be a topic of broad interest to scientists and non-scientists alike. We have argued that complexity and posetic incomparability are related. However, the incomparability of a member of a poset is a well-defined quantity leading to a possibly useful quantification of complexity. The Young Diagram Lattice was studied in detail both because of its rich (and complicated) mathematics and because of its relevance to the chemistry and physics of mixing. A numerical method was applied to this poset that led to a curve showing complexity, as a function of entropy – and that curve closely resembles the qualitative picture of complexity found elsewhere. Qualitative metaphors can be developed that may provide insights into complex systems where the precise partial ordering relation is either unknown or cannot be explicitly defined.

References

Bonchev D and Rouvray DH (2003) Ed. Complexity in Chemistry. Mathematical Chemistry Series, Ed. Bonchev D and Rouvray DH. Vol 7, Taylor and Francis, London. 210

Dashnowski A (1912) The successions of vegitation in Ohio lakes and peat deposits. Plant World 15:25-39

Dedekind R (1897) Über Zerlegungen von Zahlen durch ihre grössten gemeinsammen Teiler. In: Gesammelte Werke Bd 103-148

Huberman BH and Hogg T (1986) Complexity and Adaption. Physica 22D:376-384

Marshall AW and Olkin I (1979) Inequalities: Theory of Majorization and its Applications. Academic Press, New York

Matsen FA (1971) Spin Free Quantum Chemistry. J Phys Chem 75:1860-68

Pierce CS (1880) On the Algebra of Logic. Am Jour 3:15-57

Ruch E (1975) The Diagram Lattice as Structural Principle. Theoretica Chimica Acta (Berl.) 38:167-183

Rutherford D (1947) Substitutional Analysis. Hafner Publishing Company, New York and London

Seitz WA (2003) Thermodynamic Complexity. In: Bonchev D and Rouvray DH, (Ed.) Complexity in Chemistry Taylor and Francis: London and New York. pp 189-205

Trotter WT (1992) Combinatorics and Partially Ordered Sets. The Johns Hopkins University Press: Baltimore

Wan H and JC Wootton (2000) A global compositional complexity measure for biological sequences: AT-rich and GC-rich genomes encode less complex proteins. Computers and Chemistry 24:71-94

Young A (1900) On Quantitative Substitutional Analysis. Proc London Math Soc 33:97-146

Young A (1933) On Quantitative Substitutional Analysis (eighth paper). Proc London Math Soc 37(2):441-49

7 Historical remarks

The idea behind partial order is quite old. Famous scientists like Dedekind, Vogt, Muirhead, Hasse, Birkhoff, and Wille may be seen as pioneers of developing the mathematics of partial orders. Partial order in chemistry, biology also has a quite long tradition. Randić, Klein, Gutman, Ruch, El-Basil, Bartel and many others have published important papers about the use of partial order in chemistry. The journal "*MATCH-communications in mathematical and in computer chemistry*" publishes many papers in that context. In biology the publications of Patil, Taillie and Solomon may be mentioned.

As far as we can tell, the use of partial order in environmental systems started with the contributions of Halfon. We do not presume to write a competent chapter about the historical development of partial order in chemistry and in environmental sciences. However, in the proceeding chapter Halfon gives his view on the development of divers programs within the context of Hasse diagrams. The chapter shows the influence of Helmut Hasse, the important steps done by the Italian scientists Reggiani and Marchetti and the developments done by Halfon and in the period since 1986.

Hasse Diagrams and Software Development

Efraim Halfon

4481 Concord Place, Burlington, Ontario, Canada L7L 1J5

e-mail: info@butx.com

Abstract

This chapter describes the evolution of the use of Hasse diagrams in the environmental field and the development of the related software. Two Italian scientists, Marcello Reggiani and Roberto Marchetti, used Hasse diagrams to study the problem of model order estimation. Halfon extended the use of Hasse diagrams to ecological modelling and later to environmental chemistry. The HASSE software was initially developed in FORTRAN to be run on mainframe computers. Later Halfon in Canada and Brüggemann in Germany reprogrammed it for use on personal computers and code was added to display the diagrams interactively. Nowadays, scientific groups in Denmark, Germany, Italy and Sweden are actively developing new applications and developing new theoretical concepts.

Historical notes concerning the development of WHASSE

The volume edited by Rainer Brüggemann and Lars Carlsen presents many advances that took place in the last few years. It all started from a combination of circumstances that started from a vague idea, to Italian-Canadian collaboration, followed by German-Canadian collaboration and continuing nowadays with a fruitful international collaboration among many scientists.

In the nineteen seventies, two Italian scientists, Marcello Reggiani and Roberto Marchetti, used Hasse diagrams to study the problem of model order estimation. In the nineteen eighties Hasse diagrams were used by Halfon in ecological modelling (Halfon 1983) and later in environmental chemistry (Halfon and Brüggemann 1989, Halfon 1989). Since 1989 the collaboration with Rainer Brüggemann has led to the development of new mathematical analysis tools, new graphic tools, the establishment of regular conferences, and to this book that summarizes the international research endeavours about Hasse diagrams in chemistry and environmental sciences.

Helmut Hasse

Helmut Hasse was a German mathematician born in 1898. Hasse had over 200 publications including his book "Über die Klassenzahl abelscher Zahlkörper" that describes the graphs that took his name. Hasse left a mark on modern mathematics with discoveries including the "Hasse Local-Global Principle", the "Hasse invariant", and "Hasse's Theorem" on the number of points of an elliptic curve over a finite field. The collected mathematical papers of Hasse fill three volumes. A biography and complete references can be found in Brückner and Müller (1982), Frei (1985) and Leopoldt (1982) in the Dictionary of Scientific Biography (New York 1970-1990), in Mac Lane (1995) and on the Internet at www.geometry.net/scientists/hasse_helmut.php and http://en.wikipedia.org/wiki/Helmut_Hasse.

Hasse was a very gifted teacher, whose lectures and talks inspired an entire generation of young number theorists, far beyond the circle of his immediate students. Hasse possessed an encyclopaedic memory. His book "Über die Klassenzahl abelscher Zahlkörper" was written in the atrocious conditions of the post-war years, and he wrote it down from memory, without access to his notes or to a library. What is now known as "Hasse diagrams" in the theory of lattices and ordered sets is a legacy from the illustrations of this book (see http://en.wikipedia.org/wiki/Hasse_diagram).

Rationale on the use of partial order

Chemical compounds are ubiquitous in the environment. Some are dispersed on purpose and some are released accidentally. Some have local effects while others have a global impact. This impact has been variously quantified (see Helm 2003 and chapter by Helm, p. 285; Brüggemann and Drescher-Kaden 2003; Halfon and Brüggemann 1989; Brüggemann and Münzer 1993). For example physico-chemical properties and environmental persistence, i.e. criteria, have been combined in an index function (see Halfon 1989 or for example chapter by Brüggemann et al., p. 237). The values of these criteria are multiplied by a weight factor, to account for the relevant influence on the environment, and then all these factors are summed up to create an index. Hasse diagrams represent a different approach to the quantification of the impact of environmental chemicals.

Model order estimation

The connection of Hasse diagrams and environmental chemistry took several years to evolve: Halfon investigated the problem of developing mathematical models that would use the minimal number of state variables and parameters to describe a given ecological system (Halfon 1983, Halfon 1979). The problem, called model order estimation, can be best described as the procedure to develop a mathematical model of a given system when limited information is available about the state variables and the values of the parameters. If we develop a model with more state variables and more parameters than the solution of the problem requires, we are adding uncertainty to the model structure. This problem of model order identification is part of the larger problem of model identifiably.

The problems of identifying a mathematical model and the problem of identifying the effects of toxic contaminants on the environment are conceptually similar. On the one hand we are trying to develop a mathematical model with limited information about the system and on the other hand we are trying to identify the impact of toxic contaminants with limited knowledge about the contaminants themselves and the impact of the chemicals on the environment.

A literature review led Halfon to the field of research where Reggiani and Marchetti (1975) were active. Both engineers worked at Selenia, an industrial electronics company in Rome, Italy. Their approach was based on earlier research done a by Helmut Hasse during the 1940's and the early 1950's. Collaboration with Dr. Reggiani that lasted several years (Halfon and Reggiani 1978, Reggiani and Halfon 1979, Halfon and Reggiani 1986). Dr. Marchetti, a student at the time, moved on to other research endeavours.

Quantification of fate and impact of contaminants

The inference that the Hasse methodology that could be applied to system models (Halfon 1983) could also be extended to environmental toxicology developed quickly (Halfon and Reggiani 1986): While the index function approach was, and still is very useful, it had some weaknesses that could be overcome. One weakness is the use of the weight factor; another is the subjective formulation of the index function itself, i.e., how different criteria are combined to create the index of environmental impact. This specific weakness is common for many other multi criteria decision tools as for example ELECTRE or PROMETHEE or NAIADE etc (Roy 1990, Brans and Vincke 1985, Rauschmayer 2001). (For a comparison of Hasse diagrams with some other multi criteria decision tools, see also the chapter by Brüggemann et al., p. 237).

The addition operation within the index implies that the importance of each criterion is masked within the index: It is difficult to compute backward from an index function the importance of each criterion and the interactions with the other criteria in the index.

The publication of the Halfon and Reggiani's paper (1986) in the journal "Environmental Science and Technology" was not easy. The integration of Hasse diagrams and environmental toxicology was so innovative and unfamiliar that the editor of environmental science and technology allowed the publication of this paper even if the methodology was new and still unproven.

Development of software

Some remarks concerning the historical development can be found in the chapter by Seitz, p. 367. Here the development of the software is the main focus. The original code of the software WHASSE was originally developed in Italy by Reggiani and Marchetti (1975), as a FORTRAN program that could be executed on mainframe computers to compute Hasse diagrams. This was a good step forward since the usual development of Hasse diagrams was to draw them by hand. The automation of the development of these diagrams meant that many hypotheses could be tested in a relatively short time compared with past uses of these diagrams. In the meantime technology was developing and use of personal computers more ubiquitous.

This code was modified and extended by Halfon (Halfon and Reggiani 1986, Halfon 1989, Halfon and Reggiani 1986, Reggiani and Halfon 1979, Halfon et al. 1989) and he developed new software to visualize Hasse diagrams, i.e. to draw the circles and the lines connecting them Halfon et al. 1989. This was a difficult effort since initially it is not known how many circles will be on each line, on each level, how they will be connected and indeed whether some circles will not be connected to any others, incomparable objects. The drawing of the lines between circles is also difficult, since if we draw the lines as straight lines, they may overlap some circles making the computerized diagram very difficult to understand, thus, we had to program subroutines that would draw curved lines around the circles rather than overlapping them.

Halfon and Reggiani's paper (1986) came to the attention of Dr. Rainer Brüggemann, at the time at GSF[1] in Munich, West Germany. Brüggemann became interested in their effort and met personally at a conference at McMaster University in Hamilton, Ontario in 1986. Following this meeting Brüggemann and Halfon began a collaboration (Halfon and Brüggemann 1989) that is still continuing. In its present formulation WHASSE it includes some of the original code coupled with graphics and an interactive interface. The software is also used and being expanded in Denmark and new Italian groups.

Brüggemann is a mathematical chemist. The areas of expertise of Halfon and Brüggemann combined well and over the years they were able to extend the application of Hasse diagrams in environmental sciences especially environmental chemistry. The development of personal computers and new computer languages led to new versions of the Hasse software with more and more features. The Hasse program was converted from FORTRAN to DELPHI and the graphic subroutines properly integrated as well, Brüggemann et al. 1999. The computer program grew from a short program to one that used most space on a CDROM. Now programs written in JAVA (Pudenz 2004) and in PYTHON (Software PYTHON, see von Löwis & Fischbeck 2001) are available or under development.

More mathematical analyses were released in the public domain and were eventually published. PRORANK is a new software which is currently developed and which is mainly thought of to be used for commercial purposes, see also page 111. Another powerful program RANA was developed by Pavan et al. (2003); see also page 181. Programs with more specific tasks like Po Correlation of Sørensen et al., see page 259 are currently under development.

In 1998 the first International workshop on Hasse applications in environmental chemistry took place in Berlin with the participation of about 30 scientists. In 2004 the 6th International workshop took place in Bayreuth and was reported in local newspapers as well as in a German magazine. The contributions appeared in a special issue of MATCH (Brüggemann et al. 2005).

[1] Now: GSF -National Research Center of Health and Environment

Professor Todeschini and his team, at JRC Ispra October 2006, will organize the next international workshop.

Now besides Canada, there are different places in Germany where partial order theory and partially its software WHASSE are used. Beyond this scientific groups in Denmark and Italy are actively contributing to new applications and new theoretical concepts. Table 1 lists some activities mainly in the field of environmental chemistry.

Hasse diagrams are intrinsically very easy to create. All work can be done and crosschecked on the back of an envelope. The advent of mainframe computers and later of personal computers with their graphic capabilities has led to complex analysis and insight that were not even predictable years ago, but the basic simplicity is still there.

Table 1. Activities around WHASSE

Nation	application activity	theoretical activity	remarks
Denmark	Pesticides, monitoring, polluted sites	probability concepts, correlation	NERI Silkeborg
Denmark	Evaluation of chemicals, methods to combine QSAR modeling and Partial Ordering.	methods to apply partial orders on QSAR	Awareness Center
Germany	local water management	Combination with Cluster analysis	BTU Cottbus
Germany	regional water management	Participation of stakeholders	HU Berlin
Germany	information systems		GSF
Germany		software development	Criterion
Germany/Canada	Ecological systems	probability, graph theory	IGB/Mc Master University
Italy	Pesticides, chemicals	Characterizing quantities, integration of concepts of genetic algorithms	University Milan

References

Brans JP, Vincke PH (1985) A Preference Ranking Organisation Method (The PROMETHEE Method for Multiple Criteria Decision - Making). Management Science 31:647-656

Brückner H and Müller H (1982) Helmut Hasse (25.8.1898-26.12.1979). Mitt Math Ges Hamburg 11(1):5-7

Brüggemann R, Bücherl C, Pudenz S, Steinberg C (1999) Application of the concept of Partial Order on Comparative Evaluation of Environmental Chemicals. Acta hydrochim hydrobiol 27:170-178

Brüggemann R, Drescher-Kaden U (2003) Einführung in die modellgestützte Bewertung von Umweltchemikalien - Datenabschätzung, Ausbreitung, Verhalten, Wirkung und Bewertung, 1 Ed. Springer-Verlag Berlin

Brüggemann R, Frank H and Kerber A (2005) Proceedings of the Conference "Partial Orders in Environmental Sciences and Chemistry" (Bayreuth, 15-16 April 2004). Match - Commun Math Comput Chem 54(3):487-690

Brüggemann R, Münzer B (1993) A Graph-Theoretical Tool for Priority Setting of Chemicals. Chemosphere 27:1729-1736

Frei G (1985) Helmut Hasse (1898-1979). Expositiones Mathematicae 3(1):55-69

Halfon E (1979) Computer-based Development of large-scale Ecological Models: Problems and Prospects. In: Innis, GS and O'Neill, RV (Ed.), System Analysis of Ecosystems. pp 197-209

Halfon E (1983) Is there a best model structure? II Comparing the model structures of different fate models. Ecological Modelling 20:153-163

Halfon E (1989) Comparison of an index function and a vectorial approach method for ranking waste disposal sites. Environ Science Technol 23:600-609

Halfon E and Brüggemann R (1989) Environmental hazard ranking of chemicals spilled in the Rhine River in November 1986. Acta hydrochim hydrobiol 17:47-60

Halfon E and Reggiani MG (1978) Adequacy of ecological models. Ecological Modelling 4:41-50

Halfon E, Hodson JA and Miles K (1989) An algorithm to plot Hasse diagrams on microcomputer and Calcomp plotters. Ecol Model 47:189-197

Halfon E, Reggiani MG (1986) On Ranking Chemicals for Environmental Hazard. Environ Science Technol 20:1173-1179

Helm D (2003) Bewertung von Monitoringdaten der Umweltprobenbank des Bundes mit der Hasse-Diagramm-Technik. UWSF – Z Umweltchem Ökotox 15:85-94

Leopoldt HW (1982) Obituary: Helmut Hasse (August 25, 1898-December 26, 1979). J Number Theory 14(1):118-120

MacLane S (1995) Mathematics at Göttingen under the Nazis. Notices of the American Mathematical Society 42:10

Pavan M (2003) Total and Partial Ranking Methods in Chemical Sciences. University of Milan - Bicocca, Cycle XVI, Tutor: Prof. Todeschini pp 1-277

Pudenz S (2004) A JAVA-based software for data evaluation and decision support. In: Sørensen PB et al. (Ed.) Order Theory in Environmental Sciences - Integrative approaches, Proceedings of the 5th workshop held at NERI, 2002. NERI, Ministry of the Environment, Denmark Roskilde pp 137-147

Rauschmayer F (2001) Entscheidungshilfen im Umweltbereich - Von der monokriteriellen zur multikriteriellen Analyse. In: Meyerhoff J, Hampicke U and Marggraf R (Ed.) Jahrbuch Ökologische Ökonomik 2: Ökonomische Naturbewertung. Metropolis-Verlag Marburg pp 221-241

Reggiani MG and FE Marchetti (1975) On Assessing Model Adequacy. IEEE Trans Systems Man and Cyber. SMC-5 pp 322-330

Reggiani MG and Halfon E (1979) Comments to "Adequacy of Ecosystem Models". ISEM Journal 1:68

Roy B (1990) The outranking approach and the foundations of the ELECTRE methods. In: Bana e Costa (Ed.) Readings in Multiple Criteria Decision Aid. Springer-Verlag Berlin pp 155-183

Sørensen PB, Brüggemann R, Thomsen M and Lerche D (2005) Applications of multidimensional rank-correlation. Match - Commun Math Comput Chem 54:643-670

von Löwis M, Fischbeck N (2001) Python 2 - Einführung und Referenz der objektorientierten Skriptsprache. 2 Ed. Addison-Wesley München

Introductory References

Posets (books)

Brüggemann R, Drescher-Kaden U (2003) Einführung in die modellgestützte Bewertung von Umweltchemikalien - Datenabschätzung, Ausbreitung, Verhalten, Wirkung und Bewertung. 1 Ed. Springer-Verlag, Berlin

Davey BA, Priestley HA (1990) Introduction to Lattices and Order. Cambridge University Press, Cambridge

Erné M (1982) Einführung in die Ordnungstheorie. BI Wissenschaftsverlag Mannheim

Neggers J, Kim HS (1998) Basic Posets. Singapore: World Scientific Publishing Co

Schröder BSW (2003) Ordered Sets - An Introduction. Birkhäuser, Boston

Trotter WT (1991) Combinatorics and Partially Ordered Sets Dimension Theory; John Hopkins Series in the Mathematical Science. The J Hopkins University Press: Baltimore

Posets as tools for data exploration

Brüggemann R, Altschuh J (1991) A validation study for the estimation of aqueous solubility from n-octanol/water partition coefficients. Sci Tot Environ 109/110:41- 57

Brüggemann R, Bartel HG (1999) A Theoretical Concept to Rank Environmentally Significant Chemicals. J Chem Inf Comput Sci 39:211-217

Brüggemann R, Bücherl C, Pudenz S, Steinberg C (1999) Application of the concept of Partial Order on Comparative Evaluation of Environmental Chemicals. Acta hydrochim hydrobiol 27:170-178

Brüggemann R, Halfon E (1990) Ranking for Environmental Hazard of the Chemicals Spilled in the Sandoz Accident in November 1986. Sci Tot Environ 97/98:827-837

Brüggemann R, Halfon E (1997) Comparative Analysis of Nearshore Contaminated Sites in Lake Ontario: Ranking for Environmental Hazard. J Environ Sci Health A32(1):277-292

Brüggemann R, Halfon E, Welzl G, Voigt K, Steinberg C (2001) Applying the Concept of Partially Ordered Sets on the Ranking of Near-Shore Sediments by a Battery of Tests. J Chem Inf Comp Sc 41:918-925

Brüggemann R, Münzer B (1993) A Graph-Theoretical Tool for Priority Setting of Chemicals. Chemosphere 27:1729-1736

Brüggemann R, Münzer B, Halfon E (1994) An Algebraic/Graphical Tool to Compare Ecosystems with Respect to their Pollution - The German River "Elbe" as an Example - I: Hasse-Diagrams. Chemosphere 28:863-872

Brüggemann R, Oberemm A, Steinberg C (1997) Ranking of Aquatic Effect Tests Using Hasse Diagrams. Toxicological and Environmental Chemistry 63:125-139

Brüggemann R, Pudenz S, Voigt K, Kaune A, Kreimes K (1999) An algebraic/graphical tool to compare ecosystems with respect to their pollution. IV: Comparative regional analysis by Boolean arithmetics. Chemosphere 38:2263-2279

Brüggemann R, Schwaiger J, Negele RD (1995) Applying Hasse Diagram Technique for the evaluation of toxicological fish tests. Chemosphere 30(9):1767-1780

Brüggemann R, Zelles L, Bai QY, Hartmann A (1995) Use of Hasse Diagram Technique for Evaluation of Phospholipid Fatty Acids Distribution in Selected Soils. Chemosphere 30(7):1209-1228

Halfon E (1989) Comparison of an index function and a vectorial approach method for ranking waste disposal sites. Environ Science Technol 23:600-609

Halfon E and Brüggemann R (1989) Environmental hazard ranking of chemicals spilled in the Rhine River in November 1986. Acta hydrochim hydrobiol 17:47-60

Halfon E, Reggiani MG (1986) On Ranking Chemicals for Environmental Hazard. Environ Science Technol 20:1173-1179

Helm D (2003) Bewertung von Monitoringdaten der Umweltprobenbank des Bundes mit der Hasse-Diagramm-Technik. UWSF – Z Umweltchem Ökotox 15:85-94

Lorenz R, Brüggemann R, Steinberg C, Spieser OH (1996) Humic Material Changes effects of Terbutylazine on Behavior of Zebrafish (*Brachydanio Rerio*). Chemosphere 33(11):2145-2158

Münzer B, Brüggemann R, Halfon E (1994) An Algebraic/Graphical Tool to Compare Ecosystems with Respect to their Pollution II: Comparative Regional Analysis. Chemosphere 28:873-879

Patil GP, Taillie C (2005) Multiple indicators, partially ordered sets, and linear extensions: Multi-criterion ranking and prioritization. Environmental and Ecological Statistics 11:199-228

Schrenk C, Pflugmacher S, Brüggemann R, Sandermann HJ, Steinberg C, Kettrup A (1998) Glutathione S-Transferase Activity in Aquatic Macrophytes with Emphasis on Habitat Dependence. Ecotox Environ Saf 40:226-233

Voigt K, Gasteiger J, Brüggemann R (2000) Comparative Evaluation of Chemical and Environmental Online and CD-ROM Databases. J Chem Inf Comp Sc 40:44-49

Voigt K, Welzl G (2002) Chemical databases: an overview of selected databases and evaluation methods. Online Information Review 26:172-192

Voigt K, Welzl G, Brüggemann R (2004) Data analysis of environmental air pollutant monitoring systems in Europe. Environmetrics 15:577-596

Voigt K, Welzl G, Brüggemann R, Pudenz S (2002) Bewertungsansatz von Umweltmonitoring-Informationssystemen in Ballungsräumen. UWSF – Z Umweltchem Ökotox 14:58-64

Posets as tools to chemical investigations

Altschuh J, Brüggemann R, Behrendt H, Münzer B (1996) Relationship between Environmental Fate and Chemical Structure of Solvents. In: Gasteiger, J (Ed.) Software-Entwicklung in der Chemie 10. Gesellschaft Deutscher Chemiker, Frankfurt, pp 105-116

Babic D, Trinajstic N (1996) Möbius inversion on a poset of a graph and its acyclic subgraphs. Discrete Appl Math 67:5-11

Brüggemann R, Pudenz S, Carlsen L, Sørensen PB, Thomsen M, Mishra RK (2001) The Use of Hasse Diagrams as a Potential Approach for Inverse QSAR. SAR and QSAR Environ Res 11:473-487

Carlsen L, Lerche DB, Sørensen PB (2002) Improving the Predicting Power of Partial Order Based QSARs through Linear Extensions. J Chem Inf Comp Sc 42:806-811

Carlsen L, Sørensen PB, Thomsen M (2000) Estimation of Octanol-Water Distribution coefficients - using partial order technique. In: Sørensen PB et al. (Eds.) Order Theoretical Tools in Environmental Sciences - Proceedings of the Second Workshop, October 21st, 1999 in Roskilde, Denmark, 1 Ed. National Environmental Research Institute, Roskilde, pp 105-115

Carlsen L, Sørensen PB, Thomsen M (2001) Partial Order Ranking - based QSAR's: estimation of solubilities and octanol-water partitioning. Chemosphere 43:295-302

Carlsen L, Sørensen PB, Thomsen M, Brüggemann R (2002) QSAR's based on Partial Order Ranking. SAR and QSAR in Environmental Research 13:153-165

El-Basil S, Shalabi AS (1985) On the ordering of Kekulé Structures, spectral moments of factor graphs of benzenoid hydrocarbons. Match Commun Math Comput Chem 17:11-43

Pavan M (2003) Total and Partial Ranking Methods in Chemical Sciences. 1-277. 2003. University of Milan - Bicocca, Cycle XVI, Tutor: Prof. Todeschini. Ref Type: Thesis/Dissertation

Pudenz S, Brüggemann R, Bartel HG (2002) QSAR of Ecotoxicological Data on the Basis of Data-Driven If-Then-Rules. Ecotoxicology 11:337-342

Randić M (1990) The nature of chemical structure. J Math Chem 4:157-184

Randić M (2002) On Use of Partial Ordering in Chemical Applications. In: Voigt K and Welzl G (Eds.) Order Theoretical Tools in Environmental Sciences - Order Theory (Hasse Diagram Technique) Meets Multivariate Statistics. Shaker-Verlag, Aachen, pp 55-64

Ruch E, Gutman I (1979) The Branching Extent of Graphs. J of Combinatorics-Inform System Sciences 4:285-295

Posets and Multicriteria Decision Analysis

Colorni A, Paruccini M, Roy B (2001) A-MCD-A, Aide Multi Critere a la Decision, Multiple Criteria Decision Aiding. JRC European Commission, Ispra

Heinrich R (2001) Leitfaden Wasser - Nachhaltige Wasserwirtschaft; Ein Weg zur Entscheidungsfindung. Wasserforschung e.V. Berlin

Lerche D, Brüggemann R, Sørensen PB, Carlsen L, Nielsen OJ (2002) A Comparison of Partial Order Technique with three Methods of Multicriteria Analysis for Ranking of Chemical Substances. J Chem Inf Comp Sc 42:1086-1098

Pudenz S, Brüggemann R (2002) A New Decision Support System: METEOR. In: Voigt K and Welzl G (Eds.) Order Theoretical Tools in Environmental Sciences - Order Theory (Hasse Diagram Technique) Meets Multivariate Statistics. Shaker-Verlag, Aachen, pp 103-112

Roy B (1972) Electre III: Un Algorithme de Classements fonde sur une representation floue des Preferences En Presence de Criteres Multiples. Cahiers du Centre d'Etudes de Recherche Operationelle 20:32-43

Roy B (1980) Selektieren, Sortieren und Ordnen mit Hilfe von Prävalenzrelationen: Neue Ansätze auf dem Gebiet der Entscheidungshilfe für Multikriteria-Probleme. zfbf-Schmalenbachs Zeitschrift für betriebswirtschaftliche Forschung 32:465-497

Roy B (1990) The outranking approach and the foundations of the ELECTRE methods. In: Bana e Costa (Ed.) Readings in Multiple Criteria Decision Aid. Springer-Verlag Berlin, pp 155-183

Simon U, Brüggemann R and Pudenz S (2004) Aspects of decision support in water management - example Berlin and Potsdam (Germany) II - improvement of management strategies. Wat Res 38:4085-4092

Simon U, Brüggemann R, and Pudenz S (2004) Aspects of decision support in water management - example Berlin and Potsdam (Germany) I - spatially differentiated evaluation. Wat Res 38:1809-1816

Simon U, Brüggemann R, Mey S, Pudenz S (2005) METEOR - application of a decision support tool based on discrete mathematics. Match Commun Math Comput Chem 54(3):623-642

Posets and Linear extensions

Atkinson MD (1989) The complexity of Orders. In: Rival, I (Ed.) Algorithms and Order NATO ASI series, Series C. Mathematical and Physical Sciences, Vol 255, Kluwer Academic Publishers, Dordrecht, pp 195-230

Brüggemann R (2004) Lokale partielle Ordnungen, ein Hilfsmittel für die Bewertung? In: Wittmann J and Wieland R (Eds.) Simulation in Umwelt- und Geowissenschaften - Workshop Müncheberg 2004. Shaker-Verlag Aachen, pp 109-123

Brüggemann R, Simon U, Mey S (2005) Estimation of averaged ranks by extended local partial order models. Match Commun Math Comput Chem 54(3):489-518
Brüggemann R, Sørensen PB, Lerche D, Carlsen L (2004) Estimation of Averaged Ranks by a Local Partial Order Model. J Chem Inf Comp Sc 44:618-625
Lerche D, Sørensen PB (2003) Evaluation of the ranking probabilities for partial orders based on random linear extensions. Chemosphere 53:981-992
Lerche D, Sørensen PB, Brüggemann R (2003) Improved Estimation of the Ranking Probabilities in Partial Orders Using Random Linear Extensions by Approximation of the Mutual Probability. J Chem Inf Comp Sc 53:1471-1480
Sørensen PB et al. (2003) Probability approach applied for prioritisation using multiple criteria - Cases: Pesticides and GIS. In: Sørensen PB et al. (Eds.) Order Theory in Environmental Sciences - Integrative approaches, Proceedings of the 5th workshop held at NERI, 2002. NERI, Ministry of the Environment, Denmark, Roskilde, pp 121-136
Sørensen PB, Lerche D (2002) Quantification of the uncertainty Related to the Use of a Limited Number of Random Linear Extensions. In: Voigt K and Welzl G (Eds.) Order Theoretical Tools in Environmental Sciences - Order Theory (Hasse Diagram Technique) Meets Multivariate Statistics. Shaker-Verlag Aachen, pp 65-72
Winkler P (1982) Average height in a partially ordered set. Discr Math 39:337-341

Posac

Borg I, Shye S (1995) Facet Theory - Form and Content. Sage Publications, Thousand Oaks, California
Brüggemann R, Welzl G and Voigt K (2003) Order Theoretical Tools for the Evaluation of Complex Regional Pollution Patterns. J Chem Inf Comp Sc 43:1771-1779
Shye S (1985) Multiple Scaling - The theory and application of partial order scalogram analysis. North-Holland, Amsterdam
Systat Program http://www.spss.com/software/science/SYSTAT/, SPSS Inc., 2000

Formal Concept Analysis

Bartel HG and John PE (1999) Formalbegriffsanalytische Untersuchung von Struktur-Eigenschafts-Beziehungen an Stannaocanen und Stannatranen unter Verwendung von 119 Sn-NMR-Daten. Zeitschr für physikalische Chemie 209:141-158
Bartel HG and Brüggemann R (1998) Application of formal concept analysis to structure-activity relationships. Fresenius J Anal Chem 361:23-28

Bartel HG (1995) Formale Begriffsanalyse und Materialkunde: Zur Archäometrie altägyptischer glasartiger Produkte. Match Commun Math Comput Chem 32:27-46

Bartel HG (1996) Mathematische Methoden in der Chemie. Spektrum Akademischer Verlag, Oxford, pp 1-353

Bartel HG (1997) Ein neuer Ansatz zur formalbegriffsanalytischen Objektklassifikation und seine Anwendung auf aromatische Heterocyclische Verbindungen. Match Commun Math Comput Chem 36:185-215

Bartel HG (1998) Formal Concept Analysis in Ecotoxicology. In: Group Pragmatic Theoretical Ecology (Ed.), Proceedings of the Workshop on Order Theoretical Tools in Environmental Sciences
Berichte des IGB, Heft 6, Sonderheft I, 1998, IGB, Berlin, pp 111-117

Bartel HG (2000) Formal Concept Analysis and Chemometrics: Chemical Composition of Ancient Egyptian Bronze artifacts. Match Commun Math Comput Chem 42:25-38

Ganter B and Wille R (1986) Implikationen und Abhängigkeiten zwischen Merkmalen. In: Degens PO, Hermes HJ and Opitz O (Eds.) Die Klassifikation und ihr Umfeld. Indeks-Verlag Frankfurt, pp 171-185

Ganter B and Wille R (1996) Formale Begriffsanalyse Mathematische Grundlagen, Springer-Verlag Berlin, pp 1-286

Ganter B and Wille R (1996) Formale Begriffsanalyse Mathematische Grundlagen, Springer-Verlag Berlin, pp 1-286

Kerber A (2005) Posets and Lattices, Contexts and Concepts. Match Commun Math Comput Chem 54(3):551-560

Wille R (1996) Allgemeine Mathematik, Mathematik für die Allgemeinheit. Fachbereich Mathematik der Technischen Hochschule Darmstadt. Preprint-Nr. 1822, 1-19. Darmstadt, FB Mathematik

Wille R (1990) Concept Lattices and Conceptual Knowledge Systems. Report: TH Darmstadt. 1340, Darmstadt, TH Darmstadt, FB Mathematik

Index

MIX
Papier aus verantwortungsvollen Quellen
Paper from responsible sources
FSC® C105338

If you have any concerns about our products, you can contact us on
ProductSafety@springernature.com

In case Publisher is established outside the EU, the EU authorized representative is:
Springer Nature Customer Service Center GmbH
Europaplatz 3, 69115 Heidelberg, Germany

Printed by Libri Plureos GmbH
in Hamburg, Germany